T0178293

Springer Series in Reliability Engineering

Series Editor

Hoang Pham, Department of Industrial and Systems Engineering, Rutgers University, Piscataway, NJ, USA

More information about this series at http://www.springer.com/series/6917

Hoang Pham

Statistical Reliability Engineering

Methods, Models and Applications

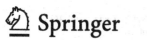 Springer

Hoang Pham
Industrial and Systems Engineering
Rutgers, The State University of New Jersey
Piscataway, NJ, USA

ISSN 1614-7839   ISSN 2196-999X   (electronic)
Springer Series in Reliability Engineering
ISBN 978-3-030-76906-2   ISBN 978-3-030-76904-8   (eBook)
https://doi.org/10.1007/978-3-030-76904-8

This Springer imprint is published by the registered company Springer Nature Switzerland AG
The registered company address is: Gewerbestrasse 11, 6330 Cham, Switzerland

In honor of my parents, who have given so much and wanted all the best for me;

To my wife, Michelle, and children, Hoang Jr., Melissa and David, for their love, patience, understanding, and support.

Preface

In today's technological world, nearly everyone depends upon the continued functioning of a wide array of complex growing range of devices used in markets ranging from driverless technology to 5G, the fifth-generation mobile network, from high-speed rail to 6G, the next-generation mobile communication technology for our everyday safety, security, mobility, and economic welfare. We expect our electric appliances, electrical power grids, hospital monitoring control, fly-by-wire next-generation aircraft, data exchange systems, and autonomous vehicle applications to function wherever and whenever we need them. When they fail, the results can be catastrophic. Internet of Everything (IoE) applications have quickly become a huge part of how we live, communicate, and do business in recent years and coming decades. All around the world, web-enabled devices are turning our world into a more switched-on place to live. As our society grows in complexity, so do the critical challenges in the area of reliability of such complex systems and devices.

In general, a system reliability is the probability that the system, including human-machine-thing in IoE applications, will not fail for a specified period of time under specified conditions. The greatest problem facing the industry today is, though given the huge large database, how to assess quantitatively and timely monitoring the reliability characteristics of such modern complex systems include IoE applications.

This book aims to present both the fundamental and the state-of-the-art methodology and methods of statistical reliability in theory and practice and recent research on *statistical reliability engineering*. It is a textbook based mainly on the author's recent research and publications as well as experience of over 30 years in this field. The topics covered are organized as follows. Chapter 1 provides a fundamental probability and statistics with which some readers are probably already familiar, but to many others may be not. It also discusses basic concepts and measures in reliability include path sets and cut sets, coherent systems, failure rate, mean time to failure, conditional reliability, and mean residual life. Chapter 2 describes most common discrete and continuous distributions such as binomial, Poisson, geometric, exponential, normal, lognormal, gamma, beta, Rayleigh, Weibull, Vtub-shaped hazard rate, etc. and its applications in reliability engineering and applied statistics. The chapter also describes some related statistical characteristics of reliability measures such as bathtub, Vtub shapes, increasing mean residual life, new better than used, etc.

Chapter 3 discusses statistical inference and common estimation techniques include the maximum likelihood, method of moments, least squared, Bayesian methods, and confidence interval estimates. It also discusses the tolerance limit estimates, goodness-of-fit tests, sequential sampling, and model selection criteria.

Chapter 4 discusses the reliability modeling and calculations for various systems including series-parallel, parallel-series, k-out-of-n, standby load-sharing, and degradable systems. It also discusses several basic mathematical reliability optimization methods and the reliability of systems with multiple failure modes. Chapter 5 discusses various system reliability estimation methods such as the maximum likelihood, uniform minimum variance unbiased, and bias-corrected estimates to estimate the reliability of some systems including the k-out-of-n system and stress-strength redundant systems. The chapter also discusses a concept of systemability and life testing cost model. Chapter 6 describes the basic of stochastic processes such as Markov process, Poisson process, nonhomogeneous Poisson process, renewal process, and quasi-renewal process, as the tools that can be applied in the maintenance modeling.

Chapter 7 discusses some basic maintenance models with various maintenance policies including age replacement, block replacement and multiple failure degradation processes, and random shocks. It also discusses the reliability and inspection maintenance modeling for degraded systems with competing failure processes. Chapter 8 aims to focus on an emergence trend in recent big data era in Industry 4.0 and the high demand of applying some statistical methods to various applications in engineering and machine learning. This chapter is devoted to the basic concepts of statistical machine learning. It first provides a brief basic linear algebra including orthogonal matrix. It further discusses the concept of Singular Value Decomposition (SVD) and its applications in the recommender systems. Finally, the chapter discusses the linear regression models with applications in machine learning aspects.

Problems are included at the end of each chapter. A few projects are also included in Chap. 4 that can be used for student group-project assignments. The detailed solutions to selected problems of each chapter are provided toward the end of the book. Appendix A contains various distribution tables. Appendix B contains some useful Laplace transform functions. The book also provides brief definitions of common glossary terms in the field of statistical reliability engineering and its related areas.

The text is suitable for a one-semester graduate course and advanced undergraduate courses in reliability engineering and engineering statistics in various disciplines including industrial engineering, systems engineering, operations research, computer science and engineering, mechanical engineering, mathematics, statistics, and business management. The instructor who wishes to use this book for a one-semester course may omit the first six sections and Sects. 10 and 11 of Chap. 1 provided that the students have had a basic probability and statistics knowledge. The book will also be a valuable reference tool for practitioners and managers in reliability engineering, data science, data analytics, applied statistics, machine learning, safety engineering, and for researchers in the field. It is also intended that the individual, after having utilized this book, will be thoroughly prepared to pursue advanced studies in reliability engineering, applied statistics, machine learning, and research in the field.

I have used an early draft version of this book as a textbook for graduate course in system reliability engineering at Rutgers University as well as a referenced reading material for a 3-day training seminar on statistical reliability modeling and its applications in machine learning. Similarly, researchers and data scientists can use the first five chapters (i.e., Chaps. 1–5), together with Chap. 8 for a 2-day seminar on statistical reliability engineering and machine learning.

I should appreciate it greatly if readers will bring to my attention any errors which they detect. I acknowledge Springer for this opportunity and professional support.

Piscataway, NJ, USA   Hoang Pham
March 2021

Contents

About the Author

Dr. Hoang Pham is a Distinguished Professor and former Chairman (2007–2013) of the Department of Industrial and Systems Engineering at Rutgers University, New Jersey. Before joining Rutgers, he was a Senior Engineering Specialist with the Boeing Company and the Idaho National Engineering Laboratory. He has been served as Editor-in-Chief, Editor, Associate Editor, Guest Editor, and board member of many journals. He is the Editor of *Springer Book Series in Reliability Engineering* and has served as Conference Chair and Program Chair of over 40 international conferences. He is the author or coauthor of 8 books and has published over 200 journal articles, 100 conference papers, and edited 17 books including *Springer Handbook in Engineering Statistics* and *Handbook in Reliability Engineering*. He has delivered over 40 invited keynote and plenary speeches at many international conferences and institutions. His numerous awards include the 2009 IEEE Reliability Society *Engineer of the Year Award*. He is a Fellow of the Institute of Electrical and Electronics Engineers (IEEE) and the Institute of Industrial Engineers (IIE).

Acronyms

AIC	Akaike's information criterion
ANOVA	Analysis of Variance
BIC	Bayesian information criterion
BT	Bathtub
BWUE	New worse than used in expectation
cdf	Cumulative distribution function
DFR	Decreasing failure rate
DMRL	Decreasing mean residual life
HBWUE	Harmonic new worse than used in expectation
HNBUE	Harmonic new better than used in expectation
IFR	Increasing failure rate
IMRL	Increasing mean residual life
LLF	Log likelihood function
ln	Natural logarithm
LOC	Lines of code
MBT	Modified bathtub
MLE	Maximum likelihood estimate
MRL	Mean residual life
MSE	Mean squared errors
MTBF	Mean time between failures
MTTF	Mean time to failure
MTTR	Mean time to repair
MVF	Mean value function
NBU	New better than used
NBUE	New better than used in expectation
NHPP	Nonhomogeneous Poisson process
NVP	N-version programming
NWU	New worse than used
PC	Pham's criterion
pdf	Probability density function
PIC	Pham's information criterion
PNZ model	Pham, Nordmann, and Zhang model

PRR	Predictive-ratio risk
rv	Random variable
SRGM	Software reliability growth model
SSE	Sum of squared errors
UBT	Upside-down bathtub
UMVUE	Uniform minimum variance unbiased estimate

Chapter 1
Basic Probability, Statistics, and Reliability

This chapter presents some fundamental elements of probability and statistics with which some readers are probably already familiar, but others may be not. Statistics is the study of how best one can describe and analyze the data and then draw conclusions or inferences based on the data available. The analysis of the reliability of a system must be based on precisely defined concepts. Since it is readily accepted that a population of supposedly identical systems, operating under similar conditions, fall at different points in time, then a failure phenomenon can only be described in probabilistic terms. Thus, the fundamental definitions of reliability must depend on the concepts from probability theory.

This chapter discusses basic definitions and measures in probability, statistics and reliability including probability axioms, basic statistics and reliability concepts and its applications in applied sciences and engineering. These concepts provide the basis for quantifying the reliability of a system. They allow precise comparisons between systems or provide a logical basis for improvement in a failure rate.

1.1 Basic of Probability

In general, probability is the branch of science concerned with the study of mathematical concepts for making quantitative inferences about uncertainty. Uncertainty is a part of the society that we are living with which indeed can also make life interesting. Uncertainty arises in every aspect of science, medicine, technology, and in our daily life in particular. Probability can be used to quantify the uncertainty of an event. A failure of an air conditioner or of an car engine can be described as a random event. Probability is an important aspect that can be used to analyze the uncertainty lifetimes of human being or lifetimes of electrical power systems.

The term *probability* refers to the number between zero and one that quantitatively measures the uncertainty in any particular event, with a probability of one indicating that the event is certain to occur and a probability of zero indicating that the event

© The Author(s), under exclusive license to Springer Nature Switzerland AG 2022
H. Pham, *Statistical Reliability Engineering*, Springer Series in Reliability Engineering,
https://doi.org/10.1007/978-3-030-76904-8_1

is certain not to occur. A *set* is merely an aggregate or collection of objects viewed as a single entity. Two *sets* X and Y are said to be equal (or identical) if and only if they consist of exactly the same elements, in which case we write $X = Y$. We shall use capital letters A, B, C... to denote sets and small letters such as a, b, c... to denote the element belonging to the set. For example, let A is the set of even positive integers, that is $A = \{2, 4, 6, ...\}$. Each even integer x is a member to the set A. We can write: $x \in A$. If an object is not a member of the set A, then we write $x \notin A$. Again, if A is the set of even positive integer, then $3 \notin A$. The theory of sets invented by the German mathematician, George Cantor in the late nineteenth century, has provided a very useful terminology that today is widely used and applicable to all the engineering, physical science and mathematics. This section will not discuss the subject in depth but only provide a few very basic concepts in order to take advantage of the set theory for topics that related to our focus in this book.

Definition 1.1 $A \cup B$ (*A union B*) is the set of elements that belong to at least one of the sets A and B, i.e., to A or B.

Definition 1.2 $A \cap B$ (*A intersection B*) is the set of elements common to both A and B.

Definition 1.3 Two sets are said to be *disjoint* or *mutually exclusive* if they have no elements in common, i.e., if

$$P(A \cap B) = 0$$
$$A \cap B = \emptyset$$

Probability theory is fundamentally concerned with specific properties of random phenomena. An *experiment* is defined as any clearly specified procedure, whereas a *trial* is a single performance of the experiment. Each possible result of an experiment that is of interset is called an *outcome*.

Example 1.1 Toss a coin twice. Let us assume "head" and "tail" as the only possible outcomes. If we denote these outcomes by H and T, respectively, the possible outcomes of the experiment are: {(H, H), (H, T), (T, H), (T, T)}.

Example 1.2 Roll two fair dice. Let us assume that the sum of the numbers on the dice is of interest. The possible outcomes of the experiment are: {2, 3, 4, 5, 6, 7, 8, 9, 10, 11, 12}.

Example 1.3 Toss a coin until it falls tails for the first time. Let us assume that the number of the toss that produced the first tail is of interest. We may obtain an infinite sequence of heads until the first tail is obtained. Thus, if a tail is obtained, we specify the outcome by recording the number of the toss that produced the first tail. The set of the possible outcomes is: {1, 2, 3, 4, 5, ...}.

The totality of outcomes associated with a real or conceptual experiment is called the sample space (denoted by S). Thus, S is a set of outcomes. An outcome is sometimes referred to as an element in the sample space.

In the Example 1.1 above, for example, the sample space can be written as

$$S = \{(H, H), (H, T), (T, H), (T, T)\}.$$

Suppose we are interested in the event (denoted by E) of heads occurring of the first toss, again in the Example 1.1. This event consists of two elements and can be written as

$$E = \{(H, H), (H, T)\}.$$

In the light of this discussion, we can define an *event* is as a set of outcomes. An event E is said to have occurred if the outcome of the experiment corresponds to an element of subset E.

A *random event* A can be characterized by the probability of the event occurring. A *random variable* is any quantity with real values that depends in a well-defined way on some process whose outcomes are uncertain. An *indicator random variable*, for example, is a random variable whose only values with nonzero probability are 0 and 1. In other words, it indicates the occurrence of a specific event by assuming the value 1 when the desired event happens and the value 0 otherwise.

The probability $P(A)$ is the likelihood or chance that A is either the case or will happen in the future. It is represented by a real number ranging from 0 to 1. $P(A)$ generally refers to a period of time T as follows:

$$P(A) = \frac{n_A}{n}$$

where n_A is the number of occurrences (chances) of event A in a period of time T and n is the number of occurrences (chances) in T. In other words, event A is a set of outcomes (a subset) to which a probability $P(A)$ is assigned. The following equations represent two main properties of random events:

$$P(A) + P(\overline{A}) = 1$$

$$P(\emptyset) = 0$$

where $\overline{A}$ is the negation of event A and $\emptyset$ is an event without outcomes i.e. a set without elements. In particular, the failure event is a random occurrence characterized by a probability function that measures the chance of the event occurring in accordance with a specific set of operating conditions.

Example 1.4 A fair coin is tossed three times.

(i)   What is the sample space?
(ii)   Find the probability of event E_1 that at least one head occurs?
(iii) Find the probability of event E_2 that at least two heads occur?

Solution

(i)   The sample space is given by

$$S = \{(H, H, H), (H, H, T), (H, T, H), (T, H, H), (T, T, H),$$
$$(T, H, T), (H, T, T), (T, T, T)\}.$$

(ii)   The probability of event E_1 that at least one head occurs is

$$E_1 = \{(H, H, H), (H, H, T), (H, T, H), (T, H, H),$$
$$(T, T, H), (T, H, T), (H, T, T)\}.$$

We assign probability 1/8 to each element in the sample space. Thus,

$$P\{E_1\} = 1/8 + 1/8 + 1/8 + 1/8 + 1/8 + 1/8 + 1/8 = 7/8$$

(iii) Similarly, the probability of event E_2 that at least two heads occur is

$$E_2 = \{(H, H, H), (H, H, T), (H, T, H), (T, H, H)\}$$

Thus,

$$P\{E_2\} = 1/8 + 1/8 + 1/8 + 1/8 = 1/2$$

Let A and B be events. The *conditional probability* is defined as the probability of A conditioned on B, or A given B, denoted $P(A|B)$:

$$P(A|B) = \frac{P(A \cap B)}{P(B)}$$

where $A \cap B$ is the *intersection* of events A and B. That is,

$$P(A \cap B) = P(A|B) \, P(B)$$

Example 1.5 A fair die is thrown and the result is known to be an even number. What is the probability that this number is divisible by 1.5?

Solution The sample space is given by $S = \{1, 2, 3, 4, 5, 6\}$. The event even numbers, denoted by B, is the subset $\{2, 4, 6\}$. The event divisible by 1.5, denoted by A, is the subset $\{3, 6\}$. Thus,

$$P(A|B) = \frac{P(A \cap B)}{P(B)} = \frac{1/6}{3/6} = \frac{1}{3}$$

We now discuss briefly the notation of independent events. It means there is no relationship between the events.

The two events A and B are said to be *statistically independent* if

$$P(A \cap B) = P(A)\, P(B).$$

If two events A and B are *statistically independent* and $P(B) > 0$ then

$$P(A|B) = P(A) \tag{1.1}$$

Intuitively, two events are independent if the knowledge of one event already happened does not influence the probability of the other happening. In general, if n events $A_1, A_2, \ldots A_n$, are mutually, statistically, independent, then

$$P\left(\bigcap_{i=1}^{n} A_i\right) = \prod_{i=1}^{n} P(A_i) \tag{1.2}$$

Example 1.6 A fair die is thrown twice. Let A, B and C denote the following events:

A   first toss is odd
B   second toss is even
C   total number of spots is equal to seven.

(i)   What is the sample space?
(ii)   Computer P(A), P(B), P(C).
(iii)   Show that A, B and C are independent in pairs.
(iv)   Show that A, B and C are not independent.

Solution

(i)   The sample space contains a total of 36 events such as $\{(1, 1), (1, 2), \ldots (6, 5), (6, 6)\}$.
(ii) $P(A) = 1/2; \quad P\{B\} = 1/2;$
       $P(C) = P\{(1, 6), (6, 1), (2, 5), (5, 2), (3, 4), (4, 3)\} = 1/6$
(iii)   Here
       $\{A \cap B\} = \{(1, 2), (1, 4), (1, 6), (3, 2), (3, 4), (3, 6), (5, 2), (5, 4), (5, 6)\}$
       $\{A \cap C\} = \{(1, 6), (3, 4), (5, 2)\}$
       $\{B \cap C\} = \{(5, 2), (3, 4), (1, 6)\}.$
       Thus,

$$P(A \cap B) = \frac{1}{4} = P(A)\, P(B)$$

$$P(A \cap C) = \frac{1}{12} = P(A)\, P(C)$$

$$P(B \cap C) = \frac{1}{12} = P(B)\, P(C).$$

We conclude that events A and B are independent; A and C are independent; and B and C are independent.

(iv)   See Problem 2.

Example 1.7 Assume that the product failures from an assembly line are 5%. Suppose that if a product that is defective, a diagnostic test indicates that it is defective 92% of the time, and if the product is good, the test indicates that it is good 96% of the time. Calculate the probability that the product is good given that the test indicates defective?

Solution Let

   A = product is good (not defective)
   A_c = product is defective
   B = a test indicates that it is good (not defective)
   B_c = a test indicates it is defective
   Then

$$P(A) = 0.95, \ P(A_c) = 0.05; \ P(B|A) = 0.96, \ P(B_c|A_c) = 0.92$$

The probability that the product is good given that the test indicates defective is:

$$
\begin{aligned}
P(A|B_c) &= \frac{P(A \cap B_c)}{P(B_c)} \\
&= \frac{P(B_c|A)P(A)}{P(B_c|A)P(A) + P(B_c|A_c)P(A_c)} \\
&= \frac{0.04 * 0.95}{(0.04 * 0.95) + (0.92 * 0.05)} = 0.4524
\end{aligned}
$$

1.2 Probability Axioms

It is commonly that we assign low probability to events that we do not expect to happen. Thus, probability is a property assigned to events in an intuitive experimental process. For example, if an event E is an impossibility, say E = A then P(A) = 0. Similarly, if an event E is a certainty, say E = D, then P(D) = 1.

   Let C be a event of the sample space ($C \subset \Omega$). A probability of event function, denoted P(C), has the following properties:

$$\text{(i)} \quad P(\Omega) = 1, \ P(C) \geq 0$$

$$\text{(ii)} \quad P\left\{ \bigcup_{i=1}^{n} C_i \right\} = \sum_{i=1}^{n} P(C_i).$$

where the events C_i have no elements in common (i.e., they are multually disjoint events). If the events A and B are not mutually exclusive, then

$$P(A \cup B) = P(A) + P(B) - P(A \cap B).$$

If $A \subset B$ then $P(A) \leq P(B)$.
Two events are *mutually (or statistically) exclusive* in the case of:

$$P(A \cap B) = 0$$
$$A \cap B = \emptyset$$

Given the events A and B we can define two new sets as follows:
The intersection of A and B is the set of all elements which belong to both A and to B—it is denoted $A \cap B$. Similarly, the union of A and B is the set of all elements which belong to A or to B or to both A and to B—it is denoted $A \cup B$.
The probability of the *union of events* A and B is

$$P(A \cup B) = P(A) + P(B) - P(A \cap B) \tag{1.3}$$

where $A \cup B$ is the *union of events* A and B.
Now considering three independent events A, B, and C:

$$P(A \cup B \cup C) = P(A) + P(B) + P(C) - P(A) \cdot P(B) - P(A) \cdot P(C)$$
$$- P(B) \cdot P(C) + P(A) \cdot P(B) \cdot P(C) \tag{1.4}$$

In the case where the events are mutually exclusive:

$$P(\bigcup_i A_i) = \sum_i P(A_i) \tag{1.5}$$

where A_i is a generic random event.
By pairwise union or intersection, we can extend the definition to a finite number of events $A_1, A_2, ..., A_n$. That is,

$$\bigcup_{i=1}^{n} A_i = A_1 \cup A_2 \cup \cdots \cup A_n$$

and

$$\bigcap_{i=1}^{n} A_i = A_1 \cap A_2 \cap \cdots \cap A_n$$

The operations of union and intersection are given by:

$$A \cup B = B \cup A \qquad \text{Commutative}$$
$$A \cap B = B \cap A \qquad \text{Commutative}$$
$$(A \cup B) \cup C = A \cup (B \cup C) \qquad \text{Associative}$$
$$(A \cap B) \cap C = A \cap (B \cap C) \qquad \text{Associative}$$
$$A \cap (B \cup C) = (A \cap B) \cup (A \cap C) \qquad \text{Distributive}$$
$$A \cup (B \cap C) = (A \cup B) \cap (A \cup C) \qquad \text{Distributive}$$

We also present the following properties, known as DeMorgan's rule:

$$(A \cup B)^c = A^c \cap B^c$$

$$(A \cap B)^c = A^c \cup B^c$$

A finite union $\bigcup_{i=1}^{n} A_i$ is an event whose elements belong to at least one of the n events. Similarly, a finite intersection $\bigcap_{i=1}^{n} A_i$ is an event whose elements belong to all the n events.

If events A_1, A_2, ..., A_n is a partition of S, then for any event C, we obtain

$$P(C) = \sum_{i=1}^{n} P(C \cap A_i) \qquad (1.6)$$

The *Venn diagram* is a useful visual representation of events. In such a diagram events are represented as regions in the plane and elements which belong to a given event are placed inside the region representing it. Frequently all the events in the diagram are placed inside a box which represents the universal set. If an element belongs to more than one event in the diagram, the two regions representing the events concerned must overlap and the element is placed in the overlapping region. In this way the picture represents the relationship between the events concerned. For example, if $A \subseteq B$ the region representing A would be enclosed inside the region representing B to ensure that every element in the region representing A is also inside that representing B (see Fig. 1.1). The problem with this Venn diagram approach is that although it aids the intuition with three or fewer events, it use is very limited for four or more events because we have to draw the events in all the combination positions as is possible that include a relationship to one another

Fig. 1.1 Venn diagram for A in B

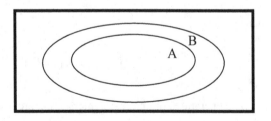

Fig. 1.2 Venn diagram for
A, B and C

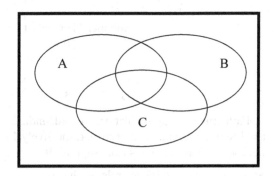

as regards intersections. This in fact is not a simple task to do with four events and
much more difficult with five. The Venn diagram in Fig. 1.2 represents the relationship
between three events A, B and C. It can be seen that the region representing $A \cap (B \cup C)$
can be obtained by the union of three elementary regions.

Let C_1 and C_2 be two events of the sample space Ω. The conditional probability
of getting an outcome in C_2 given that an outcome from C_1 is given by

$$P(C_2|C_1) = \frac{P(C_2 \cap C_1)}{P(C_1)} \tag{1.7}$$

Theorem 1.1 (Bayes' Rule) *Let $C_1, C_2, ..., C_n$ be n mutually disjoint subsets of the
sample space $\mathbb{Z}$. Let C be a subset of the union of the $C_i s$, that is*

$$C \subset \bigcup_{i=1}^{n} C_i$$

Then

$$P(C) = \sum_{i=1}^{n} P(C|C_i)P(C_i) \tag{1.8}$$

and

$$P(C_i|C) = \frac{P(C|C_i)P(C_i)}{\sum_{i=1}^{n} P(C|C_i)P(C_i)}$$

This is also known as Bayes' *rule. Equation (1.8) is known as the* Law of Total
Probability.

Example 1.8 Suppose the total inventory of a company is a collection of lots from
four different suppliers, A, B, C, and D, as indicated below:

Supplier	Percent of inventory
A	60
B	20
C	5
D	15

Furthermore, suppose that past records indicate that lots from suppliers A, B, C, and D are 5, 3, 2, and 8% defective, respectively. Find the probability that a defective item selected as random is from supplier B.

Solution Using the Bayes' rule, we have

$$P(B|defective) = \frac{(0.2)(0.03)}{(0.6)(0.05) + (0.2)(0.03) + (0.05)(0.02) + (0.15)(0.08)}$$
$$= 0.122$$

Combinatorial Methods

In real life many interesting experiments yield sample spaces with large numbers of possible outcomes. Thus, the computational procedures often are not simple, if not difficult, to obtain. To overcome such difficulty, we need to derive some combinatorial methods to provide a convenient way of facilitating the required counting. We now discuss a few of these common techniques in this subsection.

Ordered samples. Samples can be drawn from a finite population according to two procedures. First, sampling with *replacement* means the same element can be selected more than once. Second, sampling *without replacement* means an element once selected is removed from the population.

Permutation. The number of permutations of n elements chosen from a set of N elements, denotes as P(N, n), is

$$P(N, n) = \frac{N(N-1)(N-2)...(N-n+1)(N-n)!}{(N-n)!}$$
$$= \frac{N!}{(N-n)!} \quad \text{for } n \leq N \tag{1.9}$$

Example 1.9 Suppose there are 6 objects and we consider permutations of 2 objects at a time. Determine the total number of permutations?

Solution The total number of permutation of 2 objects from among 6 is equal to:

$$P(6, 2) = \frac{6!}{(6-2)!} = 30.$$

Example 1.10 A product undergoes six operations that can be performed if any sequence. In how many different ways can it be produced?

Solution The number of different ways is $N! = 6! = 720$.

Example 1.11 How many ways are there to rank n candidates for the job of corporate chief executive? If the ranking is made at random (each ranking is equally likely), what is the probability that the fifth candidate, Mr. Best, is in second place?

Solution One can easily show that the total number of rankings is n! and that Mr. Best has probability 1/n of being second. (see Problem 29.)

Combinations. Suppose we have N distinct objects and we must select n of them. In how many ways can this be done? We also assume that the order in which the choices are made is irrelevant. We call a group of n elements chosen at random from N elements (without regard to order) a *combination* of n elements from N and denote it by $C(N, n)$, that is

$$C(N, n) = \frac{P(N, n)}{n!} = \frac{N!}{n!(N - n)!} = \binom{N}{n} \tag{1.10}$$

Example 1.12 An urn contains one white, one black, one red, and one green sphere, all of equal size. Three spheres are drawn at random. In how many ways can this be done if the order in which the choice are made: (i) is relevant; (ii) is irrelevant?

In other words, we want to find the number of subsets with exactly three elements from a set of four elements, provided that the order in which the choice are made: (i) is relevant; (ii) is irrelevant?

Solution

(i)   The number of permutations of three objects from a set of four objects (order is relevant) is

$$P(4, 3) = \frac{4!}{(4 - 3)!} = 24$$

(ii) The number of combinations of three objects from a set of four objects (order is irrelevant) is

$$C(4, 3) = \frac{4!}{3!(4 - 3)!} = 4$$

Example 1.13 Suppose that 3 people are chosen randomly from among 10 to form a committee. In how many ways can such a committee be formed?

Solution In this case, the order in which the 3 people are drawn is not of importance. The solution is to obtain the number of combinations of 3 people out of 10 people, and is given by

$$C(10, 3) = \frac{10!}{3!(10 - 3)!} = 120$$

In general, the number of ways in which N distinct objects can be divided into n_i of type i (objects) such that $\sum_{i=1}^{k} n_i = N$ is given by

$$\frac{N!}{n_1! n_2! \ldots n_k!}$$

This is also called the *multinomial coefficients*.

Theorem 1.2 *If there are N distinct objects, n_i of type i where $\sum_{i=1}^{k} n_i = N$ then the number of arrangements of these objects, denoted $C(N; n_1, n_2, \ldots, n_k)$ is*

$$C(N; n_1, n_2, \ldots, n_k) = C(N, n_1) \, C(N - n_1, n_2) \ldots C(N - n_1 - \ldots - n_{k-1}, n_k)$$

$$= \frac{N!}{n_1! n_2! \ldots n_k!}$$

Example 1.14 An urn contains 20 balls of equal size: 5 red, 7 green, 2 black, and 6 white. Four balls are drawn at random (without replacement). What is the probability that all of them are of different colors?

Solution The probability that all of them are of different colors is

$$\frac{\binom{5}{1}\binom{7}{1}\binom{2}{1}\binom{6}{1}}{\binom{20}{4}} = \frac{28}{323}$$

Example 1.15 How many ways are there to form a sequence of 10 letters from 4 a's, 4 b's, 4 c's and 4 d's if each letter must appear at least twice?

Solution There are two groups of letter frequencies that sum to 10 with each letter appearing two or more times. The first group is four appearances of one letter and two appearances of each remaining letter. The second is three appearances of two letters and two appearances of the other two letters.

It is easy to show that the total number of ways to form a sequence of 10 letters from 4 a's, 4 b's, 4 c's and 4 d's if each letter must appear at least twice is 226,800. (See Problem 31.)

Fault Tree Analysis

Fault tree analysis (FTA) is a popular analytical technique for evaluating the failure probability of a system. The fault tree technique is a graphical method that represents logical relationships between events that lead to system failure. They provide a systematic mathematical framework for analyzing potential causes of failure and it roots cause analysis. From a design perspective, they allow the designer to understand the ways in which a system may fail. Fault trees comprise basic events connected by gates (such as AND gate, OR gate, priority gate etc.) in a logical path, to a top node that represents system or subsystem failure. One approach used by this standard, adopted by many industries, is largely quantitative, where FTA models an entire product, process or system, and the events and failures have a probability of an occurrence determined by analysis or test. The final result is the probability of occurrence of a top event representing probability of failure of the system. One of the advantages of FTA is that we focus on a specific hazardous state and identify each of the preconditions that need to be satisfied in order to reach such a state.

1.3 Basic Statistics

Statistics is a recognized branch of the mathematical sciences, especially tools in the theory of probability. Whenever data are collected with the purpose of studying a phenomenon or understanding the data analysis, techniques are needed to properly how to properly collect, analyze, estimate the parameters, and understand and present the data. All of such aspects are together included in the definition of statistics. The utimate goal of statistics is to gain understanding from data and make statistical inferences about population from an analysis of information contained in sample data. Any data analysis should contain following steps:

- Formulate the problem
- Define population and sample
- Collect the data
- Do descriptive data analysis
- Use appropriate statistical methods to solve the proposed problem, and
- Report the results.

Statistics is concerned with to: (1) collect data; (2) analyze data; (3) estimate the parameters from the data; and (4) understand and present data. Such aspects are often known as data collection, statistical analysis, statistical inference, and statistical findings, respectively. In general, *statistics* is the methodology for collecting, analyzing, interpreting and drawing conclusions from information. In other words, statistics is the methodology which researchers and statisticians have developed for interpreting and drawing conclusions from collected data. We will briefly discuss these aspects in this chapter, except statistical inference will be discussed in Chap. 3.

A characteristic that varies from one person or thing to another is called a variable. In other words, a variable is any characteristic that varies from one individual member of the population to another. Examples of variables for human are weight, height, sex, marital status, and eye color. The first two of these variables yield numerical information and are examples of quantitative variables; the last three yield non-numerical information and are examples of qualitative variables.

The cumulative distribution function (cdf) F is a unique function which gives the probability that a random variable X takes on values less than or equal to some value x. In other word, $F(x) = P(X \leq x)$.

The probability density function (pdf) f is the probability that X takes on the value x, that is, $f(x) = P(X = x)$.

A function f(x) is a pdf of a continuous random variable X if and only if

a. $f(x) \geq 0$ for all x
b. $\int_{-\infty}^{\infty} f(x)dx = 1$

In words, probability that X will be in area may be obtained by integrating the probability density function over the area.

If random variable is discrete, f(x) should be less than 1 because $\sum_{all\,x} f(x) = 1$. But, if random variable is continuous, the necessary condition is $\int_{-\infty}^{\infty} f(x)dx = 1$.

Example 1.16 A boy buys a local newspaper for \$0.10 and sells it for \$0.25. However, unsold newspapers cannot be returned. The boy will have to decide how many papers, denoted quantity M, that needs to order for the next day. Suppose the demand for the newspaper is a random variable X. Then the random variable Y that denotes his daily profit if the boy orders M number of papers is

$$Y_X = \begin{cases} \$0.15M & X \geq M \\ \$0.15X - \$0.10(M - X) & X < M \end{cases}$$

for $X = 0, 1, 2, 3\ldots$

In the continuous case, the pdf is the derivative of the cdf:

$$f(x) = \frac{\partial F(x)}{\partial x}$$

Example 1.17 A sample of 3 units is drawn without replacement from a batch of 15 units, of which two are bad. Let X denote the number of bad units in the sample.

(i)   Determine the probability function of X.
(ii) Determine the distribution function.

Solution

(i) $P(X = 0) = \dfrac{13}{15} \cdot \dfrac{12}{14} \cdot \dfrac{11}{13} = \dfrac{22}{35}$

Let g and b denote good and bad units, respectively. We obtain

$$P(X = 1) = P\{g, g, b\} + P\{g, b, g\} + P\{b, g, g\}$$
$$= \frac{13}{15} \cdot \frac{12}{14} \cdot \frac{2}{13} + \frac{13}{15} \cdot \frac{2}{14} \cdot \frac{12}{13} + \frac{2}{15} \cdot \frac{13}{14} \cdot \frac{12}{13}$$
$$= \frac{12}{35}$$

Similarly,

$$P(X = 2) = P\{g, b, b\} + P\{b, g, b\} + P\{b, b, g\}$$

$$= \frac{13}{15} \cdot \frac{2}{14} \cdot \frac{1}{13} + \frac{2}{15} \cdot \frac{13}{14} \cdot \frac{1}{13} + \frac{2}{15} \cdot \frac{1}{14} \cdot \frac{13}{13}$$

$$= \frac{1}{35}$$

Thus, the probability function of X is

$$P(X = x) = \begin{cases} \frac{22}{35} & x = 0 \\ \frac{12}{35} & x = 1 \\ \frac{1}{35} & x = 2 \end{cases}$$

(ii)   The probability function of X is

$$F(x) = \begin{cases} 0 & x < 0 \\ \frac{22}{35} & 0 \le x < 1 \\ \frac{34}{35} & 1 \le x < 2 \\ 1 & x \ge 2 \end{cases}$$

Example 1.18 A vending machine usually provides candy on insertion of a coin. Over a period of time it is found not to work on five 'uses' in every thousand. Suppose we use X as an indicator variable so that $x = 1$ if the machine works and $x = 0$ if no candy is obtained. Thus we have

Prob{machine works} $= 0.995$
Prob{machine does not work} $= 0.005$

or, the probability function of X can be written as

$$P(X = x) = \begin{cases} 0.995 & x = 1 \\ 0.005 & x = 0 \end{cases}$$

It is easy to see that if X is a continuous random variable with pdf f and $y = g(x)$ is a continous monotone function of x then the random variable Y has a density function h as follows:

$$h(y) = f(x) \left| \frac{dx}{dy} \right|$$

where $x = g^{-1}(y)$.

Common Measures in Statistics

Measures of central tendency and variability play a fundamental role in characterizing data. The statistics most commonly used to represent the properties of a distribution or at least to analyze the data representation fall into some common measures: the mean, standard deviation, variance, skewness, kurtosis, median, mode, and range. Measures such as the mean and the median are frequently used as measures of location or measures of central tendency because they are central or middle values. The variance and standard deviation are some popular measures of dispersion, scatter or variability. These measures are described as follows.

The mean

Let $x_1, x_2, \ldots, x_n$ be a finite set of values of the variable X. The mean or average of this set of values is equal to the total of the values divided by the number of values.

The mean or average of a finite set of n values $x_1, x_2, \ldots, x_n$ of a variable X is equal to

$$\bar{x} = \frac{\sum_{i=1}^{n} x_i}{n}$$

If X is a random variable with probability density function f, then the expected value of a random variable X, or the **mean**, is given by

$$E(X) = \begin{cases} \sum_{all\ x} x\, f(x) & \text{for X discrete} \\ \int_{-\infty}^{\infty} xf(x)dx & \text{for X continuous.} \end{cases} \tag{1.11}$$

If a new random variable Y is a function of X expressed as $Y = g(X)$, then its expectation is given by

$$E(g(X)) = \begin{cases} \sum_{all\ x} g(x)\, f(x) & \text{if g is discrete} \\ \int_{-\infty}^{\infty} g(x)f(x)dx & \text{if g is continuous.} \end{cases} \tag{1.12}$$

If a sample of n observations drawn randomly from a population, then the sample mean

$$\bar{x} = \frac{\sum_{i=1}^{n} x_i}{n} \tag{1.13}$$

is the best available estimate of the population mean μ. As n increases, $\bar{x}$ becomes a more and more precise estimate of the population mean μ.

Property 1.1 *If two random variables X_1 and X_2 are* independent, *then*

$$E(X_1 X_2) = E(X_1)E(X_2).$$

The converse of this property is not true. In other words, if $E(X_1 X_2) = E(X_1)E(X_2)$ we cannot conclude that X_1 and X_2 are independent.

Two other types of means play a role in some certain situations. They are the geometric mean *and the* harmonic mean.

Definition 1.4 The **geometric mean** of a finite set of n positive values $x_1, x_2, \ldots, x_n$ of a variable X is equal to

$$\overline{x}_{gm} = \sqrt[n]{x_1 x_2 \ldots x_n}.$$

Definition 1.5 The **harmonic mean** of a finite set of n positive values $x_1, x_2, \ldots, x_n$ of a variable X is equal to

$$\overline{x}_{hm} = \frac{n}{\sum_{i=1}^{n} \frac{1}{x_i}}.$$

The Variance

Definition 1.6 The variance of a random variable X, denoted as σ^2, with mean μ is a measure of how the values of X are spread about the mean value and is given by

$$\sigma^2 = E(X - \mu)^2 \tag{1.14}$$

It is calculated for discrete and continuous random variables, respectively, by

$$\sigma^2 = \begin{cases} \sum_{\text{all } x} (x - \mu)^2 f(x) & \text{for discrete variate} \\ \int_{-\infty}^{\infty} (x - \mu)^2 f(x) dx & \text{for continuous variate.} \end{cases} \tag{1.15}$$

If a new random variable Y is a function of X expressed as $Y = g(\mathbf{X})$, then its variance is given by

$$\sigma^2 = \begin{cases} \sum_{\text{all } x} (g(x) - E(g(x))^2 f(x) & \text{for discrete variate} \\ \int_{-\infty}^{\infty} (g(x) - E(g(x))^2 f(x) dx & \text{for continuous variate.} \end{cases} \tag{1.16}$$

The variance of the population is the mean squared deviation of the individual values from the population mean. The **standard deviation** of X, denoted by σ, is the square root of the **variance**. Of course, if the values of x are all identical, there are no differences and the estimate of the variance is zero. Likewise, if they differ only slightly from each other, the variance will be small. Obviously, the variance of any random variable is always positive, whereas the mean may be positive or negative.

Property 1.2

- *If k is an arbitrary constant, then:* $\text{Var}(kX) = k^2 \text{Var}(X)$
- *If X is a random variable, then:* $\text{Var}(X) = E(X^2) - [E(X)]^2$
- *If X and Y are independent random variables, then:*

$$\text{Var}(X + Y) = \text{Var}(X) + \text{Var}(Y)$$

- *If X and Y are two random variables, then the covariance of X and Y is*

$$
Cov(X, Y) = \begin{cases} E\{[X - E(X)][Y - E(Y)]\} \\ E(XY) - E(X)E(Y) \\ \sum_{\text{all x,y}} (x - E(x))(y - E(Y))f(x, y) \quad \text{for discrete variate} \\ \int_{-\infty}^{\infty} \int_{-\infty}^{\infty} (x - E(X)(y - E(Y))f(x, y)dxdy \text{ for continuous variate.} \end{cases}
$$

- *If X and Y are two random variables, then:*

$$
\text{Var}(X + Y) = \text{Var}(X) + \text{Var}(Y) + 2 E\{[X - E(X)][Y - E(Y)]\}
$$

If a sample of n drawn from a population with mean μ then the **sample variance** of the population is estimated by

$$
S^2 = \frac{\sum_{i=1}^{n} (x_i - \mu)^2}{n} \tag{1.17}
$$

In general μ is not known and an estimate sample mean based on the sample must be obtained. In this case, the sample variance is given by

$$
S^2 = \frac{\sum_{i=1}^{n} (x_i - \overline{x})^2}{n - 1} = \frac{\sum_{i=1}^{n} x_i^2 - \frac{\left(\sum_{i=1}^{n} x_i\right)^2}{n}}{n - 1} \tag{1.18}
$$

This gives the best estimate of the population variance from the data available. Similarly, the sample standard deviation is given by

$$
S = \sqrt{\frac{\sum_{i=1}^{n} (x_i - \overline{x})^2}{n - 1}}.
$$

The more variation there is in the observed values, the larger is the standard deviation for the variable in question. Thus the standard deviation satisfies the basic criterion for a measure of variation. This is the most common used measure of variation. However, the standard deviation does have its drawbacks. For instance, its values can be strongly affected by a few extreme observations.

Suppose a number of samples are available and it is required to estimate the variance from all samples assuming that the sample variance differ only because of sampling fluctuations, but allowing for the possibility that the sample means may be different, i.e., the samples are assumed to be drawn from population with the same variance but different means. Let n_i is the number of observations of the ith sample for $i = 1, 2, \ldots, k$. The sample variance can be obtained as follows:

$$S^2 = \frac{\sum_{j=1}^{k} \sum_{i=1}^{n_j} \left(x_{ij} - \bar{x}_j\right)^2}{N - k} \quad \text{where} \quad N = \sum_{j=1}^{k} n_j \qquad (1.19)$$

The **coefficient of variation(CV)** is the standard deviation expressed as a percentage of the mean and the expression is given by

$$CV = \frac{\sigma}{\mu} 100\%$$

The coefficient of variance for a collection of data is equal to

$$CV = \frac{S}{\bar{X}} 100\%$$

or equivalently, that

$$CV = \frac{\sqrt{\frac{\sum_{i=1}^{n} (x_i - \bar{x})^2}{n-1}}}{\frac{\sum_{i=1}^{n} x_i}{n}} \cdot 100\%$$

This coefficient of variance measure can be used to relate the spread of observations resulting from two alternative ways of measuring the same response. The main use of this measure is to compare the variability of groups of observations with widely differing mean levels.

The **range** of a set of data in a sample is the simplest of all measures of dispersion. It is simply the difference between the highest and lowest values of the data in a sample. The sample range of the variable is easy to compute. However, in using the range, a great deal of information is ignored, that is, only the largest and smallest values of the variable are considered; the other observed values are disregarded.

Example 1.19 8 participants in bike race had the following finishing times in minutes: 12, 10, 24, 31, 14, 21, 23, 15. The sample range of participants (10, 12, 14, 15, 21, 23, 24, 31) in bike race is 21 min.

Note that the range of a small sample conveys a large proportion of the information on the variation in a sample. However, the range of a large sample perhaps conveys little information since it says very little about the intermediate observations. The range perhaps can be used to calculate an approximate estimate of the standard deviation for a single set of n observations. A simple estimate of the standard deviation can be obtained for samples of equal size:

$$S = \frac{\bar{w}}{d_n} \qquad (1.20)$$

where $\overline{w}$ is the mean range and d_n is a constant depending on the size of the sample. The factors d_n are given in Table A.5 (Range values) in the Appendix A. It is interesting to note that for $n = 3\text{-}12$, d_n is very nearby equal to $\sqrt{n}$ (within less than 5%). Note that the Range method should not be used for samples of more than 12.

Example 1.20 Let mean range $\overline{w} = 25.7$ and $n = 4$. Obtain an estimate of the standard deviation σ using the mean range.

From the Table A.5, $d_n = 2.059$, the estimate of σ is

$$S = \frac{25.7}{2.059} = 12.4818$$

The **skewness coefficient** of a random variable X is a measure of the symmetry of the distribution of X about its mean value μ, and is defined as

$$S_c = \frac{E(X - \mu)^3}{\sigma^3} \tag{1.21}$$

A distribution will not in general be completely symmetrical; the frequency may fall away more rapidly on one side of the mode than on the other. When this is the case the distribution is said to be skew. Skewness is zero for a symmetric distribution, negative for a left-tailed distribution, and positive for a right-tailed distribution. In other words, skewness means lack of symmetry and measures of skewness show the extent to which the distribution departs from symmetry.

Similarly, the **kurtosis** coefficient of a random variable X is a measure of how much of the mass of the distribution is contained in the tails, and is defined as

$$K_c = \frac{E(X - \mu)^4}{\sigma^4} \tag{1.22}$$

Obviously, **kurtosis** is always positive, however, larger values represent heavier tails of the distribution. The measure of kurtosis serves to differentiate between a flat distribution curve and a sharply peaked curve.

If the data are arranged in order of magnitude, the **median** is the central point of the series, i.e., there are equal numbers of observations greater than and less than the medians. In case if n (the total number of observations) is even, it is usual to take the mean of the two central values as the median.

Definition 1.7 Suppose that n values of a variable X are ordered in increasing order of magnitude, i.e., we have $x_1, x_2, \ldots, x_n$ such that $x_{(1)} \leq x_{(2)} \leq \cdots \leq x_{(n)}$ then the median of a set of values is equal to $x_{\left(\frac{n+1}{2}\right)}$ if n is odd and $\frac{1}{2}\left(x_{\left(\frac{n}{2}\right)} + x_{\left(\frac{n}{2}\right)+1}\right)$ if n is even.

In other words, arranging the observed values of variable in a data in increasing order. If the number of observation is odd, then the sample median is the observed value exactly in the middle of the ordered list. If the number of observation is even, then the sample median is the number halfway between the two middle observed values in the ordered list. In both cases, if we let n denote the number of observations in a data set, then the sample median is at position $\frac{n+1}{2}$ in the ordered list.

Definition 1.8 The **median** of a random variable X is the value m such that the cdf

$$F(m) = \frac{1}{2}$$

The median always exists if the random variable is continuous whereas it may not always exist if the random variable is discrete.

Example 1.21 8 participants in bike race had the following finishing times in minutes: 12, 10, 24, 31, 14, 21, 23, 15. The median of participants (10, 12, 14, 15, 21, 23, 24, 31) in bike race is 18 min.

Example 1.22 The probability density function of a random variable X is

$$f(x) = \begin{cases} 2x & 0 \le x \le 1 \\ 0 & \text{otherwise.} \end{cases}$$

Obtain the median of X.

Solution The cdf of X is given by

$$F(x) = \begin{cases} 0 & x < 0 \\ x^2 & 0 \le x \le 1 \\ 1 & x > 1. \end{cases}$$

Set $F(m) = 1/2$, then the median of X is $\frac{\sqrt{2}}{2}$.

Definition 1.9 The **mode** is the value of the variate which occurs most frequently for which the frequency is a maximum.

In other words, the mode of a set of values of a variable is the most frequently occurring value in the set. The use of the mode is most often associated with discrete distributions where it is simple to count which value of the variate occurs most often.

Example 1.23 The density function of X is given by

$$f(x) = \begin{cases} \frac{2}{3}x & 0 \le x \le 1 \\ \frac{1}{3} & 1 < x \le 3. \end{cases}$$

Obtain the median of X and the mode of X.
The cdf of X is

$$F(x) = \begin{cases} 0 & x < 0 \\ \frac{x^2}{3} & 0 \le x \le 1 \\ \frac{x}{3} & 1 < x \le 3 \\ 1 & x > 3. \end{cases}$$

Since $F\left(\frac{3}{2}\right) = \frac{1}{2}$, so the median of X is equal to $\frac{3}{2}$. Similary, the mode is equal to 1 since $f(x)$ is a maximum when $x = 1$.

Which measure to choose? The mode should be used when calculating measure of center for the qualitative variable. When the variable is quantitative with symmetric distribution, then the mean is proper measure of center. In a case of quantitative variable with skewed distribution, the median is good choice for the measure of center. This is related to the fact that the mean can be highly influenced by an obsevation that falls far from the rest of the data, called an outlier.

Samples versus Population

A distinction must be drawn between the mean of a set of observed values and the mean of the underlying probability distribution or population. *Population* is the collection of all individuals or items under consideration in a statistical study. *Sample* is that part of the population from which information is collected. Often the assumption is that the population is normally distributed with mean μ and standard deviation σ. Before the characteristics of a set of data are used to estimate the corresonding properties of the population, it is necessary to know how the sample has been obtained. A *random sample* is defined as a set of observations drawn from a population in such a way that every possible possible observation has an equal chance of being drawn at every trial. In other words, the probability that any observation will have a given value is proportional to the relative frequency with which that value occurs in the population. In random values from a normal population, for example, values of x near the mean will occur more frequently than those far from the mean. The properties of individual small samples may deviate more or less widely from those of the population. These variations in the properties of small samples due only to chance causes, are known as random sampling variations. If it can be assumed that sampling is random, the observations may be used to estimate the properties of the population. However, if the sampling is not random then a bias may be introduced and the properties of the population estimated may show systematic deviation from the true properties. A *parameter* is an unknown numerical summary of the population. A *statistic* is a known numerical summary of the sample which can be used to make inference about parameters.

It is worth to note that if a distribution departs far from normality that it would not be safe to apply the common statistical tests then it would be best, by a simple transformation of the variable, to obtain an approximately normal distribution. If such a transformation is not possible, an alternative class of statistical tests, known as distribution-free methods such as the sign test and rank-sum test (Lehmann 1998), may be applicable.

One can use a special plotting paper, known as normal probability paper, which is manufactured with a probability scale transformed so that the cumulative normal distribution curve becomes a straightline. So to test whether data are normally distributed the cumulative proportional frequencies can be plotted on the normal probability paper. If the points fall close to a straight line it can be concluded that the data approximately follow a normal distribution.

As the sample size increases, its properties resemble more and more closely those of the population, provided that the sample values may be considered as randomly selected from the population. The observed mean should be regarded as an estimate of the true value which becomes better as the number of observations increases.

Given a random sample of size n from a distribution, again the sample mean and sample variance are respectively,

$$\bar{x} = \frac{1}{n} \sum_{i=1}^{n} x_i \text{ and } S^2 = \frac{1}{n-1} \sum_{i=1}^{n} (x_i - \bar{x})^2 \qquad (1.23)$$

Let μ be the population mean. $\bar{X}$ becomes a better estimate of μ as n (the number of observations) increases, and it approaches to μ as n approaches to ∞. Similarly, the sample standard deviation S (where S^2 is the sample variance) becomes a better estimate of σ as n increases where σ is the population standard deviation. In fact, the most useful measure of the spread is the standard deviation which for a sample of n observations is given by taking a square root, of the equation, S^2 above.

The **mean absolute deviation** of a set of numbers is defined as the mean of the deviations from the mean each with the absolute value:

$$MD = \frac{\sum_{i=1}^{n} |x_i - \bar{x}|}{n} \qquad (1.24)$$

where $\bar{x}$ is the sample mean. Note that when the original data vary so do the $|x_i - \bar{x}|'s$, and if all the data are the same then every $|x_i - \bar{x}|$ is equal to zero, so that the mean absolute deviation is a measure of variability. In contrast to the range, the mean absolute deviation takes into account the dispersion of the data about the mean.

The mean absolute deviation finds applications where the distribution has long tails. It is less affected by outlying results than is the standard deviation. In fact, the mean absolute deviation is of little practical use, and unless there are specific reasons for using it the standard deviation is preferable.

Sometime a set of data may contain one or more observations which fall outside the pattern exhibited by the majority of the observed values. In this case, should we assume that the **outlier** is just as valid as the other observations and calculate statistics based on all the observations or should we discard the odd value(s) as being unrepresentative? A statistic test should be applied to confirm that a suspected outlier is really as extreme as it appears. One can use the approach as follows:

An **outlier** may be regarded as significant if the outlier ratio value:

$$\frac{|\text{Extreme value} - \text{overall mean}|}{\text{overall standard deviation, s}}$$

exceeds a certain critical level. Table A.6 lists critical levels for various sample sizes when it is assumed that the other $(n - 1)$ observations are drawn form a normal distribution. In other words, the critical values in Table A.6 are based on the assumption that the non-suspect results are normally distributed. Note that these test levels should not be used when the base distribution is far from normal, i.e. if it is clearly skew.

Example 1.24 Data representing the yearly gross income (in 10^5) of six workers are given as follows:

$$0.657, \ 0.664, \ 0.655, \ 0.653, \ 0.686, \ 0.661$$

The 0.686 value is much higher than the other five incomes. We wish to test whether the 0.686 value is as an outlier?

The overall mean is

$$\text{overall mean} = (0.657 + 0.664 + 0.655 + 0.653 + 0.686 + 0.661)/6$$
$$= 0.6627$$

Similarly, the overall standard deviation is: 0.0121. The outlier ratio value is

$$\text{Outlier ratio value} = \frac{|0.686 - 0.6627|}{0.0121} = 1.9256$$

The critical level corresponding to $\alpha = 0.05$ for a sample of size 6 is 1.89 (see Table A.6). Since the outlier ratio value exceeds the critical value, therefore we can conclude that the outlying observation (0.686 value) may be regarded as significant. On this basis, we may compute the mean and standard deviation from the other five observations since the value 0.686 is as an outlier.

The **moment-generating function** (m.g.f.) of a distribution of X is definted as a function of a real variable t:

$$M(t) = E(e^{tX}). \tag{1.25}$$

It should be noted that $M(0) = 1$ for all distributions. $M(t)$ however may not exist for some $t \neq 0$.

Example 1.25 If X has a continuous distribution with pdf

$$f(x) = \lambda e^{-\lambda x} \quad \text{for } x \geq 0 \text{ and } \lambda > 0.$$

Then,

$$M(t) = E(e^{tX}) = \lambda \int_0^\infty e^{tx - \lambda x} dx = \frac{\lambda}{\lambda - t} \quad \text{for } t < \lambda.$$

This m.g.f. exists only for $t < \lambda$.

1.4 Joint Density Functions of Random LifeTimes

Assume there are n random variables X_1, X_2, ..., X_n which may or may not be mutually independent. The joint cdf, if it exists, is given by

$$P(X_1 \le x_1, X_2 \le x_2, \ldots X_n \le x_n) = \int\limits_{-\infty}^{x_n} \int\limits_{-\infty}^{x_{n-1}} .. \int\limits_{-\infty}^{x_1} f(t_1, t_2, \ldots, t_n) dt_1 \, dt_2 \ldots dt_n$$

If the n random variables are mutually statistically independent then the joint pdf can be rewritten as

$$f(x_1, x_2, \ldots, x_n) = \prod_{i=1}^{n} f(x_i).$$

Consider two random variables of the lifetime, say X and Y, in which the joint characteristic of both X and Y can be described by their joint density function f of the two variables x and y. That is,

$$\int\limits_{0}^{\infty} \int\limits_{0}^{\infty} f(x, y) \, dx \, dy = 1 \quad \text{and} \quad f(x, y) \ge 0.$$

Let F be the joint distribution function of X and Y and is defined as:

$$F(x, y) = P[X \le x, Y \le y]$$

$$= \int\limits_{0}^{x} \int\limits_{0}^{y} f(s, t) \, ds \, dt \quad \text{for } x \ge 0, y \ge 0$$

The joint pdf is simply by taking partial differentiations with respect to x and y:

$$f(x, y) = \frac{\partial^2 F(x, y)}{\partial x \, \partial y}.$$

The marginal (or simple) density function $f_X(x)$ and $f_Y(y)$ can be obtained from the $f(x,y)$ by taking integration as follows:

$$f_X(x) = \int\limits_{0}^{\infty} f(x, y) dy \quad \text{and} \quad f_Y(y) = \int\limits_{0}^{\infty} f(x, y) dx$$

and also the marginal cdf

$$F_X(x) = \int_0^x f_X(s)ds \quad \text{and} \quad F_Y(y) = \int_0^y f_Y(t)dt.$$

Similarly, if the random variables X and Y are independent, then the joint density function $f(x, y)$ is simply the product of the marginal densities $f_X(x)$ and $f_Y(y)$. That is

$$f(x, y) = f_X(x) \times f_Y(y) \quad \text{for } x \geq 0 \text{ and } y \geq 0.$$

The independence rule can also be applied to the cdf as given by:

$$F(x, y) = F_X(x) \times F_Y(y) \text{ for all } 0 \leq x \leq \infty \text{ and } 0 \leq y \leq \infty.$$

Convolutions of Density Functions

Let X and Y be two independent lifetimes with pdf $f_1(x)$ and $f_2(y)$, respectively. Let $Z = X + Y$ be the total lifetime having the pdf $g(z)$ and cdf $G(z)$. Using a convolution approach, the pdf of z can be obtain

$$g(z) = \int_0^z f_1(x) f_2(z-x) \, dx = \int_0^z f_1(z-y) f_2(y) \, dy = f_1(z) * f_2(z) \qquad (1.26)$$

where * denotes the convolutions of the distributions. Similarly, the distribution function of Z, $G(z)$, is given by

$$G(z) = P[Z \leq z]$$

$$= \iint_{x+y \leq z} f_1(x) f_2(y) \, dx \, dy$$

$$= \int_0^z f_1(x)dx \int_0^{z-x} f_2(y)dy.$$

Note that the two random variables X and Y are said to be independent if knowledge of one of the variables has no effect on the distribution of probability for the other.

Two random variables X and Y, with cdf $F_1(x)$ and $F_2(y)$ respectively, are said to be identically distributed if they have the same distribution functions, that is, $F_1(t) = F_2(t)$ for all t. Being identically distributed does not mean that $X = Y$. Rather it means that X and Y are equivalent in the sense that one should have no different to choosing between them.

As for both life times X and Y are independent, the expected $E(Z)$ and variance $V(Z)$ of total life time can be easily obtained, respectively,

$$E(Z) = E(X) + E(Y)$$

and

$$V(Z) = V(X) + V(Y).$$

Example 1.26 Suppose that X and Y are two independent life times having the pdf as follows

$$f_1(x) = \lambda e^{-\lambda x} \quad \text{for } x \geq 0$$
$$f_2(y) = \lambda e^{-\lambda y} \quad \text{for } y \geq 0$$

The density g of Z where $Z = X + Y$ is given by

$$g(z) = \int_0^z \lambda e^{-\lambda x} \lambda e^{-\lambda(z-x)} dx = \lambda^2 z \, e^{-\lambda z}$$

Therefore, the expected value of Z is

$$E(Z) = \int_0^\infty z g(z) dz = \int_0^\infty z \lambda^2 z \, e^{-\lambda z} dz = \frac{2}{\lambda}.$$

Example 1.27 Let X be a uniform r.v. over $(0, a)$ with a pdf

$$f_1(x) = \frac{1}{a} \quad \text{for } 0 < x < a$$

and Y be an exponential r.v. with the following pdf

$$f_2(x) = \lambda e^{-\lambda x} \quad \text{for } x \geq 0$$

Obtain the distribution of $(X + Y)$ under the assumption of independence?

Solution It should be noted that although $(X + Y)$ has range $(0, \infty)$ the evalution of the convolution integral using Eq. (1.26) requires separate consideration of the ranges $(0, a]$ and $[a, \infty)$.

Let $Z = X + Y$. In the first case, that is $(0, a]$, the upper limit x is correct. But if $x \geq a$ then upper limit must be replaced by a since the uniform distribution is zero for all value x larger than a. Thus, for $x \geq a$ the density of Z is given by

$$g(x) = \int_0^a \frac{1}{a} \lambda e^{-\lambda(x-t)} dt = \frac{1}{a} e^{-\lambda x} (e^{\lambda a} - 1)$$

Similarly, for $x \leq a$ the density of Z is given by

$$g(x) = \int_0^x \frac{1}{a}\lambda e^{-\lambda(x-t)}dt = \frac{1}{a}(1 - e^{-\lambda x}).$$

Thus, the density of Z is

$$g(x) = \begin{cases} \frac{1}{a}e^{-\lambda x}(e^{\lambda a} - 1) & \text{for } x \geq a \\ \frac{1}{a}(1 - e^{-\lambda x}) & \text{for } x \leq a. \end{cases}$$

Obviously, if one consider the entire range in this case $(0, \infty)$ and from the Eq. (1.26), the answer is not correct. In other word,

$$g(x) = \int_0^x \frac{1}{a}\lambda e^{-\lambda x}dx = \frac{1}{a}(1 - e^{-\lambda x}) \quad \text{for } 0 < x < \infty$$

which is not the correct answer.

Let X_i represents the lifetime of unit ith and f_i be their densities, respectively, for $i = 1, 2, \ldots, n$ and S_n be the total lifetime for n units having a density function g_n. That is

$$S_n = \sum_{i=1}^n X_i \tag{1.27}$$

We can rewrite the function S_n in a recursive form as follows

$$S_n = S_{n-1} + X_n \tag{1.28}$$

where we can easily show that the life times S_n and X_n are independent. Similarly, we can obtain as

$$g_n(z) = \int_0^z g_{n-1}(z - x)f_n(x)dx. \tag{1.29}$$

1.5   Conditional Distributions

If (X, Y) are two random variables having a joint pdf $f(x, y)$ and marginal pdf $f_X(.)$ and $f_Y(.)$, respectively, then the conditional pdf of Y given $\{X = x\}$ is defined by

$$f_{Y,X}(y|x) = \frac{f(x,y)}{f_X(x)} \tag{1.30}$$

where

$$f_X(x) = \int_{-\infty}^{\infty} f(x,y)dy$$

The conditional expectation of Y given $\{X = x\}$ is the expected value of Y with respect to the conditional pdf $f_{Y,X}(y|x)$, that is

$$E(Y|X = x) = \int_{-\infty}^{\infty} y\, f_{X,Y}(y|x)dy \tag{1.31}$$

If X and Y are independent, then

$$f(x,y) = f_X(x)\, f_Y(y)$$

Consider $E(Y|X)$ to be a random variable which is a function of X. It is interesting to compute the expected value of this function of X. That is,

$$E\{E(Y|X = x)\} = \int E(Y|X = x)\, f_X(x)dx$$

$$= \int \left\{ \int y f_{X,Y}(y|x)dy \right\} f_X(x)dx$$

$$= \iint y \frac{f(x,y)}{f_X(x)} f_X(x)\, dy\, dx$$

We can interchange the order of integration and obtain

$$E\{E(Y|X = x)\} = \int y \left\{ \int f(x,y)dx \right\} dy$$

$$= \int y f_Y(y)dy$$

$$= E(Y) \tag{1.32}$$

This result, known as the law of expected conditional mean, is very useful. Consider the following example.

Example 1.28 Let (X, N) be a pair of random variables. The conditional distribution of X given $\{N = n\}$ is the binomial $B(n, p)$. The marginal distribution of N is Poisson with mean λ. Obtain the expected value of X.

Solution $E(X) = E\{E(X|N)\} = E\{Np\} = pE(N) = p\lambda.$

Another interesting result is as follows. If (X, Y) is a pair of random variables having finite variances, then

$$V(X) = E\{V(X|Y)\} + V\{E(X|Y)\}. \tag{1.33}$$

This is known as the law of total variance.

If $X_1, X_2, \ldots, X_n$ be random variables having a joint distribution, with joint pdf $f(x_1, x_2, \ldots, x_n)$. Let $\lambda_1, \lambda_2 \ldots, \lambda_n$ be given constants. Then

$$W = \sum_{i=1}^{n} \alpha_i X_i$$

is a linear combination of the X's. It is easily shown the expected value and variance of W, respectively, that

$$E(W) = \sum_{i=1}^{n} \alpha_i E(X_i)$$

and

$$V(W) = \sum_{i=1}^{n} \alpha_i^2 V(X_i) + \sum \sum_{i \neq j} \alpha_i \alpha_j \mathrm{cov}(X_i, X_j). \tag{1.34}$$

1.6   Laplace Transformation Functions

Let X be a nonnegative life time having probability density function f. The Laplace transform of a function $f(x)$, denote f^*, is defined as

$$\ell\{f(x)\} = f^*(s) = \int_0^\infty e^{-sx} f(x) dx \quad \text{for s} \geq 0. \tag{1.35}$$

The function f^* is called the Laplace transform of the function f. The symbol ℓ in Eq. (1.35) is called the Laplace transform operator. Note that $f^*(0) = 1$. By taking a differential derivative of $f^*(s)$, we obtain

$$\frac{\partial f^*(s)}{\partial s} = -\int_0^\infty xe^{-sx}f(x)dx.$$

Substitute $s = 0$ into the above equation, the first derivative of f^*, it yields the negative of the expected value of X or the first moment of X:

$$\left.\frac{\partial f^*(s)}{\partial s}\right|_{s=0} = -E(X).$$

Similarly, the second derivative yields the second moment of X when $s = 0$, that is,

$$\left.\frac{\partial^2 f^*(s)}{\partial^2 s}\right|_{s=0} = \int_0^\infty x^2 e^{-sx}f(x)dx\Bigg|_{s=0} = E(X^2).$$

In general, it can be shown that

$$\left.\frac{\partial^n f^*(s)}{\partial^n s}\right|_{s=0} = \int_0^\infty (-x)^n e^{-sx}f(x)dx\Bigg|_{s=0} = (-1)^n\, E(X^n)$$

Note that

$$e^{-sx} = \sum_{n=0}^\infty \frac{(-sx)^n}{n!}$$

then $f^*(s)$ can be rewriten as

$$f^*(s) = \sum_{n=0}^\infty \frac{(-s)^n}{n!}\mu_n$$

where

$$\mu_n = (-1)^n \left.\frac{\partial^n f^*(s)}{\partial^n s}\right|_{s=0} = \int_0^\infty x^n e^{-sx}f(x)dx\Bigg|_{s=0} = E(X^n)$$

We can easily show that ℓ is a linear operator, that is

$$\ell\{c_1 f_1(x) + c_2 f_2(x)\} = c_1\ell\{f_1(x)\} + c_2\ell\{f_2(x)\}.$$

If $\ell\{f(t)\} = f^*(s)$, then we call $f(t)$ the inverse Laplace transform of $f^*(s)$ and write $\ell^{-1}\{f^*(s)\} = f(t)$.

A summary of some common Laplace transform functions is listed in Table B.1 in Appendix B.

Example 1.29 Use the Laplace transforms to solve the following

$$\frac{\partial f(t)}{\partial t} + 3f(t) = e^{-t} \tag{1.36}$$

with an initial condition: $f(0) = 0$. Obtain the solution $f(t)$. Here the Laplace transforms of $\frac{\partial f(t)}{\partial t}$, $f(t)$, and e^{-t} are

$$sf^*(s) - f(0), \quad f^*(s), \quad \text{and} \quad \frac{1}{s+1}$$

respectively. Thus, the Laplace transform of Eq. (1.36) is given by

$$sf^*(s) - f(0) + 3f^*(s) = \frac{1}{s+1}$$

Since $f(0) = 0$ we have

$$(s+3)f^*(s) = \frac{1}{s+1} \quad \text{or} \quad f^*(s) = \frac{1}{(s+1)(s+3)}$$

From Table B.1 in Appendix B, the inverse transform is

$$f(t) = \frac{1}{3-1}\left(e^{-t} - e^{-3t}\right) = \frac{1}{2}\left(e^{-t} - e^{-3t}\right)$$

Example 1.30 Let X be an exponential random variable with constant failure rate λ, that is, $f(x) = \lambda e^{-\lambda x}$ then we have

$$f^*(s) = \int_0^\infty \lambda e^{-sx} e^{-\lambda x} dx = \frac{\lambda}{s+\lambda}.$$

If X and Y are two independent random variables that represent life times with densities f_1 and f_2, respectively, then the total life times Z of those two X and Y, says $Z = X + Y$, has a pdf g that can be obtained as follows

$$g(z) = \int_0^z f_1(x) f_2(z-x) dx.$$

The Laplace transform of g in terms of f_1 and f_2 can be written as

$$g^*(s) = \int_0^\infty e^{-sz} g(z)dz = \int_0^\infty \int_0^z e^{-sz} f_1(x) f_2(z-x)dxdz$$

$$= \int_0^\infty e^{-sx} f_1(x)dx \int_x^\infty e^{-s(z-x)} f_2(z-x)dz$$

$$= f_1^*(s) f_2^*(s).$$

Example 1.31 If X and Y are both independent having the following pdfs: $f_1(x) = \lambda e^{-\lambda x}$ and $f_2(y) = \lambda e^{-\lambda y}$ for $x, y \geq 0$ and $\lambda \geq 0$ then we have

$$g^*(s) = f_1^*(s) f_2^*(s) = \left(\frac{\lambda}{s+\lambda}\right)^2.$$

From the Laplace transform Table B.1 in Appendix B, the inverse transform to solve for $g(z)$ is

$$g(z) = \frac{\lambda^2 t e^{-\lambda t}}{\Gamma(2)}$$

which is a special case of gamma pdf.

From Eq. (1.27), one can easily show the Laplace transform of the density function g_n of the total life time S_n of n independent life times X_i with their pdf f_i for $i = 1$, $2, \ldots, n$, that

$$g_n^*(s) = f_1^*(s) f_2^*(s)...f_n^*(s) = \prod_{i=1}^n f_i^*(s) \tag{1.37}$$

(see Problem 13).

If the pdf of n life time $X_1, X_2, \ldots, X_n$ are independent and identically distributed (i.i.d.) having a constant failure rate λ, then

$$g_n^*(s) = \left(f^*(s)\right)^n = \left(\frac{\lambda}{s+\lambda}\right)^n.$$

From the Laplace transform Table B.1 in Appendix B, we obtain the inverse transform for the solution function g as follows

$$g_n(z) = \frac{\lambda^n t^{n-1} e^{-\lambda t}}{\Gamma(n)} \tag{1.38}$$

1.7 Reliability Concepts

1.7.1 Coherent Systems

In reliability theory, both the system and its components are commonly allowed to take only two possible states or binary states: either working or failed. In a multistate system, as will be discussed in Chaps. 4 and 6, both the system and its components are allowed to experience more than two possible states, e.g. fully working, partially or degraded working or failed. A multistate system reliability model provides more flexibility for modeling of system's conditions. The reliability can be defined for a more general coherent structure. A coherent system with binary states can be simply described as follows. Consider that a system consists of n components and define

$$x_i = \begin{cases} 1 & \text{if component i is functioning} \\ 0 & \text{if component i is failed} \end{cases} \quad \text{for } i = 1, 2, ..., n$$

and

$$\phi(\mathbf{x}) = \begin{cases} 1 & \text{if the system is functioning} \\ 0 & \text{if the system is failed} \end{cases}$$

where $x = (x_1, x_2, ..., x_n)$. Binary indicators x_i and $\phi(x)$ indicate the state of component i and the state of the system, respectively. The function $\phi(x)$ is referred to as the structure function of the system. Define

$$(0_i, \mathbf{x}) = (x_1, x_2, ..., x_{i-1}, 0, x_{i+1}, ..., x_n)$$
$$(1_i, \mathbf{x}) = (x_1, x_2, ..., x_{i-1}, 1, x_{i+1}, ..., x_n).$$

The *structure function* of the system indicates that the state of the system is completely determined by the states of all components. A component is relevant if its state does affect the state of the system.

Definition 1.10 A binary system with n components is said to be coherent if:

- the structure function ϕ is non-decreasing in each argument x_i, $i = 1, 2, ..., n$;
- each component is relevant, i.e., there exists at least one vector $\mathbf{x}$ such that $\phi(1_i, \mathbf{x}) = 1$ and $\phi(0_i, \mathbf{x}) = 0$; and
- $\phi(\mathbf{0}) = 0$ and $\phi(\mathbf{1}) = 1$.

For coherent systems, when $\mathbf{x} < \mathbf{y}$ then $\phi(\mathbf{x}) \leq \phi(\mathbf{y})$.

Example 1.32 For a series system consisting of n components to function, all its components must function. Then structure function of the series system is given by

$$\phi(x) = \prod_{i=1}^{n} x_i = \min(x_1, x_2, ..., x_n)$$

For the parallel system consisting of n components to work there is at least one component to work. Then structure function of the parallel system is given by

$$\phi(x) = 1 - \prod_{i=1}^{n} (1 - x_i) = \max(x_1, x_2, ..., x_n)$$

Similarly, the k out of n system is functioning if and only if at least k components to function. Then structure function of the k out of n system is given by

$$\phi(x) = \begin{cases} 1 & if \ \sum_{i=1}^{n} x_i \geq k \\ 0 & if \ \sum_{i=1}^{n} x_i < k \end{cases}$$

We now discuss the minimal path and minimal cut sets. Let $A_1, A_2, ..., A_s$ denote the minimal path sets of a given system.

Definition 1.11 A state vector $\mathbf{x}$ is called a *path vector* (or link vector) if $\phi(\mathbf{x}) = 1$.

A path set is a set of components whose functioning ensures the functioning of the system.

Definition 1.12 A *minimal path vector* is a path vector for which the failure of any functioning components results in system failure.

Definition 1.13 A *minimal path set* is a minimal set of components whose functioning ensures the functioning of the system.

In other words, a *minimal path vector* is a path vector $\mathbf{x}$ such that $\phi(\mathbf{y}) = 0$ for any $\mathbf{y} < \mathbf{x}$ and its corresponding path set is, *minimal path set*. Let

$$\alpha_j(\mathbf{x}) = \begin{cases} 1 & \text{iff all the components of } A_j \text{ are functioning} \\ 0 & \text{otherwise} \end{cases}$$

then

$$\alpha_j(\mathbf{x}) = \prod_{i \in A_j} X_i$$

A system will function if and only iff all the components of at least one minimal path sets are functioning. That is,

$$\phi(x) = \max_{j} \alpha_j(x)$$

$$= 1 - \prod_{j=1}^{s} (1 - \alpha_j(x))$$

$$1 - \prod_{j=1}^{s} \left(1 - \prod_{i \in A_j} X_i\right). \tag{1.39}$$

Similarly, let $C_1, C_2, \ldots, C_k$ denote the minimal cut sets of a given system.

Definition 1.14 A state vector x is called a *cut vector* if $\phi(x) = 0$.
A cut set is a set of components whose failure ensures the failure of the system.

Definition 1.15 A *minimal cut vector* is a cut vector for which the repair of any failed component results in a functioning system.

Definition 1.16 A *minimal cut set* is a minimal set of components whose failure ensures the failure of the system.

In other words, a *minimal cut vector* is a cut vector x such that $\phi(y) = 1$ for any $y > x$ and its corresponding cut set is *minimal cut set*. Note that for a minimal path set to work, each component in the set must work. For a minimal cut set to fail, all components in the set must fail.
Let

$$\beta_j(x) = \begin{cases} 1 & \text{if at least one component in} \\ & \text{the jth minimal cut set is functioning} \\ 0 & \text{otherwise} \quad \text{for } j = 1, 2, \ldots, k \end{cases}$$

then

$$\beta_j(x) = \max_{i \in C_j} X_i$$

A system is not functioning (or system unreliability) if and only if all the components of at least one minimal cut set are not functioning. In other words, a system is functioning if and only if at least one component in each minimal cut sets is functioning. Thus, the structure function based on the minimal cut sets is given by

$$\phi(x) = \prod_{j=1}^{k} \beta_j(x)$$

$$= \prod_{j=1}^{k} \max_{i \in C_j} X_i$$

$$= \prod_{j=1}^{k} \left(\coprod_{i \in C_j} X_i\right)$$

$$= \prod_{j=1}^{k} \left(1 - \prod_{j=C_j}^{k} (1 - X_i)\right).$$

Let p_i is the component i reliability. That is,

$$p_i = P(X_i = 1) = E[X_i].$$

The system reliability function can be defined as

$$r(p) = P(\phi(x) = 1) = E[\phi(x)].$$

Theorem 1.3 *If ϕ is a coherent system of independent components with minimal path set $A_1, A_2, ..., A_s$ and minimal cut sets $C_1, C_2, ..., C_k$ then*

$$\prod_{j=1}^{k}\left(1 - \prod_{i \in C_j}(1 - p_i)\right) \le r(p) \le 1 - \prod_{j=1}^{s}\left(1 - \prod_{i \in A_j} p_i\right). \qquad (1.40)$$

Example 1.33 A k-out-of-n system *works* if and only if at least k of the n components work. In a k-out-of-n system, there are $\binom{n}{k}$ minimal path sets and $\binom{n}{n-k+1}$ minimal cut sets. Each minimal path set contains exactly k different components and each minimal cut set contains exactly $(n - k + 1)$ components. Thus, all minimal path sets and minimal cut sets are known. One can easily obtain the reliability of a k-out-of-n system either in term of the minimal path sets or minimal cut sets.

1.7.2   Reliability Measures

In general, a system may be required to perform various functions, each of which may have a different reliability. In addition, at different times, the system may have a different probability of successfully performing the required function under stated conditions. The term failure means that the system is not capable of performing a function when required. The term *capable* used here is to define if the system is capable of performing the required function. However, the term *capable* is unclear and only various degrees of capability can be defined. The reliability definitions given in the literature vary between different practitioners as well as researchers. The generally accepted definition is as follows.

Definition 1.17 *Reliability* is the probability that the system will perform its intended function under given conditions for a given time interval.

More specific, reliability is the probability that a product or part will operate properly for a specified period of time (design life) under the design operating conditions (such as temperature, volt, etc.) without failure. In other words, reliability may be used as a measure of the system's success in providing its function properly. Reliability is one of the quality characteristics that consumers require from the manufacturer of products.

Let T be a random variable denoting the failure time. Mathematically, reliability $R(t)$ can be defined as the probability that a system will be successful without failure during a specified time interval $(0, t)$ under given operating conditions and environment. The reliability function can be written as

$$R(t) = P(T > t) \quad t \geq 0. \tag{1.41}$$

It is also referred to as the survival function. In other words, reliability is the probability that the system is still operating at time t.

Unreliability $F(t)$, a measure of failure, is defined as the probability that the sys-tem will fail by time t:

$$F(t) = P(T \leq t) \quad \text{for } t \geq 0$$

In other words, $F(t)$ *is* the failure distribution function. If the time-to-failure random variable T has a density function $f(t)$, then

$$R(t) = \int_t^\infty f(s)ds$$

or, equivalently,

$$f(t) = -\frac{d}{dt}[R(t)] \tag{1.42}$$

The density function can be mathematically described in terms of T:

$$\lim_{\Delta t \to 0} P(t < T \leq t + \Delta t)$$

This can be interpreted as the probability that the failure time T will occur between the operating time t and the next interval of operation, $t + \Delta t$.

Consider a new and successfully tested system that operates well when put into service at time $t = 0$. The system becomes less likely to remain successful as the time interval increases. The probability of success for an infinite time interval, of course, is zero.

Thus, the system functions at a probability of one and eventually decreases to a probability of zero. Clearly, reliability is a function of mission time. For example, one can say that the reliability of the system is 0.995 for a mission time of 24 h. However, a statement such as the reliability of the system is 0.995 is meaningless because the time interval is unknown.

Properties of Reliability Functions

Reliability functions have properties as follows:

1. $R(t)$ is decreasing
2. $\lim\limits_{t \to 0} R(t) = 1 \quad \lim\limits_{t \to \infty} R(t) = 0$

Example 1.34 A computer system has an exponential failure time density function.

$$f(t) = \frac{1}{9,000} e^{-\frac{t}{9,000}} \quad t \geq 0$$

What is the probability that the system will fail after the warranty (six months or 4380 h) and before the end of the first year (one year or 8760 h)?

Solution From Eq. (1.41) we obtain

$$P(4380 < T \leq 8760) = \int\limits_{4380}^{8760} \frac{1}{9000} e^{-\frac{t}{9000}} dt$$

$$= 0.237$$

This indicates that the probability of failure during the interval from six months to one year is 23.7%.

If the time to failure is described by an exponential failure time density function, then

$$f(t) = \frac{1}{\theta} e^{-\frac{t}{\theta}} \quad t \geq 0, \, \theta > 0$$

and this will lead to the reliability function

$$R(t) = \int\limits_{t}^{\infty} \frac{1}{\theta} e^{-\frac{s}{\theta}} ds = e^{-\frac{t}{\theta}} \, t \geq 0$$

Consider the Weibull distribution where the failure time density function is given by

$$f(t) = \frac{\beta t^{\beta-1}}{\theta^{\beta}} e^{-\left(\frac{t}{\theta}\right)^{\beta}} \quad t \geq 0, \, \theta > 0, \, \beta > 0$$

Then the reliability function is

$$R(t) = e^{-\left(\frac{t}{\theta}\right)^{\beta}} \quad t \geq 0$$

Thus, given a particular failure time density function or failure time distribution function, the reliability function can be obtained directly. Section 2.2 provides further insight for specific distributions.

Definition 1.18 *Conditional reliability* is the probability of surviving a mission of length h under specified design limits given that the system has been survived up to time t.

In other words, conditional reliability $R(h|t)$ *is* the probability that a system will be successful in the interval $(t, t + h]$ given that it already survived more than t. Mathematically,

$$R(h|t) = P(T > t + h | T > t) = \frac{R(t + h)}{R(t)} \quad t \geq 0, h > 0 \tag{1.43}$$

It is obviously that the probability of prolongation by additional mission length h decreases with increasing h. Indeed the longer prolongation of life time required, the smaller probability of such an event. It is also easy to see that the probability of prolongation depends in general the time t.

Example 1.35 For the case of uniform life time density function

$$f(t) = \frac{1}{a} \quad \text{for } 0 \leq t \leq a$$

then

$$R(t) = 1 - \frac{t}{a} \quad \text{for } 0 \leq t \leq a$$

Therefore, the probability of prolongation by h, R($h|t$), is given by

$$R(h|t) = \frac{R(t + h)}{R(t)} = \frac{a - t - h}{a - t} 1 - \frac{h}{a - t} \quad 0 \leq t \leq a, 0 \leq h \leq a - t$$

It is worth noting that the reliability function $R(t)$ and conditional reliability function $R(h|t)$ are relevant when the interest is in the probability of surviving a mission of time length h. Of course, from Eq. (1.43) when $t = 0$, then the both functions are the same. One can be more interested to determine t in Eq. (1.43) so that $R(h|t)$ is greater than $R(t)$ for a given reliability function. In other words, if no such h exists for that particular t, then burn-in is not needed and the system is said to be new-better-than used (NBU) for a mission of length h.

The pth percentile point of a lifetime distribution $F(t)$ for $0 < p < 1$ is the value of t, denoted by t_p, such as $F(t) = p$. That is

$$F(t_p) = p \quad \text{or} \quad t_p = F^{-1}(p)$$

System Mean Time to Failure

Suppose that the reliability function for a system is given by $R(t)$. The expected failure time during which a component is expected to perform successfully, or the system mean time to failure (MTTF), is given by

$$MTTF = \int_0^\infty tf(t)dt \qquad (1.44)$$

Substituting

$$f(t) = -\frac{d}{dt}[R(t)]$$

into Eq. (1.44) and performing integration by parts, we obtain

$$MTTF = -\int_0^\infty td[R(t)]$$

$$= [-tR(t)] \Big|_0^\infty + \int_0^\infty R(t)dt \qquad (1.45)$$

The first term on the right-hand side of Eq. (1.45) equals zero at both limits, since the system must fail after a finite amount of operating time. Therefore, we must have $tR(t) \to 0$ as $t \to \infty$. This leaves the second term, which equals

$$MTTF = \int_0^\infty R(t)dt \qquad (1.46)$$

Thus, MTTF is the definite integral evaluation of the reliability function of the reliability function In general, if $\lambda(t)$ is defined as the failure rate function, then, by definition, MTTF is not equal to $1/\lambda(t)$.

The MTTF should be used when the failure time distribution function is specified because the reliability level implied by the MTTF depends on the underlying failure time distribution. Although the MTTF measure is one of the most widely used reliability calculalations, it is also one of the most misused calculations, it has been misinterpreted as "guaranteed minimum lifetime". Consider the results as given in Table 1.1 for a twelve-component life duration test.

Using a basic averaging technique, the component MTTF of 3660 h was estimated. However, one of the components failed after 920 h. Therefore, it is important to note that the system MTTF denotes the average time to failure. It is neither the failure

Table 1.1 Results of a twelve-component life duration test

Component	Time to failure (h)
1	4510
2	3690
3	3550
4	5280
5	2595
6	3690
7	920
8	3890
9	4320
10	4770
11	3955
12	2750

time that could be expected 50% of the time, nor is it the guaranteed minimum time of system failure.

A careful examination of Eq. (1.46) will show that two failure distributions can have the same MTTF and yet produce different reliability levels. This is illustrated in a case where the MTTFs are equal, but with normal and exponential failure distributions. The normal failure distribution is symmetrical about its mean, thus

$$R(MTTF) = P(Z \geq 0) = 0.5$$

where Z is a standard normal random variable. When we compute for the exponential failure distribution from Example 1.28, recognizing that $\theta = $ MTTF, the reliability at the MTTF is

$$R(MTTF) = e^{-\frac{MTTF}{MTTF}} = 0.368$$

Clearly, the reliability in the case of the exponential distribution is about 74% of that for the normal failure distribution with the same MTTF.

Example 1.36 *(Mixed truncated exponential distribution)* Using the Eq. (1.46), calculate the system MTTF where the random system lifetime T has a **mixed truncated exponentialdistribution** with parameters λ and m?

Solution The reliability function can be obtained as follows

$$R(t) = \begin{cases} 1 & \text{for } t < 0 \\ e^{-\lambda t} & \text{for } 0 \leq t < m \\ 0 & \text{for } t \geq m \end{cases}$$

The system MTTF is given by

$$MTTF = \int_0^m R(t)dt = \int_0^m e^{-\lambda t}dt$$
$$= \frac{1}{\lambda}\left(1 - e^{-\lambda m}\right)$$

Failure Rate Function

The probability of a system failure in a given time interval $[t_1, t_2]$ can be expressed in terms of the reliability function as

$$\int_{t_1}^{t_2} f(t)dt = \int_{t_1}^{\infty} f(t)dt - \int_{t_2}^{\infty} f(t)dt$$
$$= R(t_1) - R(t_2)$$

or in terms of the failure distribution function (or the unreliability function) as

$$\int_{t_1}^{t_2} f(t)dt = \int_{-\infty}^{t_2} f(t)dt - \int_{-\infty}^{t_1} f(t)dt$$
$$= F(t_2) - F(t_1)$$

The rate at which failures occur in a certain time interval $[t_1, t_2]$ *is* called the *failure rate*. It is defined as the probability that a failure per unit time occurs in the interval, given that a failure has not occurred prior to t_1, the beginning of the interval. Thus, the failure rate is

$$\frac{R(t_1) - R(t_2)}{(t_2 - t_1)R(t_1)}$$

Note that the failure rate is a function of time. If we redefine the interval as $[t, t + \Delta t]$, the above expression becomes

$$\frac{R(t) - R(t + \Delta t)}{\Delta t\, R(t)}$$

The rate in the above definitions is expressed as failures per unit time, when in reality, the time units might be in terms of miles, hours, etc.

In practice, we may compute the failure rate by testing or using a sample of items until all fails, recording the time of failure for each item and use the following formulas to estimate the constant failure rate λ:

$$\lambda = \frac{\text{Number of failures}}{\text{Total unit of operating hours}}$$

or

$$\lambda = \frac{\text{Number of failures}}{(\text{Units tested}) * (\text{Number of hours tested})}.$$

Example 1.37 Suppose that 15 units are tested over a 72-h period. Five units failed with one unit each failing after 5, 15, 20, 50, and 68 h. The remaining 10 units have survived at the end of the 72-h test period. Estimate the constant failure rate.

Solution The total unit operating hours, T_{total}, are:

$$T_{total} = (5 + 15 + 20 + 50 + 68) + 10(72) = 878.$$

Thus,

$$\lambda = \frac{5 \text{ failures}}{878 \text{ unit operating hours}} = 0.0057 \text{ failures per hour.}$$

In other words, on the average about 0.57% of the units would be expected to fail in a one-hour period. Furthermore, over a 72-h period, about $(0.0057 * 72) = 0.41$ or 41% of the units would be expected to fail. In the actual test given above, only 33.33% failed (i.e., $(5)(100)/15 = 33.33$).

The *hazard rate function* is, sometimes referred to as the **failure rate function,** defined as the limit of the failure rate as the interval approaches zero. Thus, the hazard rate function $h(t)$ *is* the instantaneous failure rate, and is defined by

$$
\begin{aligned}
h(t) &= \lim_{\Delta t \to 0} \frac{R(t) - R(t + \Delta t)}{\Delta t R(t)} \\
&= \frac{1}{R(t)} [-\frac{d}{dt} R(t)] \\
&= \frac{f(t)}{R(t)}
\end{aligned}
\tag{1.47}
$$

The quantity $h(t)dt$ represents the probability that a device of age t will fail in the small interval of time t to $(t + dt)$. Equivalently, the function $h(t)$ can be defined as the probability density of failure at time t, given survival up to that time. In other words, the failure rate function $h(t)$ can be defined as the probability that a device, which has survived of age t, will fail in the small time interval $(t, t + dt)$. The importance of the hazard rate function is that it indicates the change in the failure rate over the life of a population of components by plotting their hazard rate functions on a single axis. For example, two designs may provide the same reliability at a specific point in time, but the failure rates up to this point in time can differ.

The death rate, in statistical theory, is analogous to the failure rate as the force of mortality is analogous to the hazard rate function. Therefore, the hazard rate function or failure rate function is the ratio of the probability density function (pdf) to the reliability function. In reliability engineering, the survival function is often referred to as the reliability function. In actuarial science, the hazard function is known as the force of mortality.

By defintion, a mathematical reliability function is the probability that a system will be successful in the interval from time 0 to time t, given by

$$R(t) = \int_t^\infty f(s)ds = e^{-\int_0^t h(s)ds} \tag{1.48}$$

where $f(s)$ and $h(s)$ are, respectively, the failure time density and failure rate function.

Example 1.38 Consider the lifetime of a communication device, having a pdf

$$f(t) = \begin{cases} 0 & t \le t_0 \\ \frac{4(t-t_0)}{(t_1-t_0)^2} & t_0 \le t \le \frac{t_0+t_1}{2} \\ \frac{4(t_1-t)}{(t_1-t_0)^2} & \frac{t_0+t_1}{2} \le t \le t_1 \\ 0 & t_1 \le t \end{cases}$$

Then the cdf is

$$F(t) = \begin{cases} 0 & t \le t_0 \\ \frac{2(t-t_0)^2}{(t_1-t_0)^2} & t_0 \le t \le \frac{t_0+t_1}{2} \\ 1 - \frac{2(t_1-t)^2}{(t_1-t_0)^2} & \frac{t_0+t_1}{2} \le t \le t_1 \\ 1 & t_1 \le t \end{cases}$$

Mean Residual Life Function

Another important measure of reliability is the mean residual life $\mu(t)$, which is the expected remaining life beyond the present age t. Mathematically, it is defined as

$$\mu(t) = E[T - t | T \ge t]$$

$$= \int_t^\infty (s - t) \frac{f(s)}{R(t)} ds$$

$$= \frac{\int_t^\infty s f(s)ds}{R(t)} - t$$

$$= \frac{\int_t^\infty R(s)ds}{R(t)}. \tag{1.49}$$

In other words, the mean residual life (MRL) is the expected remaining life, T-t, given that the item has survived to time t. When $t = 0$, the MRL function becomes the MTTF.

The mean residual life (MRL) from Eq. (1.49) can also be written as related to the failure rate $h(t)$ through

$$\mu'(t) = \mu(t)h(t) - 1$$

We can obtain the above equation by taking the derivative of Eq. (1.49) as follows

$$\mu'(t) = \left(\frac{\int_t^\infty s f(s)ds}{R(t)} - t\right)'$$

$$= \frac{\left(\int_t^\infty R(s)ds\right)' R(t) - \left(\int_t^\infty R(s)ds\right) R'(t)}{R^2(t)}$$

$$= -\frac{R^2(t)}{R^2(t)} + \frac{f(t)\left(\int_t^\infty R(s)ds\right)}{R^2(t)}$$

$$= \mu(t)h(t) - 1.$$

In industrial reliability studies of repair, replacement, and other maintenance strategies, the mean residual life function may be proven to be more relevant than the failure rate function. Indeed, if the goal is to improve the average system life-time, then the mean residual life is the relevant measure. The function $h(t)$ relates only to the risk of immediate failure, whereas $\mu(t)$ summaries the entire residual life distribution.

For generalized Weibull distributions, the mean residual life $\mu(t)$ is generally difficult to obtain explicitly. Further analysis on the relationship between the mean residual life, and the failure rate functions for various generalizations of Weibull distributions is worth exploring.

1.8 Maintainability

Maintainability is another approach to measure a performance of any system. In other words, some applications of our daily life, such as computer security systems for example, can not tolerate any time to fail. Such failures are very costly or their continuous function is on top priorities. Therefore, a maintenance policy to perform preventive maintenance from time to time for the system to continous functioning at low costs to reduce a significant impact of system failure would be worth to carefully study. Acording to British Specifications, BS 3811:1974(1): Maintenance is a combination of any actions carried out to retain an item in or restore it to an acceptable standard. When a system fails to perform satisfactorily, repair is normally carried out to locate and correct the fault. The system is restored to operational effectiveness by making an adjustment or by replacing a component.

Maintainability is defined as the probability that a failed system will be restored to specified conditions within a given period of time when maintenance is performed according to prescribed procedures and resources. In other words, maintainability is the probability of isolating and repairing a fault in a system within a given time. Maintainability engineers must work with system designers to ensure that the system product can be maintained by the customer efficiently and cost effectively. This function requires the analysis of part removal, replacement, tear-down, and build-up of the product in order to determine the required time to carry out the operation, the necessary skill, the type of support equipment and the documentation.

There are two common types of maintenance: preventive maintenance and corrective maintenance. In preventive maintenance, the system is periodically and systematically inspected before the actual failure of the system. The maintenance action can be performed without degrading in operation or by stopping the system during a convenience time without any disruption of activities. The aim of preventive maintenance is to intentionally eliminate a severe impact when the system has failed as a less total operating cost. In corrective maintenance, it is carried out when the system has already failed in order to bring the system back to the working state as soon as possible. The time spent on all such actions often called as corrective maintenance down time.

Let T denote the random variable of the time to repair or the total downtime. If the repair time T has a repair time density function $g(t)$, then the maintainability, $V(t)$, is defined as the probability that a failed system will be back in service by time t, i.e.,

$$V(t) = P(T \leq t) = \int_0^t g(s)ds \tag{1.50}$$

For example, if $g(t) = \mu e^{-\mu t}$ where $\mu > 0$ is a constant repair rate, then

$$V(t) = 1 - e^{-\mu t}$$

which represents the exponential form of the maintainability function.

Example 1.39 Let T be the random variable of the time to repair for a system in which it has a repair time density function $g(t)$ as follows

$$g(t) = \mu_1 e^{-\mu_1 t} + \mu_2 e^{-\mu_2 t} - (\mu_1 + \mu_2)e^{-(\mu_1 + \mu_2)t}$$

The maintainability of the system is given by

$$V(t) = \int_0^t g(s)ds$$

$$= \int_0^t \left(\mu_1 e^{-\mu_1 s} + \mu_2 e^{-\mu_2 s} - (\mu_1 + \mu_2)e^{-(\mu_1+\mu_2)s} \right) ds$$

$$= 1 - \left(e^{-\mu_1 t} + e^{-\mu_2 t} - e^{-(\mu_1+\mu_2)t} \right). \tag{1.51}$$

The maintainability function given as in Eq. (1.51) above is, in fact, the cdf of a two-component parallel system where component 1 and 2 are independent and has constant failure rate μ_1 and μ_2, respectively.

Example 1.40 If T represents the random variable of the time to repair for a system where the repair time density function g(t) is given by

$$g(t) = (\mu_1 + \mu_2)e^{-(\mu_1+\mu_2)t}$$

then the system maintainability can be easily obtained as follows

$$V(t) = \int_0^t g(s)ds$$

$$= \int_0^t (\mu_1 + \mu_2)e^{-(\mu_1+\mu_2)s} ds$$

$$= 1 - e^{-(\mu_1+\mu_2)t}. \tag{1.52}$$

Similarly, the maintainability function as in Eq. (1.52) above is, easily to see later in Chap. 4, that the cdf of a two-component series system where component 1 and 2 has constant failure rate μ_1 and μ_2, respectively.

An important measure often used in maintenance studies is the mean time to repair (MTTR) or the mean downtime. MTTR is the expected value of the random variable repair time, not failure time, and is given by

$$MTTR = \int_0^\infty t g(t)dt$$

When the distribution has a repair time density given by $g(t) = \mu e^{-\mu t}$, then, from the above equation, MTTR $= 1/\mu$. When the repair time T has the log normal density function g(t), and the density function is given by

$$g(t) = \frac{1}{\sqrt{2\pi}\,\sigma t} e^{-\frac{(\ln t - \mu)^2}{2\sigma^2}} \quad t > 0$$

then it can be shown that

$$MTTR = m \, e^{\frac{\sigma^2}{2}}$$

where m denotes the median of the log normal distribution.

In order to design and manufacture a maintainable system, it is necessary to predict the MTTR for various fault conditions that could occur in the system. This is generally based on past experiences of designers and the expertise available to handle repair work.

The system repair time consists of two separate intervals: passive repair time and active repair time. Passive repair time is mainly determined by the time taken by service engineers to travel to the customer site. In many cases, the cost of travel time exceeds the cost of the actual repair. Active repair time is directly affected by the system design and is listed as follows:

1. The time between the occurrence of a failure and the system user becoming aware that it has occurred.
2. The time needed to detect a fault and isolate the replaceable component(s).
3. The time needed to replace the faulty component(s).
4. The time needed to verify that the fault has been corrected and the system is fully operational.

The active repair time can be improved significantly by designing the system in such a way that faults may be quickly detected and isolated. As more complex systems are designed, it becomes more difficult to isolate the faults.

1.9 Availability Measure

Reliability is a measure that requires system success for an entire mission time. No failures or repairs are allowed. Space missions and aircraft flights are examples of systems where failures or repairs are not allowed. Availability is a measure that allows for a system to repair when failure occurs. A repairable system once put into operation continues to operate till failure or it is stopped for a planned maintenance. After repairs or maintenance action, the system assumes to return to the working state. There are several types of availability measures as discussed in the following.

Definition 1.19 *Availability* is a point function describing the behavior of the system at a specified epoch and it does not impose any restriction on the behavior of the system before the time point.

Definition 1.20 *Pointwise availability* is the probability that the system will be able to operate with tolerance at a given instant of time.

Definition 1.21 The *point availability* is defined as the probability that the system is good at time point t and is denoted as $A(t)$. Mathematically,

$$A(t) = P\{Z(t) = 1 | Z(0) = 1\}$$

where $Z(t)$ is the state of the system.

Let

$$W(t) = \begin{cases} 1 \text{ if the system is in operating state at time t} \\ 0 \text{ if the system is in failed state at time t} \end{cases}$$

$A(t)$ can also be defined as the probability that the system functions at time t irrespective of its past history background, i.e., repairs:

$$A(t) = E\{W(t)\} = 1 \times P\{W(t) = 1\} + 0 \times P\{w(t) = 0\}$$
$$= P\{W(t) = 1\}.$$

Definition 1.22 *Interval availability* is defined as the system will be able to function within a given interval of time. More specific, for the interval of time $[t_1, t_2]$, the interval availability is defined as

$$A(t_1, t_2) = \frac{1}{t_2 - t_1} \int_{t_1}^{t_2} A(t)dt \qquad (1.53)$$

In other words, the *interval availability* in the interval $[0, T]$ is given by

$$A(T) = \frac{1}{T} \int_0^T A(t)dt \qquad (1.54)$$

Definition 1.23 The *steady state availability* of a system is defined as the probability that the system is successful at time t as t is large. Mathematically, the steady state availability is given by

$$A = A(\infty) = \lim_{T \to \infty} \frac{1}{T} \int_0^T A(t)dt \qquad (1.55)$$

The above steady state availability function can be shown as follows:

$$\text{Availability} = A \equiv A(\infty) = \frac{\text{System up time}}{\text{System up time} + \text{System down time}}$$
$$= \frac{\text{MTTF}}{\text{MTTF} + \text{MTTR}} \qquad (1.56)$$

Availability is a measure of success used primarily for repairable systems. For non-repairable systems, availability, $A(t)$, equals reliability, $R(t)$. In repairable systems, $A(t)$ will be equal to or greater than $R(t)$.

The mean time between failures (MTBF) is an important measure in repairable systems. This implies that the system has failed and has been repaired. Like MTTF and MTTR, MTBF is an expected value of the random variable time between failures. Mathematically,

$$MTBF = MTTF + MTTR$$

The term MTBF has been widely misused. In practice, MTTR is much smaller than MTTF, which is approximately equal to MTBF. MTBF is often incorrectly substituted for MTTF, which applies to both repairable systems and non-repairable systems. If the MTTR can be reduced, availability will increase, and the system will be more economical.

A system where faults are rapidly diagnosed is more desirable than a system that has a lower failure rate but where the cause of a failure takes longer to detect, resulting in a lengthy system downtime. When the system being tested is renewed through maintenance and repairs, $E(T)$ is also known as MTBF.

1.10 Mathematical Techniques

Delta Technique

Delta technique is an approach that uses the Taylor series expansion of a function of one or more random variables to obtain approximations to the expected value of the function and to its variance. For example, writing a variable x as $x = \mu + \varepsilon$ where $E(x) = \mu$ and $E(\varepsilon) = 0$, Tayfor's expansion gives

$$f(x) = f(\mu) + \varepsilon \frac{\partial f(x)}{\partial x}\Big|_{x=\mu} + \frac{\varepsilon^2}{2!} \frac{\partial^2 f(x)}{\partial x^2}\Big|_{x=\mu} \cdots + \cdots \qquad (1.57)$$

If terms involving ε^2, ε^3 etc. are assumed to be negligible then

$$f(x) \approx f(\mu) + (x - \mu)f'(\mu) \qquad (1.58)$$

so that

$$\text{var}(f(x)) \approx \left(f'(\mu)\right)^2 \text{var}(x) \qquad (1.59)$$

Beta functions. For m > 0, n > 0 (not necessarily integers)

$$\beta(m, n) \equiv \int_0^1 x^{m-1}(1 - x)^{n-1}dx = 2\int_0^{\pi/2}(\sin\theta)^{2m-1}(\cos\theta)^{2n-1}d\theta. \qquad (1.60)$$

The above transformation used was $x = \sin^2\theta$.

Gamma functions. For n > 0 (not necessarily an integer)

$$\Gamma(n) \equiv \int_0^\infty x^{n-1}e^{-x}dx = 2\int_0^\infty y^{2n-1}e^{-y^2}dy. \qquad (1.61)$$

The above transformation used was $x = y^2$. If n is a positive integer, then

$$\Gamma(n) = (n - 1)! \quad \Gamma\left(\frac{1}{2}\right) = \sqrt{\pi}.$$

Factorials of large number can be obtained using Stirling's formula as follows:

$$n! = \sqrt{2\pi n}\, n^n\, e^{-n}\left(1 + \frac{1}{12n} + \frac{1}{288n^2} + \cdots\right).$$

Also, from Eqs. (1.60) and (1.61)

$$\beta(m, n) = \frac{\Gamma(m)\,\Gamma(n)}{\Gamma(m + n)}.$$

Convex function

Definition 1.24 A function $h(x)$ is called *convex* if

$$h(\alpha x_1 + (1 - \alpha)x_2) \leq \alpha h(x_1) + (1 - \alpha)h(x_2) \qquad (1.62)$$

for all x_1 and x_2 and $\alpha \in (0, 1)$.

Definition 1.25 A function $h(x)$ is also convex if $h''(x) \geq 0$.

On the other hand, $h(x)$ is said to be *concave* if $(-h(x))$ is convex, or $h''(x) \leq 0$.

Indicator function

For any set S, the associated indicator function is

$$I_S(x) = \begin{cases} 1 & \text{if } x \in S \\ 0 & \text{if } x \notin S. \end{cases}$$

1.11   Probability and Moment Inequalities

In statistics, we often deal with situations where the exact distribution function is known. In other words, the distribution of the random variables $X_1, X_2, ..., X_n$ is assumed to be known. However, in real-world problems one often has little or no knowledge about the distribution underlying samples. In this section, we discuss common statistical approaches that aim with partial or no information about the distribution function of $(X_1, X_2, ..., X_n)$ in order to obtain bounds that can be given on the distribution of $(X_1, X_2, ..., X_n)$.

Theorem 1.4 (Chebyshev's inequality). *Let $\varepsilon > 0$ be fixed. Let X be a r.v. with mean μ and variance σ^2. Then*

$$P\{|X - \mu| \geq \varepsilon\} \leq \frac{\sigma^2}{\varepsilon^2} \qquad (1.63)$$

Corollary 1.1 *Let $\varepsilon > 0$ be fixed. Let X be a r.v. with mean μ and variance σ^2. Then*

$$P\left\{\left|\frac{X - \mu}{\sigma}\right| \geq \varepsilon\right\} \leq \frac{1}{\varepsilon^2} \qquad (1.64)$$

Theorem 1.5 (Markov's inequality) *Let $t > 0$, $a \geq 0$ be fixed. Let X be a r.v. such that $P[X \geq 0] = 1$. Then*

$$P[X \geq t] \leq \frac{E(X^a)}{t^a}. \qquad (1.65)$$

Example 1.41 A manufacturer has observed that over the course of 10 months, the numbers of products demanded were $T_1, T_2, ..., T_{10}$, respectively. If we assume that these are i.i.d. observations of demand, what bound can be put on the probability that their average $\overline{T} = \frac{\sum_{i=1}^{10} T_i}{10}$ is more than 3σ products from the average demand μ?

Solution   We obtain

$$E(\overline{T}) = \mu, \quad Var(\overline{T}) = \frac{\sigma^2}{10}.$$

From the Chebyshev's inequality,

$$P\{|\overline{T} - \mu| \geq 3\sigma\} = P\left\{\left|\frac{\overline{T} - \mu}{\sigma/\sqrt{10}}\right| \geq 3\sqrt{10}\right\} \leq \frac{1}{(3\sqrt{10})^2} = \frac{1}{90}$$

Let $X_1, X_2, ..., X_n$ and X be random variables on a probability space. We say that X_n *converges in probability* to X if

$$\lim_{n \to \infty} P(|X_n - X| < \varepsilon) = 1 \quad \text{for every } \varepsilon > 0.$$

Similarly, X_n *converges with probability 1* (or almost surely) to X if

$$P\left(\lim_{n \to \infty} X_n = X\right) = 1.$$

Moment Inequalities

The following moment inequalities are commonly used in applied statistics and probability.

Theorem 1.6 (Jensen's Inequality) *For any random variable X and a convex function* $h(x)$,

$$h(E(X)) \le E[h(X)]. \tag{1.66}$$

Example 1.42 Since the function $f(x) = x^2$ is a convex function, using Jensen's inequality we obtain

$$f[E(X)] \le E(f(X))$$

or

$$[E(X)]^2 \le E(X^2)$$

In other words, since $\text{var}(X) = E(X^2) - [E(X)]^2 \ge 0$ this implies that

$$E(X^2) \ge [E(X)]^2.$$

Theorem 1.7 (Cauchy-Schwarz Inequality) *For any two random variables X and Y,*

$$E|XY| \le \sqrt{E(X^2)E(Y^2)}. \tag{1.67}$$

Theorem 1.8 (Holder's Inequality) *For any two random variables X and Y and two positive real values a and b such that* $a + b = 1$,

$$E|XY| \le \sqrt{\left(E\left(|X|^{1/a}\right)\right)^a \left(E\left(|Y|^{1/b}\right)\right)^b}. \tag{1.68}$$

Theorem 1.9 (Liapounov's Inequality) *If one takes* $Y \equiv 1$ *in Eq. (1.68), then the Holder's inequality reduces to:*

$$E|X| \leq \left(E\left(|X|^{1/a}\right)\right)^a \ for \ 0 < a \leq 1. \tag{1.69}$$

Theorem 1.10 (Minkowski's Inequality) *For any two random variables X and Y and any real value a for $0 < a \leq 1$,*

$$\left(E\left(|X+Y|^{1/a}\right)\right)^a \leq \left(E\left(|X|^{1/a}\right)\right)^a + \left(E\left(|Y|^{1/a}\right)\right)^a. \tag{1.70}$$

1.12 Order Statistics

Let $X_1, X_2, ..., X_n$ be a random sample of size n, each with cdf $F(x)$ and pdf $f(x)$. Let $X_{(i)}$ be the ith order statistic of the sample where such characteristic values of a sample of observations that have been sorted from smallest to largest. The rth order statistic is the rth smallest value in the sample, say $X_{(r)}$. That is,

$$X_{(1)} \leq X_{(2)} \leq \cdots \leq X_{(r-1)} \leq X_{(r)} \leq \cdots \leq X_{(n)}$$

are called the *order statistics*. Imagine we are building an elevator for example and our random variable s are the weight on the elevator of various times during the day. We would want to know how much weight it can bear. In this case, we will care about the distribution of the maximum.

Let $F_{(r)}(x)$ denote the cdf of the rth order statistic $X_{(r)}$. Note that the order statistics $X_{(1)}, X_{(2)}, ..., X_{(n)}$ are neither independent nor identically distributed. Let $X_{(r)}$ be the rth order statistic. Then the cdf of $X_{(r)}$ is given by

$$
\begin{aligned}
F_{(r)}(x) &= P(X_{(r)} \leq x) \\
&= P(\text{at least } r \text{ of } X_1, X_2, ..., X_n \text{ are } \leq x) \\
&= \sum_{i=r}^{n} P(\text{exactly } i \text{ of } X_1, X_2, ..., X_n \text{ are } \leq x) \\
&= \sum_{i=r}^{n} \binom{n}{i} F^i(x)[1 - F(x)]^{n-i} \\
&= F^r(x) \sum_{j=0}^{n-r} \binom{r+j-1}{r-1} [1 - F(x)]^j \\
&= r \binom{n}{r} \int_0^{F(x)} t^{r-1}(1-t)^{n-r} dt
\end{aligned} \tag{1.71}
$$

The probability density function (pdf) of the rth order statistic $X_{(r)}$ is given by

$$f_{X_{(r)}}(x) = \frac{n!}{(r-1)!(n-r)!} f(x) F^{r-1}(x) [1 - F(x)]^{n-r} \qquad (1.72)$$

The pdf of the smallest, i.e., 1st order statistic $X_{(1)}$, value of n independent randam variables, is given by

$$f_{X_{(1)}}(x) = n f(x) [1 - F(x)]^{n-1} \qquad (1.73)$$

Similarly, the pdf of the largest, i.e., nst order statistic $X_{(n)}$, value is given by

$$f_{X_{(n)}}(x) = n f(x) [F(x)]^{n-1} \qquad (1.74)$$

Example 1.43 Assume $X_1, X_2, \ldots, X_n$ are independent variables from the uniform distribution on $[0,1]$. Then the pdf is given by

$$f(x) = \begin{cases} 1 & \text{if } 0 \leq x \leq 1 \\ 0 & \text{otherwise.} \end{cases}$$

From Eq. (1.73), the pdf of the smallest value is given by

$$f_{X_{(1)}}(x) = n f(x) [1 - F(x)]^{n-1} = n [1 - x]^{n-1}.$$

Similiarly, from Eq. (1.74), the pdf of the largest value is given by

$$f_{X_{(n)}}(x) = n f(x) [F(x)]^{n-1} = n [x]^{n-1}.$$

Note that if there are n random variables $X_1, X_2, \ldots, X_n$ which may or may not be mutually independent, then joint cdf, if it exists, is given by

$$P(X_1 \leq x_1, X_2 \leq x_2, \ldots X_n \leq x_n)$$

$$= \int_{-\infty}^{x_n} \int_{-\infty}^{x_{n-1}} \cdots \int_{-\infty}^{x_1} f(t_1, t_2, \ldots, t_n) dt_1 dt_2 \ldots dt_n \qquad (1.75)$$

If the n random variables are mutually statistically independent, then the joint pdf can be rewritten as

$$f(x_1, x_2, \ldots, x_n) = \prod_{i=1}^{n} f(x_i)$$

The Sample Distribution of the Median

Let $X_1, X_2, \ldots, X_n$ be a random sample of size n, each with cdf $F(x)$ and pdf $f(x)$. The sample median is defined as

$$
M = \begin{cases} X_{(n+1)/2} & \text{n odd} \\ (X_{(n)/2} + X_{(n/2+1)})/2 & \text{n even} \end{cases} \tag{1.76}
$$

where $X_{(1)}, X_{(2)}, \ldots, X_{(r-1)}, X_{(r)} \ldots X_{(n)}$ are the order stattistics.

Let a sample of size $n = 2m + 1$ with n large be taken from an infinite population with a density function $f(x)$ having the population median $\tilde{\mu}$. In other words, we have n independent, identically distributed random variables with density f. The sampling distribution of the median is approximately normal with mean $\tilde{\mu}$ and variance $\frac{1}{8f(\tilde{\mu})^2 m}$.

1.13 Problems

1. The probability of having various numbers of children in a family in a certain community is given by

Number of children:	0	1	2	3
Probability:	0.20	0.50	0.25	0.05

    Assume a 50–50 chance that a child born will be a boy or girl.

    (i)   Find the probability that the family has only one boy.
    (ii) What is the probability that there are two children in the family given the family has one boy?

2. A fair die is thrown twice. Let A, B and C denote the following events:

    A: first toss is odd
    B: second toss is even
    C: total number of spots is equal to seven.

    Show that A, B and C are not independent.

3. A batch of material consists of 120 units where 40 units are oversized, 20 are undersized, and 60 conform to specification. Two units are drawn at random one after the other (without replacement). What is the probability that both units will be of the same category?

4. Suppose that the probability of successful launching of a special device is 0.5. Suppose further that each trial is physically independent of preceding trials. Determine the number of trials required to make the probability of a successful launching at least 0.99.

5. Show that

$$
P\{A \cup B \cup C\} = P\{A\} + P\{B\} + P\{C\} - P\{A \cap B\}
$$

$$- P\{A \cap C\} - P\{B \cap C\} + P\{A \cap B \cap C\}$$

6. Let A, B and C be subsets of the sample space S defined by

$$S = \{1, 2, 3, 4, 5, 6\}$$
$$A = \{1, 3, 5\}; \quad B = \{1, 2, 3\}; \quad C = \{2, 4, 6\}$$

Find the events
(i) $A \cap (B \cap C)$ (ii) $A \cup (B \cup C)$

7. A fair coin is tossed until for the first time heads occur. Find the probability that the experiment ends on or before the sixth toss.

8. A consumer survey investigating the ownership of television sets and cars for a sample of 1000 families in a certain community finds the following results:

	TV	No TV
Car	115	245
No Car	380	260

What is the probability that a family has

(i)   a TV set?
(ii)   a car?
(iii) both?
(iv) either a TV set or a car or both?

9. A fair coin is tossed three successive times. What is the probability of obtaining at least two heads given that the first toss resulted in a head?

10. A group of 100 boys and 100 girls is divided into two equal groups. Find the probability p that each group will be equally divided into boys and girls.

11. Assume that the hazard rate, $h(t)$, has a positive derivative. Show that the hazard distribution

$$H(t) = \int_0^t h(x)dx$$

is strictly convex.

12. Consider the pdf of a random variable that is equally likely to take on any value *only* in the interval from a to b.

(a) Show that this pdf is given by

$$f(t) = \begin{cases} \dfrac{1}{b - a} & \text{for } a < t < b \\ 0 & \text{otherwise} \end{cases}$$

(b) Derive the corresponding reliability function $R(t)$ and failure rate $h(t)$.

(c)   Think of an example where such a distribution function would be of interest in reliability application.

13.   From Eq. (1.37), show that the Laplace transform of the density function g_n of the total life time S_n of n independent life times X_i with their pdf f_i for $i = 1, 2, ..., n$, is given by

$$g_n^*(a) = f_1^*(a) f_2^*(a)...f_n^*(a) = \prod_{i=1}^{n} f_i^*(a)$$

14.   Draw Venn diagrams to illustrate the identity below.

(a) $A \cap (B * C) = (A \cap B) * (A \cap C)$
(b) $A \cap (B \cup C) = (A \cap B) \cup (A \cap C)$

15.   One thousand new streetlights are installed in Saigon city. Assume that the lifetimes of these streetlights follow the normal distribution. The average life of these lamps is estimated at 980 burning-hours with a standard deviation of 100 h.

(a)   What is the expected number of lights that will fail during the first 800 burning-hours?
(b)   What is the expected number of lights that will fail between 900 and 1100 burning-hours?
(c)   After how many burning-hours would 10% of the lamps be expected to fail?

16.   A fax machine with constant failure rate λ will survive for a period of 720 h without failure, with probability 0.80.

(a)   Determine the failure rate λ.
(b)   Determine the probability that the machine, which is functioning after 600 h, will still function after 800 h.
(c)   Find the probability that the machine will fail within 900 h, given that the machine was functioning at 720 h.

17.   A diesel is known to have an operating life (in hours) that fits the following pdf:

$$f(t) = \frac{2a}{(t+b)^2} \quad t \geq 0$$

(a)   Determine a and b.
(b)   Determine the probability that the diesel will not fail during the first 6000 operating-hours where $a = 500$.
(c)   If the manufacturer wants no more than 10% of the diesels returned for warranty service, how long should the warranty be?

18.   The failure rate for a hydraulic component

$$h(t) = \frac{t}{t+1} \quad t > 0$$

where t is in years.

(a)   Determine the reliability function $R(t)$.
(b)   Determine the MTTF of the component.

19.   A 18-month guarantee is given based on the assumption that no more than 5% of new cars will be returned.

(a)   The time to failure T of a car has a constant failure rate. What is the maximum failure rate that can be tolerated?
(b)   Determine the probability that a new car will fail within three years assuming that the car was functioning at 18 months.

20.   Suppose that the lifetime T (in hours) of a TV tube is described by the following pdf:

$$f(t) = \begin{cases} at(b - t) & \text{for } 0 < t < b \\ 0 & \text{otherwise} \end{cases}$$

where a is a constant to be determined and b is the maximum life time.

(a)   Show that $a = \frac{6}{b^3}$.
(b)   Determine the probability that the tube will last more than k hours but less than l hours.
(c)   Determine the tube MTTF.
(d)   Obtain the probability that the tube will last an additional s hours, given it already lasted more than t hours.

21.   Show that if

$$R_1(t) \geq R_2(t) \quad \text{for all } t$$

where $R_i(t)$ *is the system reliability of the structure* i, then MTTF of the system structure 1 is always $\geq$ MTTF of the system structure 2.

22.   Assume there are 10 identical units to place on test at the same time $t = 0$ where each unit has an exponential life distribution with MTTF $\mu = 5$ h and that the units operate independently. Let N_3 be the number of failures during the third hour, in the time interval $[2, 3)$. Assume that the unit will not fail during the first two hours (in the time interval $[0, 2)$, obtain the distribution of the number of failures and expected number of failures during (a) the third time interval $[2, 3)$; and (b) the fourth interval $[3, 4)$.

23.   The lifetime of a communication system (hours) has the pdf

$$f(t) = \frac{2e^t}{(e^t+1)^2} \quad t>0$$

A random sample of $n = 30$ units from this density function is tested at the same time $t = 0$ and that the units operate independently.

(a) What is the probability that at least 2 units will fail within the first hour? Determine the expected, and the variance of the, number of failures during the first hour.

(b) What is the probability that a unit will fail during the second hour of testing, given that it has survived the first hour?

24. Consider the lifetime of a communication device, having a pdf

$$f(t) = \begin{cases} 0 & t \le t_0 \\ \frac{4(t-t_0)}{(t_1-t_0)^2} & t_0 \le t \le \frac{t_0+t_1}{2} \\ \frac{4(t_1-t)}{(t_1-t_0)^2} & \frac{t_0+t_1}{2} \le t \le t_1 \\ 0 & t_1 \le t \end{cases}$$

(a) What is the failure rate function of this communication device?

(b) Graph of this failure rate function for $t_0 = 25$ h and $t_1 = 75$ h.

25. The pdf of the truncated exponential function is give by

$$f(t) = \begin{cases} 0 & t < t_0 \\ \frac{1}{\theta}e^{-\frac{t-t_0}{\theta}} & t \ge t_0 \end{cases}$$

Show that the variance and MTTF of the life distribution are variance $\sigma^2 = \theta^2$ and $MTTF = \theta + t_0$.

26. Use the Laplace transforms to solve the following

$$\frac{\partial^2 f(t)}{\partial t^2} - 3\frac{\partial f(t)}{\partial t} + 2f(t) = 2e^{-t}$$

with an initial conditions: $f(0) = 2$ and $\frac{\partial f(0)}{\partial t} \equiv f'(0) = -1$.

27. True or false:

(a) All populations in nature are finite.

(b) A sample of observations is a collection of individual observations from a population.

(c) Parameters are sample quantities.

(d) Sample statistics provide estimates of parameters.

(e) Parameters are known when we use a sample rather than a complete enumeration.

28. How many ways are there to arrange the seven letters in the word SYSTEMS?

29. How many ways are there to rank n candidates for the job of corporate chief executive? If the ranking is made at random (each ranking is equally likely), what is the probability that the fifth candidate, Mr. Best, is in second place?

30. A committee of k people is to be chosen from a set of 7 women and 4 men. How many ways are there to form the committee if:

   (a) The committee has 5 people, 3 women and 2 men.
   (b) The committee can be any positive size but must have equal numbers of women and men.
   (c) The committee has 4 people and at least 2 are women.
   (d) The committee has 4 people and one of them must be Mr. Good.
   (e) The committee has 4 people, two of each sex, and Mr. and Mrs. Good cannot both be on the committee.

31. How many ways are there to form a sequence of 10 letters from 4 a's, 4 b's, 4 c's and 4 d's if each letter must appear at least twice?

32. The probability density function of a component is given by

$$f(t) = \begin{cases} \frac{3}{a}\left(1 - \frac{t}{a}\right)^2 & \text{for } 0 \le t \le a \\ 0 & \text{otherwise} \end{cases}$$

   Determine the following:

   a. Reliability function, $R(t)$.
   b. Failure rate function, $h(t)$.
   c. The expected life of the component, MTTF.

33. The probability density function of the lifetime of an electric motor is given by

$$f(t) = 0.25\, t e^{-0.5t} \quad t > 0$$

   where t is in years.

   (a) What is the probability of failure during the first year?
   (b) What is the probability of the motor's lasting at least 3 years?
   (c) If no more than 3% of the motors are to require warranty services, what is the maximum number of days for which the motor can be warranted?

34. Claim amounts for wind damage to insure homes are independent random variables with common density function

$$f(x) = \begin{cases} 3x^{-4} & \text{for } x > 1 \\ 0 & \text{otherwise} \end{cases}$$

   where x is the amount of a claim in thousands. Suppose 3 such claims will be made. What is the expected value of the largest of the three claims?

35. A filling station is supplied with gasoline once a week. If its weekly volume of sales in thousands of gallons is a random variable with probability density function

$$f(x) = \begin{cases} 5(1-x)^4 & 0 < x < 1 \\ 0 & \text{otherwise} \end{cases}$$

(a) what need the capacity of the tank be so that the probability of the supplies being completely used up in a given week is no more than 0.01?

(b) what is the expected volume of sales in a given week? Show your work.

36. Liquid waste produced by a factory is removed once a week. The weekly volume of waste in thousands of gallons is a continuous random variable with probability density function

$$f(x) = \begin{cases} 105x^4(1-x)^2 & \text{for } 0 < x < 1 \\ 0 & \text{otherwise.} \end{cases}$$

Assume that the storage tank has a capacity of 0.9 expressed in thousands of gallons. The cost of removing $x > 0$ units of waste at the end of the week is: $(1.25 + 0.5x)$. Additional costs $(5 + 10z)$ are incurred when the capacity of the storage tank is not sufficient and an overflow of $z > 0$ units of waste occurs during the week. What is the expected value of the weekly costs?

37. You have two light bulbs for a particular lamp. Let $X =$ the lifetime of the first bulb, and $Y =$ the lifetime of the second bulb. Suppose that X and Y are independent and that each has an exponential distribution with constant failure rate parameter $\lambda = 0.001$ per hour.

(a) What is the probability that each bulb lasts at most 1000 h?

(b) What is the probability that the total lifetime of the two bulbs is at most 2000 h?

(c) Determine the expected total lifetime of the two bulbs.

38. Claim amounts for wind damage to insure homes are independent random variables with common probability density function

$$f(x) = \begin{cases} 3x^{-4} & \text{for } x \geq 1 \\ 0 & \text{otherwise} \end{cases}$$

where x is the amount of a claim in thousands. Suppose 3 such claims will be made. What is the expected value of the smallest of the three claims?

39. Claim amounts for wind damage to insure homes are independent random variables with common probability density function

$$f(x) = \begin{cases} 3x^{-4} & \text{for } x \geq 1 \\ 0 & \text{otherwise} \end{cases}$$

where x is the amount of a claim in thousands. Suppose 5 such claims will be made. What is the expected value of the smallest of the five claims?

40.   A system consisting of one original unit plus a spare can function for a random amount of time X. If the density of X is given (in units of months) by

$$f(x) = \begin{cases} C x e^{-\frac{x}{2}} & \text{for } x > 0 \\ 0 & \text{or } x \leq 0 \end{cases}$$

where C is a constant, what is the probability that the system functions for at least 5 months?

41.   Claim amounts for wind damage to insure homes are independent random variables with common probability density function

$$f(x) = 3e^{-3x} \quad \text{for } x > 0$$

where x is the amount of a claim in thousands. Suppose 4 such claims will be made. What is the expected value of the smallest of the four claims?

42.   A certain simulation experiment is to be performed until a successful result is obtained. The trials are independent and the cost of performing the experiment is \$25,000; however, if a failure results, it costs \$5,000 to "set up" for the next trial. The experimenter would like to know the expected cost of the project. If Y is the number of trials required to obtain a successful experiment, then the cost function would be

$$C(Y) = \$25,000Y + \$5,000(Y - 1)$$

(a)   Determine the expected cost of the project.
(b)   Assume the probability of success on a single trial is 0.25. Suppose that the experimenter has a maximum of \$500,000. What is the probability that the experimental work would cost more than \$500,000?

43.   Assume that the product failures from an assembly line are 5%. Suppose that if a product that is defective, a diagnostic test indicates that it is defective 92% of the time, and if the product is good, the test indicates that it is good 96% of the time. Calculate the probability that the test will indicate that the product is defective?

44.   A filling station is supplied with gasoline once a week. If its weekly volume of sales in thousands of gallons is a random variable with probability density function

$$f(x) = \begin{cases} \frac{3}{2}(1 - x^2) & 0 \leq x \leq 1 \\ 0 & \text{otherwise} \end{cases}$$

(a) what need the capacity of the tank be so that the probability of the supplies being completely used up in a given week is no more than 0.05? Show your work.

(b) what is the expected volume of sales in a given week? Show your work.

45. The completion time X for a project has cumulative distribution function

$$
F(x) = \begin{cases}
0 & \text{for } x < 0 \\
\frac{x^3}{3} & \text{for } 0 \le x < 1 \\
1 - \frac{1}{2}(\frac{7}{3} - x)(\frac{7}{4} - \frac{3}{4}x) & \text{for } 1 \le x < \frac{7}{3} \\
1 & \text{for } x \ge \frac{7}{3}.
\end{cases}
$$

Calculate the expected completion time.

46. Claim amounts for wind damage to insure homes are independent random variables with common density function

$$
f(x) = 0.5\, x^2\, e^{-x} \text{ for } x > 0
$$

where x is the amount of a claim in thousands. What is the expected amount of the claims?

47. The scheduled length of a management meeting is b (in minutes). It is unlikely that the meeting will end before a (minutes), and it is impossible that it will last more than L since another meeting is scheduled in the same room. Assume that the cumulative distribution function F of the life time X (duration of the meeting), corresponding to the above restrictions, is given by:

$$
F(x) = \begin{cases}
0 & for\ 0 \le x \le a \\
\beta\left(\frac{x-a}{b-a}\right)^2 & for\ a \le x \le b \\
\left(\frac{1-\beta}{L-b}\right)x + \frac{\beta L - b}{L-b} & for\ b \le x \le L \\
1 & for\ L \le x < \infty
\end{cases}
$$

where $0 < \beta < 1$.

(a) Calculate the expected length of the meeting, $E(X)$.

(b) Given $a = 40$, $b = 50$, $L = 60$, and $\beta = 0.8$. Compute $E(X)$.

Reference

Lehmann EL, Casella G (1998) Theory of point estimation, 2nd ed. Springer

Chapter 2
Distribution Functions and Its Applications

In some situations, particularly when evaluating reliability of a product in the field, exact information on the failure times (Ts) is not possible to obtain, but the information on the number of failures N(a, b) during the time interval $[a, b)$ perhaps is available. The difference between the actual time til failure, T, and the number of failures, N, is that T is a continuous random variable while N is a discrete random variable. Random variables can be discrete or continuous.

Discrete—Within a range of numbers, discrete variables can take on only certain values. For example, suppose that we flip a coin and count the number of heads. Therefore, the number of heads is a discrete variable. And because the number of heads results from a random process—flipping a coin—it is a discrete random variable.

Continuous—Continuous variables, in contrast, can take on any value within a range of values. For example, suppose we randomly select an individual from a population. The age of the person selected is determined by a chance event; so age is a continuous random variable.

All probability distributions can be classified as discrete probability distributions or as continuous probability distributions, depending on whether they define probabilities associated with discrete variables or continuous variables.

This chapter describes most common distribution functions such as binomial, Poisson, geometric, exponential, normal, lognormal, student's t, gamma, Pareto, Beta, Rayleigh, Cauchy, Weibull, Pham, Vtub-shaped hazard rate, etc. and its applications and useful in engineering and applied statistics. The chapter also describes the central limit theorem and some related statistical characteristics of reliability measures including such as bathtub, Vtub shapes, increasing mean residual life, and new better than used.

© The Author(s), under exclusive license to Springer Nature Switzerland AG 2022
H. Pham, *Statistical Reliability Engineering*, Springer Series in Reliability Engineering,
https://doi.org/10.1007/978-3-030-76904-8_2

2.1 Discrete Distributions

A discrete distribution function is a distribution function for a discrete random variable. It consists of a list of the possible values the random variable can assume along with the probability of assuming each possible value. In this section, we discuss several discrete distributions such as the discrete uniform, Bernoulli, binomial, Poisson, negative binomial, geometric, logarithmic series, and power series that are commonly used to model an uncertain frequency in various applications. Examples are given to illustrate how to apply each distribution in practice.

Discrete Uniform Distribution

The simplest of all discrete probability distributions is one where the random variable assumes each of its values with an equal probability. Such a probability distribution is called a discrete uniform distribution.

If the random variable X assumes the values $x_1, x_2, ..., x_n$, with equal probabilities, then the discrete uniform distribution is given by

$$P(X = x) = \frac{1}{n} \quad \text{for } x = x_1, \ x_2, \ ..., \ x_n$$

The mean and variance of the discrete uniform distribution are

$$E(X) = \frac{\sum_{i=1}^{n} x_i}{n} \equiv \mu \ \text{ and } V(X) = \frac{\sum_{i=1}^{n} (x_i - \mu)^2}{n}$$

Example 2.1 Assume that a light bulb is select at random from a box that contain a 25-W bulb, a 40-W bulb, a 60-W bulb, a 75-W bulb, and a 100-W bulb, each occurs with probability 1/5. Thus we have a uniform probability distribution function is

$$P(X = x) = \frac{1}{5} \quad x = 25, 40, 60, 75, 100$$

Bernoulli Distribution

An experiment often consists of repeated trials, each with two possible outcomes that may be labeled success or failure. One may consider the inspection of units as they come out from the production assembly line where each test may indicate a defective or nondefective unit.

A discrete random variable X with two possible outcomes either 0 or 1 (failure or success) where p is the probability of success in each trial has the Bernoulli distribution if, for any given p

$$P(X = x) = p^x(1 - p)^{1-x} \quad x = 0 \text{ or } 1 \tag{2.1}$$

In other words, let us assume that a trial results in one the two outcomes: success (S) and failure (F) and that $P(S) = p$ and $P(F) = 1 - p$. If we define a random variable

$X = 0$ when F occurs and $X = 1$ when S occurs then

$$P(X = x) = p^x(1 - p)^{1-x} \quad x = 0 \text{ or } 1$$

Example 2.2 A vending machine usually provides snack including soft drink, coffee and cookies, on insertion of a coin. Over a period of time, it is found not to work on three uses in every thousand times on the average. If we define X as a random variable and that $x = 1$ if the machine works and $x = 0$ if it fails to work. Thus, we obtain

$$P(X = x) = \begin{cases} 0.997 & \text{for } x = 1 \\ 0.003 & \text{for } x = 0. \end{cases}$$

Binomial Distribution

The binomial distribution is one of the most widely used discrete random variable distributions in reliability and quality inspection. It has applications in reliability engineering, e.g., when one is dealing with a situation in which an event is either a success or a failure.

The binomial distribution can be used to model a random variable X which represents the number of successes (or failures) in n independent trials (these are referred to as Bernoulli trials), with the probability of success (or failure) being p in each trial. The pdf of the binomial distribution, $b(n, p)$, is given by

$$P(X = x) = \binom{n}{x} p^x (1 - p)^{n-x} \quad x = 0, 1, 2, \ldots, n$$

$$\binom{n}{x} = \frac{n!}{x!(n - x)!} \tag{2.2}$$

where n = number of trials.

x = number of successes.

p = single trial probability of success.

In fact, a binomial random variable can be considered as a sum of a Bernoulli random variables, that is, as the number of successes for example, in n Bernoulli trails.

The expected value and variance of the binomial random variable X are, respectively

$$E(X) = np$$

and

$$V(X) = np(1 - p)$$

The coefficient of skewness is given by

$$S_c = \frac{E(X - \mu)^3}{\sigma^3} = \frac{np(1 - p)(1 - 2p)}{\sqrt{(np(1 - p))^3}} = \frac{1 - 2p}{\sqrt{np(1 - p)}}$$

If $p = 0.5$ then $S_c = 0$ and the distribution is symmetric. If $p < 0.5$, the distribution is positively skewed, and it is negatively skewed if $p > 0.5$.

The coefficient of kurtosis is

$$K_c = 3 - \frac{6}{n} + \frac{1}{np(1 - p)}$$

The reliability function, $R(k)$, (i.e., at least k out of n items are good) is given by

$$R(k) = \sum_{x=k}^{n} \binom{n}{x} p^x (1 - p)^{n-x} \tag{2.3}$$

Example 2.3 Suppose in the production of lightbulbs, 90% are good. In a random sample of 20 lightbulbs, what is the probability of obtaining at least 18 good lightbulbs?

Solution The probability of obtaining at least 18 good lightbulbs is given by

$$R(18) = \sum_{x=18}^{20} \binom{20}{x} (0.9)^x (0.1)^{20-x} = 0.667$$

Example 2.4 A life insurance company is interested to predict in the coming year on the number of claims that it may receive on a particular town due to the severe weather such as floods. All of the policies in the town are written on floods. There are currently 250 active policies in the town. The company estimates that the probability of any flood occurs in the coming year is 0.002. Determine the probability that the insurance company receives more than 4 claims in this town out of 250 active policies assuming that the claims are independent.

Solution Let N be the number of claims in the coming year. Then $N \sim$ binomial $(250, p = 0.002)$. Then the desired probability is given by

$$P(N > 4) = 1 - \sum_{n=0}^{4} \binom{250}{n} (0.002)^n (0.998)^{250-n} = 1.6664 \ 10^{-4}$$

Example 2.5 Assume there are 10 identical units to place on test at the same time $t = 0$ where each unit has an exponential life distribution with a mean time to failure MTTF $\mu = 5$ h and that the units operate independently.

(a) Determine the expected and the variance of number of failures during the first hour, i.e. interval [0, 1).
(b) What is the expected and the variance of the number of failures during the interval [1, 2).

Solution

(a) Let N_1 be the number of failures during the first hour, in the time interval [0, 1). The probability of failure of a unit in the time interval [0, 1) is

$$q_1 = 1 - e^{-\frac{1}{5}} = 0.1813$$

The distribution of N_1 that the probability of exactly i failures in [0, 1) is

$$P(N_1 = i) = \binom{10}{i}(0.1813)^i(0.8187)^{10-i}$$

The expected number of failures (N_1) during the first hour is

$$E(N_1) = n\,q_1 = 10\,(0.1813) = 1.813$$

The variance of N_1 is

$$\begin{aligned} V(N_1) &= n\,q_1\,(1 - q_1) \\ &= 10\,(0.1813)(0.8187) = 1.4843 \end{aligned}$$

Similarly, let N_2 be the number of failures during the second hour, in the time interval [1, 2). Assume that the unit will not fail during the first hour, the probability of failure of a unit in the time interval [1, 2) is

$$\begin{aligned} q_2 &= P(1 < T \le 2) = P(T \le 2) - P(T \le 1) \\ &= (1 - e^{-\frac{2}{5}}) - (1 - e^{-\frac{1}{5}}) = 0.1484 \end{aligned}$$

The distribution of N_2 that the probability of exactly i failures during the time interval [1, 2) is

$$P(N_2 = i) = \binom{10}{i}(0.1484)^i(0.8516)^{10-i}$$

The expected number of failures (N_2) during the interval [1, 2) is

$$E(N_2) = n\,q_2 = 10\,(0.1484) = 1.484$$

The variance of N_2 is

$$V(N_2) = n\, q_2\, (1 - q_2)$$
$$= 10\, (0.1484)(0.8516) = 1.2638$$

(b)   Similarly, one can easily obtain the distribution of the number of failures and expected number of failures during the third hour, the fourth hour (see Problem 22 in Chap. 1).

Example 2.6 An electronic manufacturer produces air cooling units for computer systems. There is a 3% chance that any given cooling unit is defective, independent of the other units produced. Determine the probability that more than 2 units in a shipment of 100 are defective.

Solution Let X be the number of defectives in a shipment of 100, then X has a binomial distribution with n = 100 and p = 0.97, i.e., $X \sim b(100, 0.97)$ where p is the probability of a non-defective unit. The desired probability is given by

$$P(X > 2) = 1 - P(X \le 2)$$
$$= 1 - \sum_{i=0}^{2} \binom{100}{i}(0.03)^i\,(0.97)^{100-i}$$
$$= 0.5802$$

Negative Binomial Distribution

A negative binomial distribution can be used to model the number of failures before the rth success in repeated mutually independent Bernoulli trials, each with probability of success p. In other words, the random variable X that represents the number of failures before the rth success, each with probability of failure $(1 - p)$, has the negative binomial distribution is given by

$$P(X = x) = \binom{x + r - 1}{r - 1} p^r (1 - p)^x \quad x = 0,\ 1,\ 2,\ \dots \tag{2.4}$$

In fact, if x is a given positive integer and $(X + r)$ is the number of independent Bernoulli trials required to get r successes when the probability of success in a single trial is p, then the random variable X has the probability distribution given in Eq. (2.4). Applications include acceptance sampling in quality control and modeling of demand for a product.

When $r = 1$, the negative binomial distribution becomes geometric distribution. When r is a positive integer number, the negative binomial distribution function is also known as the Pascal distribution. From Eq. (2.4), the mean and the variance of the negative binomial distribution are

$$E(X) = \frac{r\,(1 - p)}{p} \tag{2.5}$$

and

$$V(X) = \frac{r\,(1-p)}{p^2}$$

respectively. We also can consider the number Y of Bernoulli trials up to and including the rth success. In this case, from Eq. (2.4)

$$P(Y = n) = P(X = n - r) = \binom{n-1}{r-1} p^r (1-p)^{n-r} \quad n = r,\, r+1,\, r+2, \ldots$$

$$(2.6)$$

Both X and Y are referred to as having a negative binomial distribution.

Example 2.7 A given event occurs (success) with probability p and failure with probability $(1 - p)$ and independent trials are performed. Let X be r fewer than the number of trials needed to produce r successes. That is, X is the number of failures encountered on the way to the first r successes, and therefore, $X = x$ means we need $(x + r)$ trials in general. Then it is easy to obtain that

$$P(X = x) = p \binom{x+r-1}{r-1} p^{r-1} q^x \quad x = 0, 1, 2, \ldots$$

or

$$P(X = x) = \binom{x+r-1}{r-1} p^r q^x \quad x = 0, 1, 2, \ldots$$

Since the last trial must be a success and there are $\binom{x+r-1}{r-1}$ possible ways to order with $(r - 1)$ successes. Here X is negative binomial with r. The geometric distribution is a special case of negative binomial, that is when $r = 1$.

Example 2.8 An entrepreneurer of start-up company calls potential customers to sale a new product. Assume that the outcomes of consecutive calls are independent and that on each call he has 20% chance of making a sale. His daily goal is to make 5 sales and he can make only 25 calls in a day.

(a) What is the probability that he achieves his goal in 15 trials?
(b) What is the probability that he achieves his goal in between 15 and 20 trials?

(c)   What is the probability that he does not achieve his daily goal?

Solution

(a)   We wish to find the probability that his fifth sale (success) will be at trial 15.
      Note that in this case, the number of successes $r = 5$ is fixed. Thus the number
      of calls is random, and we have the negative binomial distribution. Here p
      $= 0.20$. Let Y be the number of trials up to and including the fifth success.
      Therefore, from the negative binomial distribution below

$$P(Y = n) = \binom{n-1}{r-1} p^r (1-p)^{n-r} \quad n = 1, 2, \ldots$$

$$\text{where } \binom{n}{r} = \frac{n!}{r!(n-r)!}$$

we can obtain

$$P(Y = 15) = \binom{14}{4}(0.2)^5(0.8)^{10} = 0.0344.$$

(a)   see Problem 27.
(b)   Let S_n be the number of successes in n trials. Here the number of trials is fixed.
      (i.e., $n = 20$). We can then use the binomial distribution. The entrepreneurer
      does not achieve his daily goal if $S_n \leq 2$. Thus,

$$P(S_{20} \leq 2) = \sum_{x=0}^{2} P(S_{20} \leq x)$$

$$= \binom{20}{0}(0.2)^0(0.8)^{20} + \binom{20}{1}(0.2)^1(0.8)^{19}$$

$$+ \binom{20}{2}(0.2)^2(0.8)^{18}$$

$$= 0.1969$$

Truncated Binomial Distribution

The probability distribution

$$P(X = x) = \frac{\binom{n}{x} p^x (1-p)^{n-x}}{1 - (1-p)^n} \quad x = 1, 2, \ldots, n$$

$$\binom{n}{x} = \frac{n!}{x!(n-x)!} \tag{2.7}$$

is known as the truncated binomial distribution, truncated below $x = 1$, where $n =$ number of trials; $x =$ number of successes; and $p =$ single trial probability of success.

This is the conditional distribution of the number of successes when it is known that there are no failures. The mean of the truncated binomial distribution is given by

$$E(X) = \frac{np}{1 - (1-p)^n} \tag{2.8}$$

Compound Binomial Distribution

If the parameter p of the binomial distribution

$$P(X = x) = \binom{n}{x} p^x (1-p)^{n-x} \quad x = 0, 1, 2, ..., n$$

$$\binom{n}{x} = \frac{n!}{x!(n-x)!} \tag{2.9}$$

behaves as a random variable with the density function

$$f(p) = \frac{1}{B(a,b)} p^{\alpha-1}(1-p)^{\beta-1} \quad 0 \le p \le 1 \tag{2.10}$$

where a and b are positive-valued parameters and

$$B(a,b) = \int_0^1 y^{a-1}(1-y)^{b-1}dy$$

then the probability distribution of the number of successes X in n trials is given by

$$P(X = x) = \frac{\binom{n}{x} B(a+x, n+b-x)}{B(a,b)} \quad x = 0, 1, 2, ..., n \tag{2.11}$$

The mean and the variance of the compound binomial distribution can be obtained as follows:

$$E(X) = \frac{na}{a+b} \tag{2.12}$$

and

$$V(X) = \frac{n\,a\,b}{(a+b)(a+b+1)}\left(1 + \frac{n}{a+b}\right) \tag{2.13}$$

respectively.

Multinomial Distribution

A random variable $X = (X_1, X_2, ..., X_k)$ has the multinomial distribution if for any $p_i \geq 0$ for $i = 1, 2, ..., k$ and $\sum_{i=1}^{k} p_i = 1$

$$P[X_1 = x_1, X_2 = x_2, ..., X_k = x_k] = \begin{cases} \dfrac{n!}{x_1! x_2! ... x_k!} p_1^{x_1} p_2^{x_2} ... p_k^{x_k} \\ 0 \quad \text{otherwise} \end{cases} \tag{2.14}$$

where $x_i = 0, 1, ..., n$ and $\sum_{i=1}^{k} x_i = n$.

Example 2.9 Suppose an experiment of a new product can result in exactly m possible outcomes and their corresponding probabilities are $s_1, ..., s_m$ and $p_1, ..., p_m$, respectively. Suppose the experiment is, independently, repeated n times and let X_i be the number of outcomes of type s_i for $i = 1, 2, ..., m$, then it has the multinomial distribution as follows

$$P[X_1 = s_1, X_2 = s_2, ..., X_k = s_m] = \begin{cases} \dfrac{n!}{s_1! s_2! ... s_m!} p_1^{s_1} p_2^{s_2} ... p_m^{s_k} \\ 0 \quad \text{otherwise} \end{cases}$$

where $s_i = 0, 1, ..., n$ and $\sum_{i=1}^{m} s_i = n$.

Poisson Distribution

Although the Poisson distribution can be used in a manner similar to the binomial distribution, it is used to deal with events in which the sample size is unknown. A Poisson random variable is a discrete random variable distribution with probability density function given by

$$P(X = x) = \frac{\lambda^x e^{-\lambda}}{x!} \quad \text{for } x = 0, 1, 2, ... \tag{2.15}$$

where λ is a constant failure rate and x is the number of events. In other words, $P(X = x)$ is the probability of exactly x failures occur.

A Poisson random variable X, i.e. $X \sim$ Poisson (λ), can interpret as follows: It is the number of occurrences of a rare event during some fixed time period in which the expected number of occurrences of the event is λ and individual occurrences of the event are independent of each other. A Poisson distribution is used to model a Poisson process and also is useful for modeling claim frequency on individual policies such as auto insurance, household insurance, customer services. A Poisson random variable has mean and variance both equal to λ where λ is called the parameter of the distribution. The skewness coefficient is

$$S_c = \frac{1}{\sqrt{\lambda}}$$

and the kurtosis coefficient is

$$K_c = 3 + \frac{1}{\lambda}$$

The reliability Poisson distribution by time t, $R(k)$ (the probability of k or fewer failures) can be defined as follows

$$R(k) = \sum_{x=0}^{k} \frac{(\lambda t)^x e^{-\lambda t}}{x!} \tag{2.16}$$

This distribution can be used to determine the number of spares required for the reliability of standby redundant systems during a given mission.

Example 2.10 The number of X-ray machine failures per week in a laboratory has a Poisson distribution with parameter $\lambda = 2$. Currently facilities at this laboratory can repair two machines per week. Failures in excess of two will have to repair by an outside specialist contractor. Determine

(a) the probability of having X-ray machines repaired by a specialist contractor on any given week?
(b) the expected number of X-ray machines repaired by the specialist contractor per week?

Solution The probability of having exactly x number of X-ray machine failures per week is

$$P(X = x) = \frac{\lambda^x e^{-\lambda}}{x!} \quad \text{for } x = 0, 1, 2, \dots.$$

(a) The probability of having X-ray machines repaired by a specialist contractor is

$$P(X \geq 3) = 1 - P(X \leq 2)$$

$$= 1 - \sum_{k=0}^{2} \frac{2^k e^{-2}}{k!} = 0.3232$$

(b)   The expected number of X-ray machines repaired by the specialist contractor
is

$$\sum_{k=3}^{\infty} (k-2)P(X=k) = \sum_{k=3}^{\infty} (k-2)\frac{2^k e^{-2}}{k!} = 0.54$$

Example 2.11 A nuclear plant is located in an area susceptible to both high winds
and earthquakes. From the historial data, the mean frequency of large earthquakes
capable of damaging important plant structures is one every 50 years. The corresponding frequency of damaging high winds is once in 25 years. During a strong
earthquake, the probability of structure damage is 0.1. During high winds, the damage
probability is 0.05. Assume that earthquakes and high winds can be described by
independent Poisson random variables and that the damages caused by these events
are independent. Let us exercise the following questions:

(a)   What is the probability of having strong winds but not large earthquakes during
a 10-year period?
(b)   What is the probability of having strong winds and large earthquakes in the
10-year period?
(c)   What is the probability of building damage during the 10-year period?

Solution To begin with question (a), let the random variable X and Y represent the
number of earthquakes and the number of occurrences of high winds, respectively.
We assume that the two random variables are statistically independent. The mean of
X and Y are, respectively, given by

$$\lambda_X = \frac{1}{50 years}(10 years) = 0.2$$

and

$$\lambda_Y = \frac{1}{25 year}(10 years) = 0.4$$

The conditional damage probabilities are given as follows:

$$P(damage/earthquake) = 0.1$$

and

$$P(damage/wind) = 0.05$$

Let the event

$A = \{\text{strong winds and no earthquakes}\}$.
$B = \{\text{strong winds and large earthquakes}\}$.
$C = \{\text{building damage}\}$.

Assuming that winds and earthquakes are independents, the probability of having strong winds but not earthquakes during the 10-year period can be written as

$$P(A) = P(winds)\, P(no\ earthquakes)$$
$$= [1 - P(no\ winds)]P(no\ earthquakes)$$

Therefore, we obtain

$$P(A) = (1 - e^{-0.4})(e^{-0.2}) = 0.27$$

(b) The probability of having strong winds and earthquakes during the 10-year period can be obtained

$$P(B) = P(winds)\, P(earthquakes)$$
$$= [1 - P(no\ winds)][1 - P(no\ earthquakes)]$$
$$= (1 - e^{-0.4})(1 - e^{-0.2}) = 0.06$$

(c) We assume that there will not be multiple occurrences of earthquakes and high winds during the 10-year period. Therefore, the probability of building damage can be written as

$$P(C) = P(damage/earthquakes)\, P(earthquakes)$$
$$+ P(damage/wind)P(wind)$$
$$- P(damage/earthquakes\ and\ wind)P(earthquake\ and\ wind)$$
$$= P(damage/earthquakes)\, P(earthquakes)$$
$$+ P(damage/wind)P(wind)$$
$$- P(damage/earthquakes)P(damage/wind)P(earthquake\ and\ wind)$$
$$= (1 - e^{-0.2})(0.1) + (1 - e^{-0.4})(0.05) - (0.05)(0.1)(0.06)$$
$$= 0.0343$$

Lemma 2.1 *An independent sum of Poisson random variables is also Poisson.*

For example, if $X_1 \sim$ Poisson (λ_1), $X_2 \sim$ Poisson (λ_2), and X_1 and X_2 are independent, then $X_1 + X_2 \sim$ Poisson $(\lambda_1 + \lambda_2)$.

Suppose that for each time t, the number of arrivals in the interval $[0, t]$ is Poisson (λt). Then the waiting time until the first arrival is Exponential (λ) and the waiting time until the nth arrival is Gamma (n, λ).

Example 2.12 The major accident of hospitalization tends to vary with the snow storms over the winter season of a given year. Data from an ambulance emergency center suggests that one should expect 2 hospitalization in October, 3 hospitalization in November, 4 hospitalizations in December, and 6 hospitalizations in January. Assume that hospitalizations are independent. Determine the probability that there are more than 10 hospitalization over a period from October to January.

Solution Let X_1 be the number of hospitalizations in October, X_2 be the number of hospitalizations in November, X_3 be the number of hospitalizations in December, and X_4 be the number of hospitalizations in January. Let N be the total number of hospitalizations over a period from October to January, then.

$$X_1 \sim \text{Poisson (2)}, X_2 \sim \text{Poisson (3)}, X_3 \sim \text{Poisson (4)}, X_4 \sim \text{Poisson (6)}.$$

and so,

$$N = X_1 + X_2 + X_3 + X_4 \sim Poisson \ (2 + 3 + 4 + 6 = 15).$$

Then we obtain

$$P(N > 10) = 1 - P(N \leq 10)$$

$$= 1 - \sum_{n=0}^{10} \frac{(15)^n \, e^{-15}}{n!} = 0.883$$

Truncated Poisson Distribution

The probability function

$$P(X = x) = \frac{\lambda^x}{(e^\lambda - 1)x!} \quad \text{for } x = 1, 2, \dots. \tag{2.17}$$

is called the truncated Poisson density function below $x = 1$ where λ is the constant failure rate and x is the number of events. The mean of the truncated Poisson distribution is

$$E(X) = \frac{\lambda}{(1 - e^{-\lambda})} \tag{2.18}$$

Compound Poisson Distribution

Let $X_k, k = 1, 2, \dots$ be i.i.d. random variables and let N be a non-negative integer-value random variable, independent of the X_k. Denote

$$S_N = \sum_{i=1}^{N} X_i \qquad (2.19)$$

and $S_0 = 0$.

If N is a Poisson random variable, then S_N is called a compound Poisson random variable.

Theorem 2.1 *Let $E(X_1) = \mu$ and $Var(X_1) = \sigma^2$. If N is Poisson with parameter λ then $S_N = \sum_{i=1}^{N} X_i$ has mean and variance.*

$$E(S_N) = \mu\lambda \qquad (2.20)$$

and

$$V(S_N) = \lambda(\mu^2 + \sigma^2). \qquad (2.21)$$

Geometric Distribution

Geometric distribution plays important roles in many applications including finances, manufacturing quality control, stock market, insurance, biology, and reliability. Consider a sequence of independent trials in which each having the same probability for success, say p. Let N be a random variable that counts for the number of trials until the first success. This distribution is called the geometric distribution. It has a pdf given by

$$P(N = n) = p(1 - p)^{n-1} \qquad n = 1, 2, \dots \qquad (2.22)$$

For the sake of brevity, N is called a geometric random variable. In fact, a geometric random variable N can be written in terms of a sequence $\{X_i\}$ for $i = 1, 2, \dots$ of Bernouli independent identically random variable defined as follows:

$$N = \min\{i : X_i = 1\} \qquad (2.23)$$

where

$$X_i = \begin{cases} 1 & \text{with probability p} \\ 0 & \text{with probability } 1 - p \end{cases} \qquad i = 1, 2, \dots$$

In other words, the right-hand side of Eq. (2.23) is the waiting time for the first success in the Bernoulli trials with probability p of success at each trial. This waiting time is equal to n, if first $(n - 1)$ trials are not successful and the nth trial is successful. From Eq. (2.22), the corresponding cdf is given by

$$F(n) = P(N \leq n) = p \sum_{i=1}^{n} (1 - p)^{i-1} = 1 - (1 - p)^n \quad n = 1, 2, \ldots$$

Thus,

$$P(N > n) = (1 - p)^n \quad n = 1, 2, \ldots$$

The expected value and variance are, respectively

$$E(N) = \frac{1}{p} \tag{2.24}$$

and

$$V(N) = \frac{1 - p}{p^2}$$

The coefficient of skewness is given by

$$S_c = \frac{E(X - \mu)^3}{\sigma^3} = \frac{2 - p}{\sqrt{1 - p}}$$

Example 2.13 At busy time a telephone exchange is very near capacity, so callers have difficult placing their calls. It may be of interest to know the number of attempts necessary in order to gain a connection. Suppose that we let $p = 0.05$ be the probability of a connection during busy time. We are interested in knowing the probability that 4 attempts are necessary for a successful call. What is the expected number of calls necessary to make a connection?

Solution The probability that four attempts are necessary for a successful call is:

$$P(X = 4) = p(1 - p)^3 = (0.05)(1 - 0.05)^3 = 0.0429$$

The expected number of calls necessary to make a connection is

$$E(X) = \frac{1}{p} = \frac{1}{0.05} = 20$$

Application—Software Debugging: Suppose we have to test a complex software program for the purpose of detecting and correcting errors. During debugging, we load specific data into the program and force the program to run along the prescribed path. If after several attempts, the program run correctly then it is considered as debugged. But programmers know that almost all large programs are not free of errors. Let the program to be debugged contain a fault. How long does it take to detect this error? Suppose that the probability of detecting the error during any attempt is equal to p. Let X_i be the time needed to obtain the ith output for given inputs. Given that all X_i are i.i.d. random variable, the total debugging time T can be defined as

$$T = Y_N = \sum_{i=1}^{N} X_i$$

where N is a geometric random variable such that

$$P(N = n) = p(1 - p)^{n-1} \quad n = 1, 2, \ldots$$

and the desired time T can be represented as a geometric sum.

Example 2.14 The probability of failure of a medical device when it is tested is equal to 0.05. Assume independence of tests. What is the probability that failure occurs in less than 15 tests?

Solution The required probability that failure occurs in less than 15 tests is

$$
\begin{aligned}
P(N \le 14) &= 1 - (1 - p)^{14} \\
&= 1 - (0.95)^{14} \\
&= 0.5123
\end{aligned}
$$

It should be noted that the geometric distribution is the only discrete distribution having the *memoryless property*. In other words, mathematically, it is

$$P(N \ge (n + s)|N > s) = P(N \ge n) \quad n = 1, 2, \ldots \tag{2.25}$$

This memoryless property can be shown as follows. We know that

$$P(N > n) = P(N \ge n + 1) = (1 - p)^n$$

Thus,

$$
\begin{aligned}
P(N \ge (n + s)|N > s) &= \frac{P(N \ge n + s)}{P(N > s)} \\
&= \frac{(1 - p)^{n+s-1}}{(1 - p)^s} \\
&= (1 - p)^{n-1} \\
&= P(N \ge n).
\end{aligned}
$$

The Pascal Distribution

The random variable X for the Pascal distribution represents the number of trials until the rth success where p is the probability of success of each trial. The Pascal distribution of X is given by

$$P(X = n) = \binom{n-1}{r-1} p^r (1-p)^{n-r} \quad n = r, r+1, r+2, \ldots \quad (2.26)$$

The Pascal distribution is the probability distribution of a certain number of successes and failures in a series of independent and identically distributed Bernoulli trials. For n Bernoulli trials with success probability p, the Pascal distribution gives the probability of r successes and $(n-r)$ failures, with a success on the last trial. In other words, the Pascal distribution is the probability distribution of the number of Bernoulli trials before the rth success in a Bernoulli process, with probability p of successes on each trial. A Bernoulli process is a discrete time process, and so the number of trials, failures, and successes are integers. The mean and the variance of the Pascal distribution are

$$E(X) = \frac{r}{p} \quad (2.27)$$

and

$$V(X) = \frac{r(1-p)}{p^2} \quad (2.28)$$

respectively. When $r = 1$ the Pascal distribution becomes geometric distribution. It should be noted that the Pascal distribution and the negative binomial distribution, logically, are in the same logical form.

Hypergeometric Distribution

The hypergeometric distribution is obtained as follows. A sample of size n is drawn, without replacement, from a population of size N composed of k number of successes and $(N - k)$ number of failures. Let X denote a random variable that represents the number of successes in the sample. Then X is a hypergeometric random variable with pdf as follows

$$P(X = x) = \frac{\binom{k}{x}\binom{N-k}{n-x}}{\binom{N}{n}} \quad x = 0, 1, 2, \ldots, n \quad (2.29)$$

Let $p = \frac{k}{N}$ denote the probability of success prior to selecting the first unit. The expected value and variance of the hypergeometric random variable X are, respectively

$$E(X) = \frac{nk}{N} = np$$

and

$$V(X) = n\frac{k}{N}\left(1 - \frac{k}{N}\right)\frac{(N-n)}{(N-1)} = np(1-p)\frac{(N-n)}{(N-1)}$$

Thus the expected number of successes in a sample of size n is np regardless of whether we sample with or without replacement. The variance in the case of sampling without replacement is smaller than the corresponding binomial variance (see Eq. 2.2) by the factor $\frac{N-n}{N-1}$. The behavior of the variances of binomial and hypergeometric distributions is different as the sample size, n, increases. In the binomial distribution, each new unit of the sample contributes the same amount to variance, so the latter grows linearly with n. In sampling without replacement, variance changes in proportion to the product $n(N-n)$; hence it initially grows to reach a maximum when sampling exhausts half of the population and then declines to zero when $n = N$. Variance for $n = 1$ is the same as for $n = N - 1$.

The following theorem connects the binomial and hypergeometric distributions in the situation when the successes come from two sources where each follows the binomial distribution and the total number of successes is fixed.

Theorem 2.2 *Let X and Y be independent random variables with binomial distribution $X \sim b(m, p)$ and $Y \sim b(n, p)$. Then*

$$P(X = k|X + Y = s) = \frac{\binom{m}{k}\binom{n}{s-k}}{\binom{m+n}{s}}. \tag{2.30}$$

If $\frac{n}{N} < 0.1$, the hypergeometric distribution can be approximated by the binomial $b\left(n, \frac{k}{N}\right)$, where the expected of the hypergeometric random variable is

$$E(X) = \frac{nk}{N}$$

Example 2.15 Suppose there are 500 items in the lot and their manufacturer claims that no more than 5% are defective. Then when the shipment is received at an assembly plant, a quality control manager asks his staff to take a random sample of size 20 from it without replacement and will accept the lot if there are at most

1defective item in the sample. If let say 5% are defective in the entire lot, then it consists of 25 defective items. What is the probability that the manager will accept the lot?

Solution Let X be the number of defective items in a sample of size 20, then it has a hypergeometric distribution and it probability function is as follows

$$P(X = x) = \frac{\binom{25}{x}\binom{475}{20-x}}{\binom{500}{20}} \quad x = 0, 1, 2, ..., 20$$

The probability that the manager will accept the lot, means that there will be at most 1 defective item in a random sample of size 20, is as follows:

$$P(X \le 1) = \frac{\binom{25}{0}\binom{475}{20}}{\binom{500}{20}} + \frac{\binom{25}{1}\binom{475}{19}}{\binom{500}{20}}$$

$$= 0.3511941 + 0.3850813 = 0.7362754$$

Logarithmic Series Distribution

The probability mass function for this distribution is

$$P(X = x) = -\frac{\theta^x}{x \log(1-\theta)} \quad x = 1, 2, ..., \infty \tag{2.31}$$

where $0 < \theta < 1$ is a positive-valued parameter, is known as the logarithmic series distribution. The logarithmic series distribution is useful in the analysis of various kinds of data. A description of some of its applications can be found in Johnson and Kotz (1970).

The mean and the variance of this distribution are, respectively,

$$E(X) = \frac{\theta}{-(1-\theta) \log(1-\theta)} \tag{2.32}$$

and

$$V(X) = \frac{\theta}{-(1-\theta)^2 \log(1-\theta)} \tag{2.33}$$

Power Series Distribution

The power series distribution, which was introduced by Kosambi (1949) represents a family of discrete distributions, such as Poisson, binomial and negative binomial. Its probability mass function is given by

$$P(X = x) = \frac{a_x \, \theta^x}{f(x)} \quad x = 0, 1, 2,... \tag{2.34}$$

where

$$a_x \geq 0, \ \theta > 0 \text{ and } f(\theta) = \sum^{\infty} a_x \, \theta^x.$$

This function is defined provided that θ falls inside the interval of convergence of the series in Eq. (2.34) above. It can be seen that the logarithmic series distribution and Poisson distribution are special cases of the power series distribution.

2.2 Continuous Distributions

The continuous distribution function is a distribution function for a continuous random variable where it can take any value in any interval of real numbers or the entire real values domain. In this section, we discuss various continuous distributions—the exponential, the gamma, truncated exponential, Weibull, normal, log normal, chi square, Student's t, beta, log log, that commonly use to model the lifetime of system, products and in statistics, reliability engineering, and industrial management applications.

Exponential Distribution

The exponential distribution plays an essential role in reliability engineering because it has a constant failure rate. It represents the distribution of a random variable giving the time interval between independent events. This distribution has been used to model the lifetime of electronic and electrical components and systems. This distribution is appropriate when a used component that has not failed is as good as a new component—a rather restrictive assumption. Therefore, it must be used diplomatically since numerous applications exist where the restriction of the memoryless property may not apply.

A random variable T is said to have an exponential distribution if its probability density function (pdf) has the form

$$f(t) = \lambda e^{-\lambda t} \ t \geq 0, \ \lambda > 0 \tag{2.35}$$

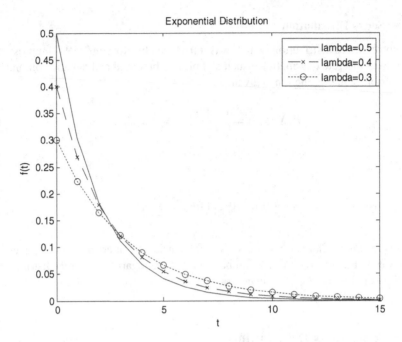

Fig. 2.1 Exponential density function for various values of λ

where $\lambda > 0$ is a parameter of the distribution, also called the constant failure rate. If a random variable T has this distribution, we write $T \sim \text{Exp}(\lambda)$. Figure 2.1 shows the exponential pdf for various values of λ. The reliability function is given by

$$R(t) = \int_t^\infty \lambda e^{-\lambda s} ds = e^{-\lambda t} \quad t \geq 0 \qquad (2.36)$$

An alternate common used of the pdf of an exponential failure time density function is given by

$$f(t) = \frac{1}{\theta} e^{-\frac{t}{\theta}} \quad t \geq 0, \ \theta > 0 \qquad (2.37)$$

and this will lead to the reliability function

$$R(t) = \int_t^\infty \frac{1}{\theta} e^{-\frac{s}{\theta}} ds = e^{-\frac{t}{\theta}} \quad t \geq 0 \qquad (2.38)$$

where $\theta = 1/\lambda > 0$ is a scale parameter of the distribution and $\lambda \geq 0$ is a constant failure rate. The pth percentile point of the exponential distribution with MTTF θ is

$$t_p = -\theta \ln(1 - p).$$

Example 2.16 A certain manufacturing plant has five production lines, all making use of a specific bulk product. The amount of product used in one day can be modeled as having an exponential distribution with a mean of 6 (measurements in thousands), for each of the five lines. If the lines operate independently, find the probability that exactly three of the five lines use more than 6 thousands on a given day.

Solution Let T be the amount used in a day. The probability that any given line uses more than 6 thousands is

$$P(T > 6) = \int_6^\infty \frac{1}{6} e^{-\frac{x}{6}} \, dx = e^{-1} = 0.3679$$

The probability that exactly three of the five lines use more than 6 thousands on a given day is

$$P(\text{exactly 3 use more than 6}) = \binom{5}{3}(0.3679)^3 (1 - 0.3679)^2 = 0.199$$

The hazard function or failure rate for the exponential density function is constant, i.e.,

$$h(t) = \frac{f(t)}{R(t)} = \frac{\frac{1}{\theta}e^{-\frac{1}{\theta}}}{e^{-\frac{1}{\theta}}} = \frac{1}{\theta} = \lambda$$

The failure rate for this distribution is λ, a constant, which is the main reason for this widely used distribution. Because of its constant failure rate property, the exponential is an excellent model for the long flat "intrinsic failure" portion of the bathtub curve. Since most parts and systems spend most of their lifetimes in this portion of the bathtub curve, this justifies frequent use of the exponential (when early failures or wear out is not a concern). The exponential model works well for inter arrival times. When these events trigger failures, the exponential lifetime model can be used.

The mean and variance of an exponentially distributed r.v. T with constant rate λ are given, respectively,

$$E(T) = \frac{1}{\lambda} \quad \text{and} \quad V(T) = \frac{1}{\lambda^2}.$$

We note that the mean and standard deviation of the exponential distribution are equal. We will now discuss some properties of the exponential distribution that are useful in understanding its characteristics, when and where it can be applied.

Property 2.1 (Memoryless property) *The exponential distribution is the only continuous distribution satisfying.*

$$P\{T \geq t\} = P\{T \geq t + s | T \geq s\} \quad \text{for } t > 0, s > 0 \tag{2.39}$$

This result indicates that the conditional reliability function for the lifetime of a component that has survived to time s is identical to that of a new component. This term is the so-called "used-as-good-as-new" assumption. It is worth to note that the exponential distribution and geometric distribution are the only momoryless probability distribution functions. In other words, if time to failure of a unit follows the exponential distribution, it is the lifetime of an item that does not age.

Example 2.17 Suppose the life of a certain security alarm system is exponential distribution with constant failure rate $\lambda = 0.00008$ failures per hour.

(a) What is the probability that a security system will last at least 10,000 h?
(b) What is the probability that a security system will last at least 10,000 h given that it has already survived 5,000 h?

Solution

(a) The probability that a security system will last at least 10,000 h is

$$P\{T \geq 10,000\} = e^{-\lambda t} \, e^{-(0.00008)(10,000)} = e^{-0.8} = 0.4493$$

(b) The probability that a security system will last at least 10,000 h given that it has already survived 5,000 h is

$$P\{T \geq 10,000 | T \geq 5,000\} = P\{T \geq 5,000\}$$
$$= e^{-\lambda t} = e^{-(0.00008)(5,000)} = 0.6703$$

Property 2.2 *If T_1, T_2, ..., T_n, are independently and identically distributed exponential random variables (r.v.'s) with a constant failure rate λ, then*

$$2\lambda \sum_{i=1}^{n} T_i \sim \chi^2(2n) \tag{2.40}$$

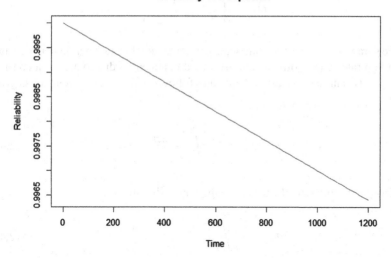

Fig. 2.2 Reliability function versus time

where $\chi^2(2n)$ is a chi-squared distribution with degrees of freedom $2n$. This result is useful for establishing a confidence interval for λ in Chap. 3.

Example 2.18 A manufacturer performs an operational life test on ceramic capacitors and finds they exhibit constant failure rate with a value of 3×10^{-6} failure per hour. What is the reliability of a capacitor at 1000 h?

Solution The reliability of a capacitor at 1000 h is

$$R(t = 1000) = e^{-\lambda t} = e^{-3(10^{-6})(1000)} = 0.997$$

The resulting reliability plot is shown in Fig. 2.2.

Property 2.3 *If T_1, T_2, ..., T_n are independently exponentially distributed random variables (r.v.'s) with constant failure rate $\lambda_1, \lambda_2, ..., \lambda_n$, and let $Y = \min\{T_1, T_2, ..., T_n\}$ then Y is also exponentially distributed with constant parameter $\lambda = \sum_{i=1}^{n} \lambda_i$.*

It is easy to show that if T_1, T_2, ..., T_n, are independently exponentially distributed random variables (r.v.'s) with constant failure rate $\lambda_1, \lambda_2, ..., \lambda_n$, and if $Y = \max\{T_1, T_2, ..., T_n\}$ then Y is not exponentially distributed (see Problem 6).

Lower-bound Truncated Exponential Distributio

The lower-bound truncated exponential distribution is an exponential distribution starting at t_0. The model can be used when no unit fails before the time t_0. The pdf of a lower-bound truncated exponential function is give by

$$f(t) = \begin{cases} 0 & t < t_0 \\ \frac{1}{\theta} e^{-\frac{t-t_0}{\theta}} & t \geq t_0 \end{cases} \qquad (2.41)$$

The parameters θ and t_0 are the scale parameter and location parameter. Note that the failure rate of this lower-bound truncated exponential distribution is a constant for all $t \geq t_0$. This can be easily obtained as follows. The reliability function is given by

$$R(t) = \int_t^\infty f(x)dx = \int_t^\infty \frac{1}{\theta} e^{-\frac{x-t_0}{\theta}} dx = e^{-\frac{t-t_0}{\theta}} \qquad (2.42)$$

Thus, the failure rate of truncated exponential distribution is

$$h(t) = \frac{f(t)}{R(t)} = \frac{\frac{1}{\theta} e^{-\frac{t-t_0}{\theta}}}{e^{-\frac{t-t_0}{\theta}}} = \frac{1}{\theta}. \qquad (2.43)$$

The variance and MTTF of the life distribution are (see Problem 25 in Chap. 1):

$$\text{variance } \sigma^2 = \theta^2$$

and

$$MTTF = \theta + t_0.$$

Upper-bound Truncated Exponential Distribution

For many applications, it is desired to consider to truncate a distribution on an upper tail. Let us assume that the exponential density be truncated at time T. The truncated density function is given by

$$f(t) = \begin{cases} 0 & t > T \\ \frac{\lambda e^{-\lambda t}}{1 - e^{-\lambda T}} & 0 < t \leq T \end{cases} \qquad (2.44)$$

The reliability function is

$$R(t) = \int_t^T f(x)dx = \int_t^T \frac{\lambda e^{-\lambda x}}{1 - e^{-\lambda T}} dx = \frac{e^{-\lambda t} - e^{-\lambda T}}{1 - e^{-\lambda T}}. \qquad (2.45)$$

Obviously as for the upper-bound truncated exponential distribution, the failure rate is not a constant for all $t > T$. To see this, the failure rate of this upper-bound truncated exponential distribution is

$$h(t) = \frac{f(t)}{R(t)} = \frac{\frac{\lambda e^{-\lambda t}}{1-e^{-\lambda T}}}{\frac{e^{-\lambda t}-e^{-\lambda T}}{1-e^{-\lambda T}}} = \frac{\lambda}{1-e^{-\lambda(T-t)}}. \tag{2.46}$$

and is a function of t.

Mixed Truncated Exponential Distribution

The mixed distribution is one whose distribution function resembles both the continuous on an interval and the discrete on another interval. A random variable X with a continuous distribution function

$$F(x) = \begin{cases} 1 - e^{-\lambda x} & \text{for } 0 \le x < a \\ 1 & \text{for } x \ge a \end{cases} \tag{2.47}$$

and has a point mass of size $e^{-\lambda a}$ at $x = a$ is said to have a mixed truncated exponential distribution with parameters λ and a.

The mixed distribution has both discrete and continuous parts. If f_X denotes the density for the continuous part and p_X denotes the mass function for the discrete part, then the expected value of a random variable X can be obtained as follows:

$$E(X) = \underbrace{\int x \, f_x(x) dx}_{\text{continuous part}} + \underbrace{\sum x \, p_X(x)}_{\text{discrete part}} \tag{2.48}$$

Note that in this context, $\int_{\text{continuous part}} f_x(x)dx + \int_{\text{discrete part}} p_X(x) = 1$ where the integral is calculated over the continuous part and the sum is calculated over the discrete part.

Note that the distribution is mixed with a point mass of size $e^{-\lambda a}$ at $x = a$ and a continuous distribution on the interval $0 \le x < a$, and using Eq. (2.48), we can obtain the expected value of X as follows

$$E(X) = \underbrace{\int x \, f_x(x) dx}_{\text{continuous part}} + \underbrace{\sum x \, p_X(x)}_{\text{discrete part}}$$

$$= \int_0^a x \, \lambda e^{-\lambda x} dx + a \, e^{-\lambda a}$$

$$= \frac{1}{\lambda}\left(1 - e^{-\lambda a}\right) \tag{2.49}$$

From the distribution function (see Eq. 2.47), the reliability function is given by

$$R(t) = \begin{cases} 1 & \text{for } t < 0 \\ e^{-\lambda t} & \text{for } 0 \leq t < a \\ 0 & \text{for } t \geq a \end{cases}$$

The system MTTF is given by

$$MTTF = \int_0^a e^{-\lambda t} dt = \frac{1}{\lambda}\left(1 - e^{-\lambda a}\right) \qquad (2.50)$$

which is the same as with the expected value discussed here. Similarly, the variance of a mixed truncated exponential distribution with parameters λ and a is

$$V(X) = \frac{1}{\lambda^2}\left(1 - 2\lambda a e^{-\lambda a} - e^{-2\lambda a}\right). \qquad (2.51)$$

Detail is left to the reader. See Problem 12.

Uniform Distribution

This represents the distribution of a random variable whose values are completely uncertain. It is used to represent when one knows nothing about the system. Let us denote X be a random variable having a uniform distribution on the interval (a, b) where $a < b$. The pdf is given by

$$f(x) = \begin{cases} \frac{1}{b-a} & a \leq x \leq b \\ 0 & \text{otherwise.} \end{cases} \qquad (2.52)$$

The mean and the variance are, respectively,

$$E(X) = \frac{a+b}{2} \qquad (2.53)$$

and

$$V(X) = \frac{(b-a)^2}{12} \qquad (2.54)$$

The cdf is

$$F(x) = \frac{x-a}{b-a} \quad \text{for } a \leq x \leq b$$

If a and b are unknown parameters, they are estimated respectively by the smallest and the largest observations in the sample.

Normal Distribution

Normal distribution, also known as Gaussian distribution, plays an important role in classical statistics owing to the *Central Limit Theorem*, and, in fact, is the most commonly used of the theoretical distribution. In production engineering, the normal distribution primarily applies to measurements of product susceptibility and external stress. This two-parameter distribution is used to describe systems in which a failure results due to some wearout effect for many mechanical systems. The normal distribution takes the well-known bell shape curve. This distribution is symmetrical about the mean (expected value) and the spread is measured by variance. The larger the value, the flatter the distribution. The pdf is given by

$$f(t) = \frac{1}{\sigma\sqrt{2\pi}} e^{-\frac{1}{2}(\frac{t-\mu}{\sigma})^2} \quad -\infty < t < \infty \qquad (2.55)$$

where μ is the location parameter (is also the mean) and σ is the scale parameter (is also the standard deviation of the distribution, and σ is the variance. We will denote a normal distribution by $N(\mu, \sigma^2)$. Most of the common statistical tests are based on the assumption that the data are normally distributed, an assumption which cannot often be verified, but since in most cases the quantities required are averages, the tests are of general applicability.

The pth percentile point of the normal distribution $N(\mu, \sigma^2)$ is

$$t_p = \mu + z_p\,\sigma.$$

The cumulative distribution function (cdf) is

$$F(t) = \int_{-\infty}^{t} \frac{1}{\sigma\sqrt{2\pi}} e^{-\frac{1}{2}(\frac{s-\mu}{\sigma})^2} ds$$

The reliability function is

$$R(t) = \int_{t}^{\infty} \frac{1}{\sigma\sqrt{2\pi}} e^{-\frac{1}{2}(\frac{s-\mu}{\sigma})^2} ds$$

There is no closed form solution for the above equation. However, tables for the standard normal density function are readily available (see Table A.1 in Appendix A) and can be used to find probabilities for any normal distribution. If

$$Z = \frac{T - \mu}{\sigma} \qquad (2.56)$$

is substituted into the normal pdf, we obtain

$$f(z) = \frac{1}{\sqrt{2\pi}} e^{-\frac{z^2}{2}} \quad -\infty < Z < \infty$$

This is a so-called standard normal pdf, with a mean value of 0 and a standard deviation of 1. The standardized cdf is given by

$$\Phi(t) = \int_{-\infty}^{t} \frac{1}{\sqrt{2\pi}} e^{-\frac{1}{2}s^2} ds \tag{2.57}$$

where Φ is a standard normal distribution function. Thus, for a normal random variable T, with mean μ and standard deviation σ,

$$P(T \leq t) = P\left(Z \leq \frac{t - \mu}{\sigma}\right) = \Phi\left(\frac{t - \mu}{\sigma}\right)$$

where Φ yields the relationship necessary if standard normal tables are to be used.

It should be noted that the coefficient of kurtosis in the normal distribution is 3. The hazard function for a normal distribution is a monotonically increasing function of t. This can be easily shown by proving that $h'(t) \geq 0$ for all t. Since

$$h(t) = \frac{f(t)}{R(t)}$$

then

$$h'(t) = \frac{R(t) f'(t) + f^2(t)}{R^2(t)} \geq 0$$

One can try this proof by employing the basic definition of a normal density function f.

Example 2.19 A component has a normal distribution of failure times with $\mu = 2{,}000$ h and $\sigma = 100$ h. Obtain the reliability of the component at $1{,}900$ h.

Note that the reliability function is related to the standard normal deviate z by

$$R(t) = P(Z > \frac{t - \mu}{\sigma})$$

where the distribution function for Z is given by Eq. (2.47). For this particular application,

$$R(1,900) = P\left(Z > \frac{1,900 - 2,000}{100}\right) = P(Z > -1).$$

From the standard normal table in Table A.1 in Appendix A, we obtain

$$R(1,900) = 1 - \Phi(-1) = 0.8413.$$

Example 2.20 A part has a normal distribution of failure times with $\mu = 40{,}000$ cycles and $\sigma = 2000$ cycles. Find the reliability of the part at 38,000 cycles.

Solution The reliability at 38,000 cycles

$$\begin{aligned}
R(38000) &= P\left(z > \frac{38000 - 40000}{2000}\right) \\
&= P(z > -1.0) \\
&= \Phi(1.0) = 0.8413
\end{aligned}$$

The resulting reliability plot is shown in Fig. 2.3.

The normal distribution is flexible enough to make it a very useful empirical model. It can be theoretically derived under assumptions matching many failure mechanisms. Some of these are corrosion, migration, crack growth, and in general, failures resulting from chemical reactions or processes. That does not mean that the normal is always the correct model for these mechanisms, but it does perhaps explain why it has been empirically successful in so many of these cases.

The normal distribution is flexible enough to make it a very useful empirical model. It can be theoretical derived under assumptions matching many failure mechanisms. Some of these are: corrosion, migration, crack growth, and in general, failures resulting from chemical reactions or processes. That does not mean that the normal

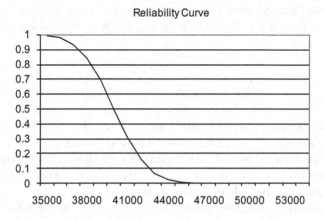

Fig. 2.3 Normal reliability plot versus time

is always the correct model for these mechanisms, but it does perhaps explain why it has been empirically successful in so many of these cases.

Truncated Normal Distribution

Truncated normal distribution plays an important role in modeling the distribution of material strength and mechanical systems. The pdf of this truncated normal is given by

$$f(t) = \frac{e^{-\frac{1}{2}\left(\frac{t-\mu}{\sigma}\right)^2}}{\sigma\sqrt{2\pi}\left(1 - \Phi\left(\frac{t_0-\mu}{\sigma}\right)\right)} \quad t_0 \le t < \infty \tag{2.58}$$

If $t_0 > 0$, the truncated normal distribution can be used in modeling the distribution of material strength.

The reliability function is given by

$$R(t) = \begin{cases} \frac{1-\Phi\left(\frac{t-\mu}{\sigma}\right)}{1-\Phi\left(\frac{t_0-\mu}{\sigma}\right)} & \text{for } t \ge t_0 \\ 1 & \text{for } 0 < t < t_0 \end{cases} \tag{2.59}$$

The failure rate is

$$h(t) = \frac{e^{-\frac{1}{2}\left(\frac{t-\mu}{\sigma}\right)^2}}{\sigma\sqrt{2\pi}\left[1 - \Phi\left(\frac{t-\mu}{\sigma}\right)\right]} \quad \text{for } t \ge t_0 \tag{2.60}$$

Log-normal Distribution

The log-normal lifetime distribution is a very flexible model that can empirically fit many types of failure data. This distribution, with its applications in mechanical reliability engineering, is able to model failure probabilities of repairable systems, the compressive strength of concrete cubes, the tensile strength of fibers as well as to model the uncertainty in failure rate information. The log-normal density function is given by

$$f(t) = \frac{1}{\sigma t\sqrt{2\pi}}e^{-\frac{1}{2}\left(\frac{\ln t-\mu}{\sigma}\right)^2} \quad t \ge 0 \tag{2.61}$$

where μ and σ are parameters such that $-\infty < \mu < \infty$, and $\sigma > 0$. Note that μ and σ are not the mean and standard deviations of the distribution.

The relationship to the normal (just take natural logarithms of all the data and time points and you have "normal" data) makes it easy to work with many good software analysis programs available to treat normal data.

Mathematically, if a random variable X is defined as $X = \ln T$, then X is normally distributed with a mean of μ and a variance of σ^2. In other word, if $X \sim N(\mu, \sigma^2)$ then $T = e^X \sim$ lognormal with parameters μ and σ That is,

$$E(X) = E(\ln T) = -\mu$$

and

$$V(X) = V(\ln T) = \sigma^2$$

Since $T = e^X$, the mean of the log-normal distribution can be found by using the normal distribution. Consider that

$$E(T) = E(e^X) = \int_{-\infty}^{\infty} \frac{1}{\sigma\sqrt{2\pi}} e^{[x - \frac{1}{2}(\frac{x-\mu}{\sigma})^2]} dx$$

and by rearrangement of the exponent, this integral becomes

$$E(T) = e^{\mu + \frac{\sigma^2}{2}} \int_{-\infty}^{\infty} \frac{1}{\sigma\sqrt{2\pi}} e^{-\frac{1}{2\sigma^2}[x - (\mu + \sigma^2)]^2} dx$$

Thus, the mean of the log normal distribution is

$$E(T) = e^{\mu + \frac{\sigma^2}{2}}$$

Proceeding in a similar manner,

$$E(T^2) = E(e^{2X}) = e^{2(\mu + \sigma^2)}$$

thus, the variance for the log normal is

$$V(T) = e^{2\mu + \sigma^2}[e^{\sigma^2} - 1]$$

The coefficient of skewness of this distribution is

$$S_c = \frac{e^{3\sigma^2} - 3e^{\sigma^2} + 2}{\left(e^{\sigma^2} - 1\right)^{\frac{3}{2}}}$$

It is interesting that the skewness coefficient does not depend on μ and grows very fast as the variance σ^2 increases. The cumulative distribution function for the log normal is

$$F(t) = \int_0^t \frac{1}{\sigma s \sqrt{2\pi}} e^{-\frac{1}{2}(\frac{\ln s - \mu}{\sigma})^2} ds$$

and this can be related to the standard normal deviate Z by

$$F(t) = P(T \leq t) = P(\ln T \leq \ln t)$$

$$= P\left(Z \leq \frac{\ln t - \mu}{\sigma}\right)$$

Therefore, the reliability function is given by

$$R(t) = P\left(Z > \frac{\ln t - \mu}{\sigma}\right) \tag{2.62}$$

and the hazard function would be

$$h(t) = \frac{f(t)}{R(t)} = \frac{\Phi\left(\frac{\ln t - \mu}{\sigma}\right)}{\sigma t R(t)}$$

where Φ is a cdf of standard normal distribution function.

Example 2.21 The failure time of a certain component is log normal distributed with $\mu = 5$ and $\sigma = 1$. Find the reliability of the component and the hazard rate for a life of 50 time units.

Solution Substituting the numerical values of μ, σ, and t into Eq. (2.61), we compute

$$R(50) = P\left(Z > \frac{\ln 50 - 5}{1}\right) = P(Z > -1.09)$$

$$= 0.8621$$

Similarly, the hazard function is given by

$$h(50) = \frac{\Phi\left(\frac{\ln 50 - 5}{1}\right)}{50(1)(0.8621)} = 0.032 \text{ failures/unit.}$$

Thus, values for the log normal distribution are easily computed by using the standard normal tables.

Example 2.22 The failure time of a part is log normal distributed with $\mu = 6$ and $\sigma = 2$. Find the part reliability for a life of 200 time units.

Solution The reliability for the part of 200 time units is Fig. 2.4.

$$R(200) = P\left(Z > \frac{\ln 200 - 6}{2}\right) = P(Z > -0.35)$$

$$= 0.6368$$

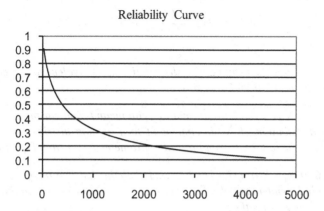

Fig. 2.4 Log normal reliability plot versus time

The log normal lifetime model, like the normal, is flexible enough to make it a very useful empirical model. Figure 2.4 shows the reliability of the log normal vs time. It can be theoretically derived under assumptions matching many failure mechanisms. Some of these are: corrosion, migration, crack growth, and in general, failures resulting from chemical reactions or processes. That does not mean that the log normal is always the correct model for these mechanisms, but it does perhaps explain why it has been empirically successful in so many of these cases.

Chi-Square (χ^2) **Distribution**

The chi-square distribution is in fact from a combination of a modified function between normal distribution and gamma distribution. The chi-square probability density function of a random variable T is given by:

$$ f(t) = \frac{(t)^{\frac{n}{2}-1} e^{-\frac{t}{2}}}{2^{n/2}\, \Gamma\left(\frac{n}{2}\right)} \quad t \geq 0,\ n = 1, 2, \dots \tag{2.63}$$

where the parameter n is called the degrees of freedom (d.f.) of chi-square, i.e., $T \sim \chi_n^2$. The mean and the variance of chi-square r.v. T with n d.f. are, respectively,

$$ E(T) = n $$

and

$$ V(T) = 2n. $$

Property 2.4 *As from the gamma pdf*

$$f(x) = \frac{\beta^\alpha t^{\alpha-1}}{\Gamma(\alpha)} e^{-t\beta}$$

if $\beta = \frac{1}{2}$ and $\alpha = \frac{1}{2}$ then the gamma distribution is equivalent to the chi-square distribution with $n = 1$ from Eq. (2.63).

Property 2.5 *If X is a standard normal r.v. with mean 0 and variance 1, i.e., N(0, 1), then the distribution of X^2 is equivalent to a gamma distribution with $\beta = \frac{1}{2}$ and $\alpha = \frac{1}{2}$ or chi-square distribution with parameter $n = 1$ from Eq. (2.63).*

Property 2.6 *If $X_i \sim N(0, 1)$ for $i = 1, 2, ..., n$ then the distribution of the sum of the squares of n independent standard normal r.v. (T) is the chi-square. In other words, $T = \sum_{i=1}^{n} X_i^2 \sim \chi_n^2$ with the pdf*

$$f(t) = \frac{(t)^{\frac{n}{2}-1} e^{-t/2}}{2^{n/2} \Gamma\left(\frac{n}{2}\right)} \quad \text{for } t \geq 0$$

See Problem 20 in details.

Property 2.7 *If $T_i \sim \chi_{n_i}^2$ are independent chi-square random variable with n_i d.f. for $i = 1, 2, ... k$ then the sum of these random variables of chi-square has chi-square distribution with n degrees of freedom where n is the sum of all degrees of freedom. In other words, $\sum_{i=1}^{k} T_i \sim \chi_n^2$ where $n = \sum_{i=1}^{k} n_i$ is the d.f.*

Example 2.23 Four random variables have a chi-square distribution with 3, 7, 9, and 10 d.f. (i.e., $T_i \sim \chi_{n_i}^2$). Determine $P\left(\sum_{i=1}^{k} T_i \leq 19.77\right)$?

Solution From Table A.3 in the Appendix, since $\sum_{i=1}^{k} T_i \sim \chi_{29}^2$ the probability that

$$P\left(\sum_{i=1}^{k} T_i \leq 19.77\right) = 0.90.$$

Student's t Distribution

The Student's t probability density function of a random variable T is given as follows:

$$f(t) = \frac{\Gamma\left(\frac{r+1}{2}\right)}{\sqrt{\pi r}\, \Gamma\left(\frac{r}{2}\right)\left(1 + \frac{t^2}{r}\right)^{\frac{r+1}{2}}} \quad \text{for } -\infty < t < \infty \qquad (2.64)$$

In other words, if a random variable T is defined as

$$T = \frac{W}{\sqrt{\frac{V}{r}}} \qquad (2.65)$$

where W is a standard normal random variable and V is distributed as chi-square with r degrees of freedom, and W and V are statistically independent, then T is Student's t distributed with parameter r is referred to as the degrees of freedom (see Table A.2 in Appendix A). The mean and variance of the Student's t distributed r.v. T with parameter r are given, respetively,

$$E(T) = 0$$

and

$$V(T) = \frac{r}{r-2}.$$

The F Distribution

Let we define the random variable F is as follows

$$F = \frac{\frac{U}{r_1}}{\frac{V}{r_2}} \tag{2.66}$$

where U is distributed as chi-square with r_1 degrees of freedom, V is distributed as chi-square with r_2 degrees of freedom, and U and V are statistically independent, then the probability density function of F is given by

$$f(t) = \frac{\Gamma\left(\frac{r_1+r_2}{2}\right) \left(\frac{r_1}{r_2}\right)^{\frac{r_1}{2}} (t)^{\frac{r_1}{2}-1}}{\Gamma\left(\frac{r_1}{2}\right) \Gamma\left(\frac{r_2}{2}\right) \left(1 + \frac{r_1 t}{r_2}\right)^{\frac{r_1+r_2}{2}}} \quad \text{for t} > 0 \tag{2.67}$$

The F distribution is a two parameter, r_1 and r_2, distribution which are the degrees of freedom of the underlying chi-square random variables (see Table A.3 in Appendix A). The mean and variance of the F distribution with parameters, r_1 and r_2, are given, respetively,

$$E(T) = \frac{r_2}{r_2-2} \quad \text{for } r_2 > 2$$

and

$$V(T) = \frac{2r_2^2(r_1+r_2-2)}{r_1(r_2-4)(r_2-2)} \quad \text{for } r_2 > 4$$

It is worth to note that if T is a random variable with a t distribution and r degrees of freedom, then the random variable T^2 has an F distribution with parameters $r_1 =$

1and $r_2 = r$. Similarly, if F is F-distributed with r_1 and r_2 degrees of freedom, then the random variable Y, defined as

$$Y = \frac{r_1 F}{r_2 + r_1 F}$$

has a beta distribution with parameters $r_1/2$ and $r_2/2$.

Weibull Distribution

The Weibull distribution (Weibull 1951) is a popular distribution for modeling failure times because it reflects the physics of failure and it models either increasing or decreasing failure rates simply. The Weibull is the preferred distribution for the time effects due to fatigue, erosion and wear.

The exponential distribution is often limited in applicability owing to the memoryless property. The Weibull distribution is a generalization of the exponential distribution and is commonly used to represent fatigue life, ball bearing life, laser diodes, and vacuum tube life. It is extremely flexible and appropriate for modeling component lifetimes with fluctuating hazard rate functions and to represent various types of engineering applications such as electrical devices. It is also used as the distribution for product properties such as strength (electrical or mechanical) and resistance. It is used to describe the life or roller bearings, electronic components, ceramics, and capacitors. The Weibull distribution is also used in biological and medical applications to model the time to occurrence of tumors as well as human life expectancy.

The well known Weibull bathtub curve describes the failure rate for most electrical devices. In the initial stage, the failure rate starts off high and decreases over time. After this in the random failure stage, the failure rate is constant. The last stage is the wearout period in which the failure rate steadily increases. The lifetime of equipment using semiconductors, for example, is much shorter than the devices themselves which usually do not reach the wearout stage. Recent improvements in laser diode technology have extended average lifetime to a level of typical semiconductor devices. More importantly, the failure rate in the initial stage has been reduced by upgrading materials, improvements in processing, and the use of screening technology.

Many engineers prefer to model lifetime data by the two-parameter Weibull distribution with shape parameter β and scale parameter θ while others prefer the three-parameter Weibull distribution with shape parameter β, scale parameter θ, and location parameter γ.

The three-parameter Weibull has the following probability density function:

$$f(t) = \frac{\beta(t-\gamma)^{\beta-1}}{\theta^\beta} e^{-(\frac{t-\gamma}{\theta})^\beta} \quad t \geq \gamma \geq 0 \tag{2.68}$$

where $\beta > 0$ is the shape parameter, $\theta > 0$ is the scale parameter, $\gamma \geq 0$ is the location parameter.

The corresponding reliability function $R(t)$ and hazard function $h(t)$ are, respectively:

$$R(t) = e^{-(\frac{t-\gamma}{\theta})^\beta} \text{ for } t \geq \gamma \geq 0, \beta > 0, \theta > 0 \tag{2.69}$$

and

$$h(t) = \frac{\beta(t-\gamma)^{\beta-1}}{\theta^\beta} \quad t \geq \gamma \geq 0, \beta > 0, \theta > 0 \tag{2.70}$$

The scale-parameter, θ, has the same units as T, for example, second, minutes, hours, days, or cycle. The scale parameter indicates the characteristic life of the product, the time by which roughly 63.2% of the units are expected to fail. The shape parameter β is a unit less number. For most products and material, β is in the range 0.5–5. The parameter β determines the shape of the distribution, and θ determines the spead.

Often the location parameter γ is not used. Its value can be set to zero to obtain the two-parameter Weibull distribution. The latter can describe the failure of an element throughout its lifecycle. The probability density function for a Weibull distribution with two parameters (i.e., $\gamma = 0$) is given by

$$f(t) = \frac{\beta t^{\beta-1}}{\theta^\beta} e^{-(\frac{t}{\theta})^\beta} \quad t \geq 0, \beta > 0, \theta > 0 \tag{2.71}$$

where β and θ are the shape parameter and the scale parameter, respectively. The corresponding reliability function for the two-parameter Weibull distribution is given by:

$$R(t) = e^{-(\frac{t}{\theta})^\beta} \quad t \geq 0 \tag{2.72}$$

Furthermore, the MTTF for the two-parameter Weibull distribution is given by

$$MTTF = \theta \, \Gamma\left(1 + \frac{1}{\beta}\right)$$

where Γ is the gamma function. The hazard function of the two-parameter Weibull distribution is given by

$$h(t) = \frac{\beta \, t^{\beta-1}}{\theta^\beta} \quad t > 0, \beta > 0, \theta > 0 \tag{2.73}$$

It can be shown that the failure rate function is decreasing for $\beta < 1$, increasing for $\beta > 1$, and constant when $\beta = 1$. A decreasing failure rate would imply infant mortality and that the failure rate of a given product decreases as time progresses. A

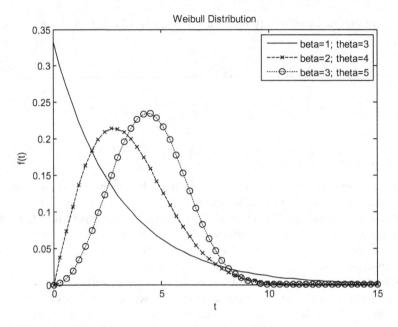

Fig. 2.5 Weibull pdf for various values of θ and β where $\gamma = 0$

constant failure rate implies that products are failing from random events. As for an increasing failure rate, it would imply that, as wear-out, products will likely to fail as time progresses.

For many products, such as computer processors, the hazard function graphical representation resembles a bathtub, with a decreasing hazard rate early during product life, accounting for infant mortality, a constant and low hazard rate accounting for the product lifetime, and then increasing hazard function, accounting for rapid wear-out, during the later age of the product. Figures 2.5, 2.6, 2.7 and 2.8 show the pdf of two-parameter Weibull distribution for various values of θ and β (here $\gamma = 0$), and various values of θ and $\beta = 2$ and $\gamma = 0$, respectively. Figure 2.9 illustrates a typical bathtub curve function.

When $\beta = 2$, and $\theta = \sqrt{2}\,\gamma$, the Weibull distribution becomes a Rayleigh distribution as the pdf of Rayleigh distribution is given by

$$f(t) = \frac{t}{\gamma^2} e^{-\frac{t^2}{2\gamma^2}} \quad t \geq 0$$

Example 2.24 The failure time of a certain component has a Weibull distribution with $\beta = 4$, $\mu = 2000$, and $\sigma = 1000$. Find the reliability of the component and the hazard rate for an operating time of *1500* h.

Solution A direct substitution into Eq. (2.69) yields

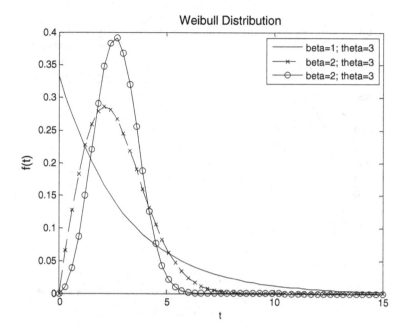

Fig. 2.6 Weibull pdf for various values of β where $\theta = 3.1$ and $\gamma = 0$

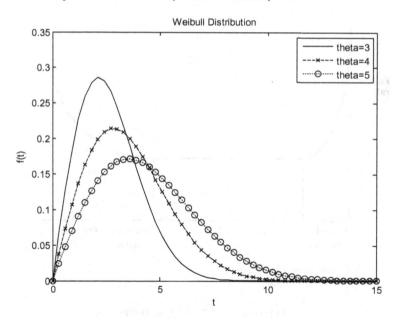

Fig. 2.7 Weibull pdf for various values of theta where $\beta = 2$ and $\gamma = 0$

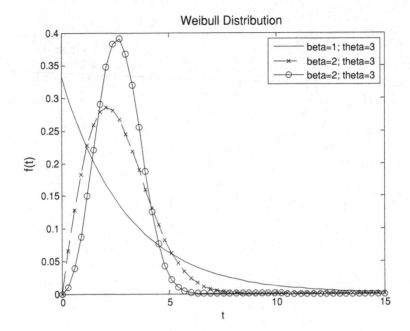

Fig. 2.8 Weibull pdf for various values of beta where $\theta = 3$ and $\gamma = 0$

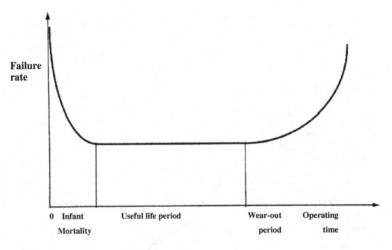

Fig. 2.9 Typical bathtub curve illustration

$$R(1500) = e^{-\left(\frac{1500-1000}{2000}\right)^4} = 0.996$$

Using Eq. (2.70), the desired hazard function is given by

$$h(1500) = \frac{4(1500 - 1000)^{4-1}}{(2000)^4} = 3.13 \times 10^{-5} \text{ failures/hour}$$

Note that the Rayleigh and exponential distributions are special cases of the Weibull distribution at $\beta = 2$, $\gamma = 0$, and $\beta = 1$, $\gamma = 0$, respectively. For example, when $\beta = 1$ and $\gamma = 0$, the reliability of the Weibull distribution function in Eq. (2.69) reduces to

$$R(t) = e^{-\frac{t}{\theta}}$$

and the hazard function given in Eq. (2.70) reduces to $1/\theta$, a constant. Thus, the exponential is a special case of the Weibull distribution. Similarly, when $\beta = 2$ and $\theta = \sqrt{2}\,\alpha$, the two-parameter Weibull distribution becomes the *Rayleigh* density function as follows:

$$f(t) = \frac{t}{\alpha^2} e^{-\frac{t^2}{2\alpha^2}} \quad t \geq 0 \tag{2.74}$$

The Weibull distribution again is widely used in engineering applications. It was originally proposed for representing the distribution of the breaking strength of materials. The Weibull model is very flexible and also has theoretical justification successfully in many applications as a purely empirical model.

Gamma Distribution

The gamma distribution can be used as a failure probability function for components whose distribution is skewed. Useful also for representation of times required to perform a given task or accomplish a certain goal but more general. The failure density function for a gamma distribution is

$$f(t) = \frac{t^{\alpha-1}}{\beta^\alpha \Gamma(\alpha)} e^{-\frac{t}{\beta}} \quad t \geq 0, \alpha, \ \beta > 0 \tag{2.75}$$

where α is the shape parameter and β is the scale parameter. In this expression, $\Gamma(\alpha)$ is the gamma function, which is defined as

$$\Gamma(\alpha) = \int_0^\infty t^{\alpha-1} e^{-t} dt \quad \text{for } \alpha > 0$$

Hence,

$$R(t) = \int_t^\infty \frac{1}{\beta^\alpha \Gamma(\alpha)} s^{\alpha-1} e^{-\frac{s}{\beta}} \, ds$$

When α is an integer, it can be shown by successive integration by parts that

$$R(t) = e^{-\frac{t}{\beta}} \sum_{i=0}^{\alpha-1} \frac{\left(\frac{t}{\beta}\right)^i}{i!} \tag{2.76}$$

and

$$h(t) = \frac{f(t)}{R(t)} = \frac{\frac{1}{\beta^\alpha \Gamma(\alpha)} t^{\alpha-1} e^{-\frac{t}{\beta}}}{e^{-\frac{t}{\beta}} \sum_{i=0}^{\alpha-1} \frac{\left(\frac{t}{\beta}\right)^i}{i!}}$$

The mean and variance of the gamma random variable are, respectively,

$$\text{Mean (MTTF)} \quad E(T) = \alpha \beta$$

and

$$\text{Variance} \quad V(T) = \alpha \beta^2.$$

The gamma density function has shapes that are very similar to the Weibull distribution. When $\alpha = 1$, the gamma distribution is that of an exponential distribution with the constant failure rate $1/\beta$. In general, if α is a positive integer, Eq. (2.74) becomes

$$f(t) = \frac{t^{\alpha-1}}{\beta^\alpha (\alpha - 1)!} e^{-\frac{t}{\beta}} \quad t \geq 0, \alpha \text{ is an integer, } \beta > 0$$

and is known as the *Erlang* distribution.

The gamma distribution can also be used to model the time to the nth failure of a system if the underlying failure distribution is exponential.

Suppose $X_1, X_2, \ldots, X_n$ are independent and identically distributed random variables. Furthermore, suppose that these random variables follows the exponential distribution with parameter θ ($= 1\beta$ as in Eq. 2.74) then the distribution of a new random variable T where $T = X_1 + X_2 + \ldots + X_n$ is gamma distribution (or known as Erlang distribution) with parameters θ and n (see Problem 10). The pdf of T is given by (see Problem 10):

$$f(t) = \frac{\theta^n t^{n-1}}{\Gamma(n)} e^{-t\theta} \quad t > 0$$

Example 2.25 The time to failure of a component has a gamma distribution with α = 3 and β = 5. Determine the reliability of the component and the hazard rate at 10 time-units.

Solution Using Eq. (2.75), we compute

$$R(10) = e^{-\frac{10}{5}} \sum_{i=0}^{2} \frac{(\frac{10}{5})^i}{i!} = 0.6767$$

The hazard rate is given by

$$h(10) = \frac{f(10)}{R(10)} = \frac{0.054}{0.6767} = 0.798 \text{ failures/unit time}$$

Other form of Gamma distribution. The other form of the gamma probability density function can be written as follows:

$$f(t) = \frac{\beta^\alpha t^{\alpha-1}}{\Gamma(\alpha)} e^{-t\beta} \quad t > 0 \qquad (2.77)$$

This pdf is characterized by two parameters: shape parameter α and scale parameter β. When $0 < \alpha < 1$, the failure rate monotonically decreases; when $\alpha > 1$, the failure rate monotonically increase; when $\alpha = 1$ the failure rate is constant.

The mean, variance and reliability of the gamma random variable are, respectively

$$\text{Mean (MTTF)} \quad E(T) = \frac{\alpha}{\beta}.$$

$$\text{Variance} \quad V(T) = \frac{\alpha}{\beta^2}.$$

$$\text{Reliability} \quad R(t) = \int_t^\infty \frac{\beta^\alpha x^{\alpha-1}}{\Gamma(\alpha)} e^{-x\beta} dx.$$

The coefficient of skewness of this gamma distribution is given by

$$S_c = \frac{2}{\sqrt{\alpha}}$$

Example 2.26 A mechanical system time to failure is gamma distribution with α = 3 and $1/\beta = 120$. What is the system reliability at 280 h?

Solution The system reliability at 280 h is given by

$$R(280) = e^{\frac{-280}{120}} \sum_{k=0}^{2} \frac{(\frac{280}{120})^2}{k!} = 0.85119$$

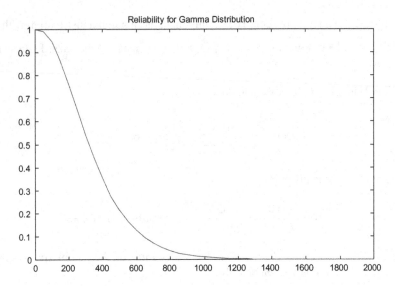

Fig. 2.10 Gamma reliability function versus time

and the resulting reliability plot is shown in Fig. 2.10.

Example 2.27 Consider a communication network of having 10 stations where the repair time for each station, in case a station is failed, follows the exponential distribution with $\beta = 19$ h. If 10 stations have failed simultaneously and one repair person is available, what is the probability that communication failure will last more than 45 min?

Solution This problem can be formulated using a gamma distribution (see Eq. 2.77) with $\alpha = 10$ and $\beta = 19$. Thus (45 min is same as 0.75 h),

$$P(T > 0.75) = 1 - P(T \leq 0.75)$$

$$= 1 - \int_0^{.75} \frac{(19)^{10}t^9}{(10-1)!} e^{-19t} dt \approx 0.11$$

The gamma model is a flexible lifetime model that may offer a good fit to some sets of failure data. It is not, however, widely used as a lifetime distribution model for common failure mechanisms. A common use of the gamma lifetime model occurs in Bayesian reliability applications.

Property 2.8 *Let X_i follows gamma distributed with α_i is the shape parameter and common scale parameter β for $i = 1, 2, ..., k$ and is denoted as $X_i \sim G(\alpha_i, \beta)$. If X_1, $X_2,..., X_n$ are independent random variables then the distribution of T, where $T = X_1 + X_2 + ... + X_k$, is also gamma distributed, as $T \sim G(\alpha, \beta)$. with scale parameter β and the shape parameter is $\alpha = \sum_{i=1}^k \alpha_i$.*

Proof As exercise, see Problem 11.

Erlang Distribution

Erlang distribution is a special case of the gamma distribution. It represents the distribution of a random variable which is itself the result of the sume of exponential component processes. In fact, if $\alpha = n$, an integer, the gamma distribution becomes Erlang distribution. The failure density function for Erlang distribution is

$$f(t) = \frac{t^{n-1}}{\beta^n (n-1)!} e^{-\frac{t}{\beta}} \quad t \geq 0, \text{ n is an integer, } \beta > 0 \tag{2.78}$$

Hence, the reliability of the Erlang distribution is given by:

$$R(t) = e^{-\frac{t}{\beta}} \sum_{i=0}^{n-1} \frac{(\frac{t}{\beta})^i}{i!}$$

Example 2.28 Consider an electric power system consisting of n similar diesel generators connected in a standby structure. When generator 1 (primary unit) fails, standby unit 1 (generator 2) starts operating automatically, and so on until all n generators fail. Assume that each generator has an exponential life distribution wth constant rate λ or MTTF β and that all the generators operate independently. Determine the distribution function of the system life length.

Let T_i be the life length of generator i and a random variable T be the sum of the life length of all n generators, that is

$$T = \sum_{i=1}^{n} T_i$$

Thus, the distribution of T is the Erlang distribution with a pdf:

$$f(t) = \frac{t^{n-1}}{\beta^n (n-1)!} e^{-\frac{t}{\beta}} \quad t \geq 0, \beta > 0$$

Beta Distribution

The two-parameter Beta density function, $f(t)$, is given by

$$f(t) = \frac{\Gamma(\alpha + \beta)}{\Gamma(\alpha) \Gamma(\beta)} t^{\alpha-1}(1-t)^{\beta-1} \quad 0 \leq t \leq 1, \alpha > 0, \beta > 0 \tag{2.79}$$

where α and β are the distribution parameters. This two-parameter beta distribution has commonly used in many reliability engineering applications and also an important role in the theory of statistics. This is a distribution of broad applicability restricted to bounded random variables. It is used to represent the proportion of non-conforming items in a set. Note that the beta-distributed random variable takes on values in the interval (0, 1), so the beta distribution is a natural model when the random variable represents a probability. Likewise, when $\alpha = \beta = 1$, the beta distribution reduces to the uniform distribution.

The mean and variance of the beta random variable are, respectively,

$$E(T) = \frac{\alpha}{\alpha + \beta}$$

and

$$V(T) = \frac{\alpha\beta}{(\alpha + \beta + 1)(\alpha + \beta)^2}$$

The coefficient of skewness of beta distribution is given by

$$S_c = \frac{2(\beta - \alpha)\sqrt{(\alpha + \beta + 1)}}{(\alpha + \beta + 2)\sqrt{\alpha\beta}}.$$

Pareto Distribution

The Pareto distribution was originally developed to model income in a population. Phenomena such as city population size, stock price fluctuations, and personal incomes have distributions with very long right tails. The probability density function of the Pareto distribution is given by

$$f(t) = \frac{\alpha k^\alpha}{t^{\alpha+1}} \quad k \le t \le \infty \tag{2.80}$$

The mean, variance and reliability of the Pareto distribution are, respectively,

Mean $E(T) = k/(\alpha - 1)$ for $\alpha > 1$.

Variance $V(T) = \alpha k^2 / [(\alpha - 1)^2(\alpha - 2)]$ for $\alpha > 2$.

Reliability $R(t) = \left(\dfrac{k}{t}\right)^\alpha$.

The Pareto and log normal distributions have been commonly used to model the population size and economical incomes. The Pareto is used to fit the tail of the distribution, and the log normal is used to fit the rest of the distribution.

Rayleigh Distribution

The Rayleigh distribution is a flexible lifetime function that can apply to many degradation process failure modes. The Rayleigh probability density function is

$$f(t) = \frac{t}{\sigma^2} \exp\left[\frac{-t^2}{2\sigma^2}\right] \tag{2.81}$$

where σ is the scale parameter. The mean, variance, and reliability of Rayleigh function are, respectively,

$$\text{Mean } E(T) = \sigma\left(\frac{\pi}{2}\right)^{\frac{1}{2}}$$

$$\text{Variance } V(T) = \left(2 - \frac{\pi}{2}\right)\sigma^2$$

$$\text{Reliability } R(t) = e^{-\frac{t^2}{2\sigma^2}}.$$

Example 2.29 Rolling resistance is a measure of the energy lost by a tire under load when it resists the force opposing its direction of travel. In a typical car, traveling at sixty miles per hour, about 20% of the engine power is used to overcome the rolling resistance of the tires. A tire manufacturer introduces a new material that, when added to the tire rubber compound, significantly improves the tire rolling resistance but increases the wear rate of the tire tread. Analysis of a laboratory test of 150 tires shows that the failure rate of the new tire is linearly increasing with time (hours). It is expressed as.

$$h(t) = 0.5 \times 10^{-8}t$$

Determine the reliability of the tire after one year.

Solution The reliability of the tire after one year (8760 h) of use is

$$R(1 year) = e^{-\frac{0.5}{2} \times 10^{-8} \times (8760)^2} = 0.8254$$

Figure 2.11 shows the resulting reliability function.

Pham Distribution

A two-parameter distribution with a Vtub-shaped hazard rate curve was developed by Pham (2002) known as *loglog distribution with Vtub-shaped* or *Pham distribution* (Cordeiro et al. 2016).

Note that the loglog distribution with Vtub-shaped and Weibull distribution with bathtub-shaped failure rates are not the same. As for the bathtub-shaped, after the infant mortality period, the useful life of the system begins. During its useful life, the system fails as a constant rate. This period is then followed by a wear out period during which the system starts slowly increases with the on set of wear out. For the Vtub-shaped, after the infant mortality period, the system starts to experience at a relatively low increasing rate, but not constant, and then increasingly more failures due to aging. Practitioners can debate among these two shapes such as the bathtub shaped and Vtub shaped.

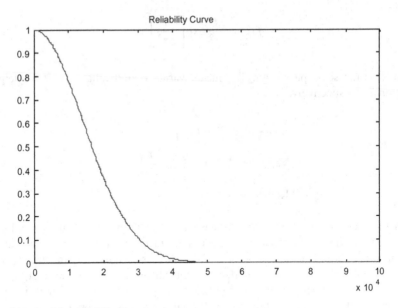

Fig. 2.11 Rayleigh reliability function versus time

The behavior of those with Vtub shaped failure rate can be observed in many practical applications and systems including human life expectancy, medical devices and electronic system products. For example, suppose that you buy a new car and let T be the time of the first breakdown. Typically the failure rate function h(t) is initially high and then decline to a constant. It remains at this level for several years, eventually beginning to increase. The reason is that early failures are typically caused by hidden faults of material and electrical devices such as engines, undetected in manufacture control. If they do not show up immediately, then there are probability of faults and risk of failure remains constant for some time. This would well represent the bathtub-shaped. On the other hand, often users may have experiences due to some minor problems after a few months or so. In this case, the Vtub-shaped can be represented the failure rate. The later case, which is failure rate begins to increase, is due to the wearing out of various parts.

The Pham pdf is given as follows (Pham 2002):

$$f(t) = \alpha \; \ln(a) \; t^{\alpha-1} \; a^{t^\alpha} \; e^{1-a^{t^\alpha}} \quad \text{for} \quad t > 0, \; a > 1, \; \alpha > 0 \qquad (2.82)$$

The corresponding cdf and reliability functions are given by

$$F(t) = \int_0^t f(x)dx = 1 - e^{1-a^{t^\alpha}}$$

and

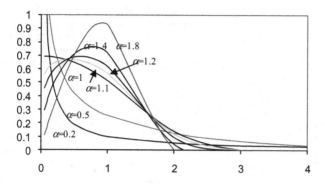

Fig. 2.12 Probability density function for various values α with a $= 2$

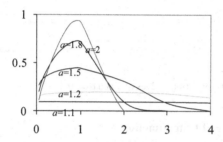

Fig. 2.13 Probability density function for various values a with $\alpha = 1.5$

$$R(t) = e^{1-a^{t^{\alpha}}} \qquad (2.83)$$

respectively. Figures 2.12 and 2.13 describe the density functions for various values of a and α, respectively.

The corresponding failure rate of the Pham distribution is given by

$$h(t) = \alpha \ \ln a \ t^{\alpha-1} \ a^{t^{\alpha}} \qquad (2.84)$$

Figure 2.14 describes the failure rate curve for various values of a and α.

It is worth to note that for given a and α, the Pham distribution F (or Loglog distribution) is decreasing failure rate (DFR) for $t \leq t_0$ and increasing failure rate (IFR) for $t \geq t_0$ where

$$t_0 = \left(\frac{1-\alpha}{\alpha \ln a} \right)^{\frac{1}{\alpha}}.$$

(See Problem 13.)

When $\alpha \geq 1$, the Loglog distribution is an IFR. (See Problem 14.)

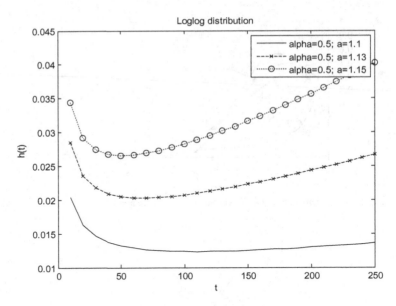

Fig. 2.14 Failure rate $h(t)$ for various values of a and $\alpha = 0.5$

Two-Parameter Hazard Rate Function

This is a two-parameter function that can have increasing and decreasing hazard rates. The hazard rate, $h(t)$, the reliability function, $R(t)$, and the pdf are, respectively, given as follows

$$h(t) = \frac{n\lambda t^{n-1}}{\lambda t^n + 1} \quad \text{for } n \geq 1, \lambda > 0, t \geq 0$$

$$R(t) = e^{-\ln(\lambda t^N + 1)}$$

$$(2.85)$$

and

$$f(t) = \frac{n\lambda t^{n-1}}{\lambda t^n + 1} e^{-\ln(\lambda t^n + 1)} \quad n \geq 1, \lambda > 0, t \geq 0 \qquad (2.86)$$

where $n = $ shape parameter; $\lambda = $ scale parameter.

Three-Parameter Hazard Rate Function

This is a three-parameter distribution that can have increasing and decreasing hazard rates. The hazard rate, $h(t)$, is given as

$$h(t) = \frac{\lambda(b+1)[\ln(\lambda t + \alpha)]^b}{(\lambda t + \alpha)} \quad b \geq 0, \lambda > 0, \alpha \geq 0, t \geq 0 \qquad (2.87)$$

The reliability function $R(t)$ for $\alpha = 1$ is

$$R(t) = e^{-[\ln(\lambda t + \alpha)]^{b+1}}$$

The probability density function $f(t)$ is

$$f(t) = e^{-[\ln(\lambda t + \alpha)]^{b+1}} \frac{\lambda(b+1)[\ln(\lambda t + \alpha)]^{b}}{(\lambda t + \alpha)}$$

where b = shape parameter; λ = scale parameter, and α = location parameter.

Vtub-Shaped Failure Rate Function

Pham (2019) present a distribution function that characterizes by a Vtub-shaped failure rate function. As for the bathtub-shaped, during its useful life period the failure rate is constant. For the Vtub-shaped, after the infant mortality period the system will begin to experience a rather low increasing failure rate but never be at a constant rate due to aging. The Vtub-shaped failure rate function $h(t)$ is defined as (Pham 2019):

$$h(t) = at \ln(bt) + \frac{a}{b} \quad \text{for } t > 0,\, a > 0,\, b > 0 \tag{2.88}$$

respectively. Figure 2.15 shows the vtub-shaped of failure rate $h(t)$ for $a = 0.75$ and various values of b (i.e. 0.35, 0.45, 0.55).

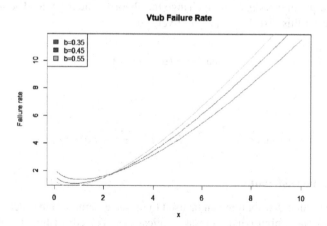

Fig. 2.15 A vtub-shaped failure rate for $a = 0.75$ and various values of b (i.e. 0.35, 0.45, 0.55)

The corresponding probability density function $f(t)$ and reliability function $R(t)$ are as follows:

$$f(t) = \left(at\ln(bt) + \frac{a}{b}\right) e^{-\{at[\frac{t}{2}\ln(bt) - \frac{t}{4} + \frac{1}{b}]\}} \quad \text{for } t > 0, a > 0, b > 0 \quad (2.89)$$

and

$$R(t) = e^{-at[\frac{t}{2}\ln(bt) - \frac{t}{4} + \frac{1}{b}]} \quad \text{for } t > 0, a > 0, b > 0$$

respectively.

Triangular Distribution

A triangular distribution (sometime called a triangle distribution) is a continuous probability distribution shaped like a triangle. It is defined by

a: the minimum value.

c: the peak value (the height of the triangle).

b: the maximum value, where $a \leq c \leq b$.

The triangular probability density function is given by

$$f(t) = \begin{cases} 0 & t < a \\ \frac{2(t-a)}{(b-a)(c-a)} & a \leq t \leq c \\ \frac{2(b-t)}{(b-a)(b-c)} & c \leq t \leq b \\ 0 & t > b \end{cases} \quad (2.90)$$

The three parameters a, b and c change the shape of the triangle. The mean and the variance for this distribution, respectively, are:

$$\text{mean} = \frac{1}{3}(a + b + c)$$

and

$$\text{variance} = \frac{1}{18}\left(a^2 + b^2 + c^2 - ab - ac - bc\right).$$

Extreme-value Distribution

The extreme-value distribution can be used to model external events such as floods, tornado, hurricane, high winds, in risk applications. The cdf of this distribution is given by

$$F(t) = e^{-e^{-t}} \quad \text{for } -\infty < t < \infty \quad (2.91)$$

Cauchy Distribution

The Cauchy distribution can be applied in analyzing communication systems in which two signals are received and one is interested in modeling the ratio of the two signals. The Cauchy probability density function is given by

$$f(t) = \frac{1}{\pi(1 + t^2)} \quad \text{for } -\infty < t < \infty \tag{2.92}$$

It is worth to note that the ratio of two standard normal random variables is a random variable with a Cauchy distribution. Note that a Student's t-distribution with $r = 1$ degree of freedom (see Eq. 2.64) becomes a Cauchy distribution.

Logistic Distribution

The logistic distribution is a flexible function that has the form of logistic function. It is widely used in the field of reliability modeling, especially population growth. The logistic two parameter distribution function whose pdf is given as

$$f(t) = \frac{b(1 + \beta)e^{-bt}}{\left(1 + \beta e^{-bt}\right)^2}; \quad t \geq 0, \quad 0 \leq b \leq 1, \quad \beta \geq 0 \tag{2.93}$$

The reliability function of logistic distribution can be easily obtained

$$R(t) = \frac{(1 + \beta)e^{-bt}}{1 + \beta e^{-bt}}$$

Half logistic distribution. Another type of logistic distribution known as *half logistic distribution* can be defined, which is a one parameter continuous probability distribution. The pdf of a half logistic distribution is given as

$$f(t) = \frac{2be^{-2bt}}{\left(1 + e^{-bt}\right)^2}; \quad t \geq 0, \quad 0 \leq b \leq 1, \tag{2.94}$$

and the reliability function is given as

$$R(t) = \frac{2}{1 + e^{bt}}$$

2.3 Characteristics of Failure Rate Functions

This section discusses briefly properties of some common related Weibull lifetime distributions.

Notation

T	Lifetime random variable.
$f(t)$	Probability density function (pdf) of T.
$F(t)$	Cumulative distribution function (cdf).
$h(t)$	hazard rate function.
$H(t)$	Cumulative failure rate function $[H(t) = \int_0^t h(x)dx]$.
$R(t)$	Reliability function $[= 1 - F(t)]$.
$\mu(t)$	Mean residual life defined by $E(T - t \mid T > t)$.
μ	Mean lifetime.

Acronym

IFR	Increasing failure rate.
DFR	Decreasing failure rate.
BT	Bathtub.
MBT	Modified bathtub.
UBT	Upside-down bathtub.
MRL	Mean residual life.
IMRL	Increasing mean residual life.
DMRL	Decreasing mean residual life.
NBU	New better than used.
NWU	New worse than used.
NBUE	New better than used in expectation.
BWUE	New worse than used in expectation.
HNBUE	Harmonic new better than used in expectation.
HBWUE	Harmonic new worse than used in expectation.

In the context of reliability modelling, some well-known interrelationships between the various quantities such as pdf, cdf, failure rate function, cumulative failure rate function, and reliability function, for a continuous lifetime T, can be summarized as

$$h(t) = \frac{f(t)}{1 - F(t)} = \frac{f(t)}{R(t)}$$

$$H(t) = \int_0^t h(x)dx$$

$$R(t) = e^{-H(t)}$$

A life distribution can be classified according to the shape of its $h(t)$, or its $\mu(t)$. For convenience sake, the life classes are often given by their abbreviations IFR, DFR, BT, MBT, etc.

Definition 2.1 The distribution F is IFR (DFR) if the failure rate $h(t)$ is nondecreasing (nonincreasing) in t. That is,

$$h(t_2) \geq h(t_1) \quad \text{for all } t_2 \geq t_1 \geq 0$$

Definition 2.2 The distribution F is BT (UBT) if the failure rate $h(t)$ has a bathtub (upside-down bathtub) shape.

Definition 2.3 F is MBT (modified bathtub) if $h(t)$ is first increasing, then followed by a bathtub curve.

Definition 2.4 The distribution F has a *Vtub-shaped* failure rate if there exists a change point t_0 such that the distribution F is DFR for $t < t_0$ and IFR for $t > t_0$ and $h'(t_0) = 0$.

Definition 2.5 The distribution F has an *upside-down Vtub-shaped* failure rate if there exists a change point t_0 such that the distribution F is IFR for $t < t_0$ and DFR for $t > t_0$ and $h'(t_0) = 0$.

Definition 2.6 The distribution F is said to be IMRL (DMRL) if the mean residual life function $\mu(t)$ is an increasing (decreasing) function of t. That is,

$$\mu(t_2) \geq \mu(t_1) \quad \text{for all } t_2 \geq t_1 \geq 0$$

where

$$\mu(t) = \frac{\int_t^\infty R(x)dx}{R(t)}.$$

Birnbaum and Esary (1966a) introduced the concept of an increasing failure rate average (IFRA) distribution.

Definition 2.7 The distribution F is said to be IFRA (DFRA) if

$$\frac{\int_0^t h(x)dx}{t} \quad \text{is increasing (decreasing) in t.} \tag{2.95}$$

Definition 2.8 The distribution F is said to be NBU (NWU) if.

$$R(x + y) \leq (\geq) R(x)R(y)$$

Definition 2.9 The distribution F is said to be NBUE (NWUE) if.

$$\int_t^\infty R(x)dx \leq (\geq) \mu R(t) \text{ where } \mu = \int_0^\infty R(x)dx.$$

Definition 2.10 The distribution F is said to be HNBUE (HNWUE) if

$$\int\limits_{t}^{\infty} R(x)dx \ \le (\ge) \ \mu e^{-\frac{t}{\mu}} \ \text{where} \ \mu = \int\limits_{0}^{\infty} R(x)dx.$$

It is well-known that if the distribution F is IFR (DFR) then F is DMRL (IMRL) (see Birnbaum and Esary (1966b)).

That is,

F is IFR $\Rightarrow$ F is DMRL.

F is DFR $\Rightarrow$ F is IMRL.

Indeed, it can be easily to show that.

F is IFR $\Rightarrow$ F is IFRA $\Rightarrow$ F is NBU $\Rightarrow$ F is NBUE $\Rightarrow$ F is HNBUE.

F is DFR $\Rightarrow$ F is DFRA $\Rightarrow$ F is NWU $\Rightarrow$ F is NWUE $\Rightarrow$ F is HNWUE.

Theorem 2.3 *Given a and α, the Pham distribution F (or Loglog distribution, see Sect. 2.2) has a Vtub-shaped failure rate when*

$$t_0 = \left(\frac{1-\alpha}{\alpha \ln a} \right)^{\frac{1}{\alpha}}.$$

(Also see Problem 13).

Corollary 2.1 *For $\alpha \ge 1$, the Pham distribution F is a DMRL.*

Proof

When $\alpha \ge 1$, Pham distribution F is IFR. This implies that F is a DMRL.

For more details concerning these and other ageing classes, see Pham (2006). Table 2.1 offers a brief but by no means exhaustive summary of common lifetime distributions with two or more parameters and their characteristics of failure rate functions.

2.4 The Central Limit Theorem

The central limit theorem (CLT) is one of the most important theorem in reliability application and probability in general.

Theorem 2.4 (Central Limit Theorem). *Assume that there are n independent random variables X_1, X_2,..., X_n drawn from the same distribution having mean μ and variance σ^2. Define the sample mean as follows:*

$$\bar{X}_n = \frac{1}{n}(X_1 + X_2 + ... + X_n)$$

Table 2.1 Reliability functions for some parametric distributions that can be considered as generalized Weibull distributions (Pham and Lai 2007)

Author	Reliability function	Characteristics
Gompertz	$R(t) = \exp\left\{ \frac{\theta}{\alpha}\left(1 - e^{\alpha t}\right)\right\}$, $\theta > 0, -\infty < \alpha < \infty; t \geq 0$	IFR if $\alpha > 0$ DFR if $\alpha < 0$
Weibull	$R(t) = \exp(-\lambda t^{\alpha})$, $\lambda, \alpha > 0; t \geq 0$	Exponential if $\alpha = 1$ IFR if $\alpha > 1$ DFR if $\alpha < 1$
Smith and Bain (Exponential power model)	$R(t) = \exp\left\{1 - e^{(\lambda t)^{\alpha}}\right\}$, $\alpha, \lambda > 0; t \geq 0$	BT if $0 < \alpha < 1$ Special cases of Chen's model
Xie and Lai	$R(t) = \exp\left\{-(t/\beta_1)^{\alpha_1} - (t/\beta_2)^{\alpha_2}\right\}$ $\alpha_1, \alpha_2, \beta_1, \beta_2 > 0; t \geq 0$	IFR if $\alpha_1, \alpha_2 > 1$ DFR if $\alpha_1, \alpha_2 < 1$ BT if $\alpha_1 < 1, \alpha_2 > 1$
Chen	$R(t) = \exp\left\{-\lambda\left[e^{t^{\beta}} - 1\right]\right\}, \lambda, \beta > 0; t \geq 0$	BT if $\beta < 1$ IFR if $\beta \geq 1$ Exponential power if $\lambda = 1$
Pham	$R(t) = \exp\left\{1 - a^{t^{\alpha}}\right\}; \alpha > 0, a > 1; t \geq 0$	DFR for $t \leq t_0$ IFR for $t \geq t_0$ $t_0 = \left(\frac{1-\alpha}{\alpha \ln a}\right)^{\frac{1}{\alpha}}$
Xie, Tang and Goh	$R(t) = \exp\left\{\lambda\beta\left[1 - e^{(t/\beta)^{\alpha}}\right]\right\}$ $\alpha, \beta, \lambda > 0; t \geq 0$	Chen's model if $\beta = 1$ IFR if $\alpha \geq 1$ BT if $0 < \alpha < 1$
Lai et al. (2003), Gurvich et al. (1997)	$R(t) = \exp\left\{-at^{\alpha}e^{\lambda t}\right\}$ $\lambda \geq 0, \alpha, a > 0; t \geq 0$	Weibull if $\lambda = 0$: Exponential if $\beta = 0, \alpha = 1$ IFR if $\alpha \geq 1$ BT if $0 < \alpha < 1$
Nadarajah and Kotz (2005)	$R(t) = \exp\left\{-at^b\left(e^{ct^d} - 1\right)\right\}$ $a, d > 0; b, c \geq 0; t \geq 0$	Xie, Tang & Goh's model if $b = 0$ Chen's model if $b = 0, c = 1$
Bebbington et al. (2006)	$R(t) = \exp\left\{-\left(e^{\alpha t - \beta/t}\right)\right\}; \alpha, \beta > 0; t \geq 0$	IFR iff $\alpha\beta > (27/64)$ MBT if $\alpha\beta < (27/64)$
Hjorth (1980)	$R(t) = \frac{\exp(-\delta t^2/2)}{(1+\beta t)^{\theta/\beta}}$, $\delta, \beta, \theta > 0; t \geq 0$	BT shape if $0 < \delta < \theta\beta$
Slymen and Lachenbruch (1984)	$R(t) = \exp\left\{-\exp\left[\alpha + \frac{\beta(t^{\theta} - t^{-\theta})}{2\theta}\right]\right\}$, $t \geq 0$	BT if $2\{(\theta + 1)t - \theta - 2 - (\theta - 1)t\theta - 2\}$ $\times (t\theta - 1 - t - \theta - 1) - 2$ bounded
Phani (1987)	$R(t) = \exp\left\{-\lambda \frac{(t-a)^{\beta_1}}{(b-t)^{\beta_2}}\right\}$ $\lambda > 0, \beta's > 0, 0 \leq a \leq t \leq b < \infty$	Kies (1958) [9] if $\beta_1 = \beta_2 = \beta$ IFR if $\beta \geq 1$ BT if $0 < \beta < 1$
Mudholkar and Srivastava (1993) (Exponentiated Weibull)	$R(t) = 1 - \left[1 - \exp(-t/\beta)^{\alpha}\right]^{\theta}, t \geq 0;$ $\alpha, \beta > 0, \theta \geq 0$	Weibull if $\theta = 1$ Exponential if $\alpha = \theta = 1$ DFR if $\alpha, \theta < 1$ IFR if $\alpha, \theta > 1$ BT of IFR if $\alpha > 1, \theta < 1$: UBT of DFR if $\alpha < 1, \theta > 1$:
Mudholkar et al. (1996)	$R(t) = 1 - \left[1 - \left(1 - \lambda\left(\frac{t}{\beta}\right)^{\alpha}\right)^{1/\lambda}\right]$, $\alpha, \beta > 0; t \geq 0$	BT for $\alpha < 1, \lambda > 0$ IFR for $\alpha \geq 1, \lambda \geq 0$ DFR for $\alpha \leq 1, \lambda \geq 0$ UBT for $\alpha > 1, \lambda > 0$ Exponential for $\alpha = 1, \lambda = 0$
Marshall and Olkin (1997)	$R(t) = \frac{\nu \exp[-(t/\beta)^{\alpha}]}{1 - (1-\nu)\exp[-(t/\beta)^{\alpha}]}$ $\alpha, \beta, \nu > 0; t \geq 0.$	IFR if $\nu \geq 1, \alpha \geq 1$ DFR if $\nu \geq 1, \alpha \leq 1$ MBT $\alpha = 2, \nu = 0.1$ or 0.05

(continued)

Table 2.1 (continued)

Author	Reliability function	Characteristics
Nikulin and Haghighi (2006)	$R(t) = \exp\left\{1 - (1 + (t/\beta)^\alpha)^\theta\right\}$, $t \geq 0;\ \alpha, \beta > 0, \theta \geq 0$	Can achieve various shapes, See Sect. 2.4 below

If $\bar{X}_n$ is the sample mean of n i.i.d. random variables, then if the population variance σ^2 is finite, the sampling distribution of $\bar{X}_n$ is approximately normal as $n \to \infty$. Mathematically, if $X_1, X_2,..., X_n$ is a sequence of i.i.d. random variables with

$$E(X_1) = \mu \text{ and } V(X_1) = \sigma^2$$

then

$$\lim_{n \to \infty} P\left\{\frac{\bar{X}_n - \mu}{\frac{\sigma}{\sqrt{n}}} \leq z\right\} = \Phi(z) \qquad (2.96)$$

where $\Phi(z)$ is the cdf of $N(0,1)$.

It can easily show that the mean and variance of $\bar{X}_n$ are μ and variance $\frac{\sigma^2}{n}$, respectively. Let Z_n denote the standardized $\bar{X}_n$, that is

$$Z_n = \frac{\bar{X}_n - \mu}{\sigma / \sqrt{n}}$$

It follows from the Central Limit Theorem that if n is large enough, Z_n has approximately a standard normal distribution, i.e. $N(0,1)$ distribution. This implies that the cdf of Z_n is given by

$$P(Z_n \leq z) = \frac{1}{\sqrt{2\pi}} \int_{-\infty}^{z} e^{-s^2/2} ds.$$

Similary, $\bar{X}_n$ has approximately a normal $N\left(\mu, \sigma^2/n\right)$ distribution as n is large enough. It can be shown that $\sum_{i=1}^{n} X_i$ has approximately a normal $N\left(n\mu, n\sigma^2\right)$ distribution. In most cases, a sample size of 30 or more gives an adequate approximate normality.

Note that when n is large and p is small, the Poisson distribution can be used to approximate the binomial distribution, $b(n, p)$ where the pdf of binomial distribution is

$$P(X = x) = \binom{n}{x} p^x (1 - p)^{n-x}$$

Normal Approximations

When n is large and p is not small, the Central Limit Theorem (above) allows us to use the normal distribution to approximate the binomial $b(n, p)$ distribution. In fact, let 's consider that there are n independent Bernoulli random variables X_1, X_2,..., X_n drawn from the same distribution each with p, the probability of success. Then the sum of n Bernoulli random sample is the binomial distribution. Mathematically, using the Central Limit Theorem we can write as follows:

$$Z_n = \frac{\sum_{i=1}^{n} X_i - np}{\sqrt{np(1 - p)}} \sim N(0, 1) \text{ as } n \text{ is large} \tag{2.97}$$

In practice, one can use the Central Limit Theorem for the approximation of the binomial distribution $b(n, p)$ by a normal distribution if: (1) values of n and p satisfy $np \geq 10$ and $n(1 - p) \geq 10$ or (2) the values of p are between 0.4 and 0.75, $np \geq 5$ and $n(1 - p) \geq 5$.

How large should n be to get a good approximation? A general rule is

$$n > \frac{9}{p(1 - p)}$$

Example 2.30 The probability of a defective unit in a manufacturing process is $q = 0.06$. That means, the probability of a succesful unit is $p = 0.94$. Using the normal approximation, obtain the probability that the number of defective units in a sample of 1000 units is between 50 and 70.

Solution The probability that the number of defective units in a sample of 1000 units is between 50 and 70 is given by using the normal approximation:

$$P(50 \leq X \leq 70) \simeq \Phi\left(\frac{70.5 - 60}{\sqrt{56.4}}\right) - \Phi\left(\frac{49.5 - 60}{\sqrt{56.4}}\right) \quad \text{as } nq = 60 \text{ is large}$$
$$= 0.8379$$

In general, for a < b we can obtain

$$P(a \leq X \leq b) \simeq \Phi\left(\frac{b + \frac{1}{2} - np}{\sqrt{np(1 - p)}}\right) - \Phi\left(\frac{a - \frac{1}{2} - np}{\sqrt{np(1 - p)}}\right)$$

where X is binomial $b(n,p)$ with the expected of X is np for large n.

2.5 Problems

1. An operating unit is supported by $(n - 1)$ identical units on cold standby. When it fails, a unit from standby takes its place. The system fails if all n units fail. Assume that units on standby cannot fail and the lifetime of each unit follows the exponential distribution with failure rate λ.

    (a) What is the distribution of the system lifetime?
    (b) Determine the reliability of the standby system for a mission of 100 h when $\lambda = 0.0001$ per hour and $n = 5$.

2. Assume that there is some latent deterioration process occurring in the system. During the interval $[0, a - h]$ the deterioration is comparatively small so that the shocks do not cause system failure. During a relatively short time interval $[a - h, a]$, the deterioration progresses rapidly and makes the system susceptible to shocks. Assume that the appearance of each shock follows the exponential distribution with failure rate λ. What is the distribution of the system lifetime?

3. Consider a series system of n Weibull components. The corresponding lifetimes $T_1, T_2,..., T_n$ are assumed to be independent with pdf

$$f(t) = \begin{cases} \lambda_i^\beta \beta t^{\beta-1} e^{-(\lambda_i t)^\beta} & \text{for } t \geq 0 \\ 0 & \text{otherwise} \end{cases}$$

where $\lambda > 0$ and $\beta > 0$ are the scale and shape parameters, respectively.

    (a) Show that the lifetime of a series system has the Weibull distribution with pdf

$$f_s(t) = \begin{cases} \left(\sum_{i=1}^{n} \lambda_i^\beta \right) \beta t^{\beta-1} e^{-\left(\sum_{i=1}^{n} \lambda_i^\beta \right) t^\beta} & \text{for } t \geq 0 \\ 0 & \text{otherwise} \end{cases}$$

    (b) Find the reliability of this series system.

4. The failure rate function, denoted by $h(t)$, is defined as

$$h(t) = -\frac{d}{dt} \ln[R(t)]$$

Show that the constant failure rate function implies an exponential distribution.

5. Show that if a random variable X is geometric for some p

$$P(X = x) = \begin{cases} (1 - p)p^x & \text{for } x = 0, 1, ... \\ 0 & \text{otherwise} \end{cases}$$

Then it has the memoryless property for any positive integers i and j

$$P(X \geq i + j/X \geq i) = P(X \geq j).$$

6. Show that if $T_1, T_2, ..., T_n$, are independently exponentially distributed random variables (r.v.'s) with constant failure rate $\lambda_1, \lambda_2, ..., \lambda_n$, and if. $Y = \max\{T_1, T_2, ..., T_n\}$ then Y is not exponentially distributed.

7. The time to failure T of a unit is assumed to have a log normal distribution with pdf

$$f(t) = \frac{1}{\sqrt{2\pi}} \frac{1}{\sigma t} e^{-\frac{(\ln t - \mu)^2}{2\sigma^2}} \quad t > 0$$

Show that the failure rate function is unimodal.

8. A diode may fail due to either open or short failure modes. Assume that the time to failure T_0 caused by open mode is exponentially distributed with pdf

$$f_0(t) = \lambda_0 e^{-\lambda_0 t} \quad t \geq 0$$

and the time to failure T_1 caused by short mode has the pdf

$$f_s(t) = \lambda_s e^{-\lambda_s t} \quad t \geq 0$$

The pdf for the time to failure T of the diode is given by

$$f(t) = p\, f_0(t) + (1 - p)f_s(t) \quad t \geq 0$$

(a) Explain the meaning of p in the above pdf function.
(b) Derive the reliability function $R(t)$ and failure rate function $h(t)$ for the time to failure T of the diode.
(c) Show that the diode with pdf $f(t)$ has a decreasing failure rate (DFR).

9. Events occur according to an NHPP in which the mean value function is

$$m(t) = t^3 + 3t^2 + 6t \quad t > 0.$$

What is the probability that n events occur between times $t = 10$ and $t = 15$?

10. Show that if X_i is exponentially distributed with parameter θ then $T = X_1 + X_2 + \ldots + X_n$, is Erlang distribution with parameters θ and n where the pdf is given.
by

$$f(t) = \frac{\theta^n t^{n-1}}{\Gamma(n)} e^{-t\theta} \quad t > 0$$

11. Let X_i follows gamma distributed with α_i is the shape parameter and common scale parameter β for $i = 1, 2, \ldots, k$ and is denoted as $X_i \sim G(\alpha_i, \beta)$. If X_1, $X_2, \ldots, X_n$ are independent random variables, show that the distribution of T, where $T = X_1 + X_2 + \ldots + X_k$, is also gamma distributed, as $T \sim G(\alpha, \beta)$. with scale parameter β and the shape parameter is $\alpha = \sum_{i=1}^{k} \alpha_i$

12. Show that the variance of a mixed truncated exponential distribution (see Eq. 2.47) with parameters λ and a as

$$Var(X) = \frac{1}{\lambda^2} \left(1 - 2\lambda a e^{-\lambda a} - e^{-2\lambda a}\right).$$

13. Show that for any given a and α, the Pham distribution F (see Sect. 2.2) is decreasing failure rate (DFR) for $t \leq t_0$ and increasing failure rate (IFR) for $t \geq t_0$ where

$$t_0 = \left(\frac{1 - \alpha}{\alpha \ln a}\right)^{\frac{1}{\alpha}}.$$

14. Show that when $\alpha \geq 1$, the Pham distribution is an IFR.
15. A physician treats 10 patients with elevated blood pressures. The drug administered has a probability of 0.9 of lowering the blood pressure in an individual patient. What is the probability that at least 8 of the 10 patients respond successfully to treatment?
16. Phone calls arrive at a business randomly throughout the day according to the Poisson model at the rate of 0.5 calls per minute. What is the probability that in a given five minute period not more than three calls arrive?
17. Suppose an electrical firm manufactures N fuses a week k of which are good and $(N - k)$ of which are defective. A sample of size n is inspected weekly, x of which are good and $(n - x)$ are defective. Determine the probability that exactly x number of good fuses in the sample of size n.
18. In exploring for oil, an energy company has been recording the average number of drilling excavations prior to locating a productive oil well. In 25 experiences,

it was found that 4.5 drillings, on the average, were required before a well was found. Let N be a random variable that counts for the number of drillings necessary to find a productive well. Define p is the probability that a drilling is a productive. Assuming that all drillings are mutually independent, obtain the probability that exactly n number of drillings necessary in order to find a productive well.

19. Suppose an item fails according to the exponential failure density function.

$$f(x) = \lambda e^{-\lambda x} \quad \lambda > 0, x \geq 0.$$

Let $N(t)$ be the number of failures of a component during a given time interval $(0, t)$. Determine the probability of $N(t)$.

20. Show that the pdf of the sum of squares of n independent standard normal random variables $X_i, i = 1, 2, ..., n$ (i.e., $X_i \sim N(0, 1)$ is the chi-square given in Eq. (2.63).

21. The number of seeds (N) produced by a certain kind of plant has a Poisson distribution with parameter λ. Each seed, independently of how many there are, has probability p of forming into a developed plant. Find the mean and variance of the number of developed plants.

22. A system has m locations at which the same component is used, each component having an exponentially distributed failure time with mean $\frac{1}{\lambda}$. When a component at any location fails it is replaced by a spare. In a given mission the component at location k is required to operate for time t_k. What is the minimum number of components, n, (i.e., m originals and $(n-m)$ spares) required to ensure a successful mission with probability at least 0.96? Compute the minimum number of components (n) when $m = 3$, $t_k = k$, and $\lambda = 0.5$.

23. It has been observed that the number of claims received by an insurance company in a week is Poisson distributed with constant failure parameter λ where the value of the parameter is obtained by statistical estimation. What is the probability that a week will pass without more than two claims? Obtain this probability when $\lambda = 10$ per week.

24. The number of accidents per month in a factory has been found to have a geometric distribution with parameter p (probability of an accident at each month), where the parameter is estimated by statistical methods. Let random variable X is defined as the number of accidents in a given month. What is the probability that there will be more than two accidents on a month? Calculate when $p = 0.1$.

25. There are n workers in an assembly line factory and it is known that the number of workers independently reporting for work on a given day is binomially distributed with parameters n and p where p is the probability of an individual who presents for work. What is the probability that there will be at least two absentees on a given day selected at a random? Determine this probability when $p = 0.95$ and $n = 20$.

26. Show that the number of failures before n successes in Bernoulli trials is negative binomial with parameters n and p where p is the probability of success at each trial.

27. An entrepreneurer of start-up company calls potential customers to sale a new product. Assume that the outcomes of consecutive calls are independent and that on each call he has 20% chance of making a sale. His daily goal is to make 5 sales and he can make only 25 calls in a day. What is the probability that he achieves his goal in between 15 and 20 trials?

28. Prove Theorem 2.1.

29. A class consists of 10 girls and 12 boys. The teacher selects 6 children at random for group projects. Is it more likely that she chooses 3 girls and 3 boys or that she chooses 2 girls and 4 boys?

30. Suppose the life of automobile batteries is exponentially distributed with constant failure rate parameter $\lambda = 0.0001$ failures per hour.

    (a) What is the probability that a battery will last more than 5,000 h?
    (b) What is the probability that a battery will last more than 5,000 h given that it has already survived 3,000 h?

31. The probability of failure of an equipment when it is tested is equal to 0.2. What is the probability that failure occurs in less than four tests? Assume independence of tests.

32. An electric network has N relays. If a relay is tripped the repair time is a random variable that obeys the exponential distribution with constant failure rate parameter $\lambda = 19/h$. If one repair man is available and ten relays have tripped simultaneously, what is the probability that power failure will last more than $\frac{3}{4}$ h?

33. An instrument which consists of k units operates for a time t. The reliability (the probability of failure-free performance) of each unit during the time t is p. Units fail independently of one another. When the time t passes, the instrument stops functioning, a technician inspects it and replaces the units that failed. He needs a time s to replace one unit.

    (a) Find the probability P that in the time $2\,s$ after the stop the instrument will be ready for work.
    (b) Calculate P when $k = 10$, $p = 0.9$, $t = 25$ h, and $s = 20$ min.

34. Two types of coins are produced at a factory: a fair coin and a biased one that comes up heads 55 percent of the time. A quality manager has one of these coins but does not know whether it is a fair coin or a biased one. In order to ascertain which type of coin he has, he shall perform the following statistical test: He shall toss the coin 1000 times. If the coin lands on heads 525 or more times, then he shall conclude that it is a biased coin, whereas, if it lands heads less than 525 times, then he shall conclude that it is the fair coin. If the coin is actually fair, use the Normal approximation to find the probability that he shall reach a false conclusion?

35. Consider a system of two independent components with failure time T_i for component i, i = 1, 2, with corresponding probability density function f_i and distribution function F_i.

    (a) Regardless of the connections between the two components, what is the probability of component 1 failure before component 2 failure?
    (b) If T_i has an exponential distribution with constant failure rate λ_i, i = 1, 2, what is the probability of component 1 failure before component 2 failure?
    (c) Calculate (b) when $\lambda_1 = 0.05$ failures per hour and $\lambda_2 = 0.03$ failures per hour.

36. Suppose there are 100 items in the lot and their manufacturer claims that no more than 2% are defective. When the shipment is received at an assembly plant, a quality control manager asks his staff to take a random sample of size 10 from it without replacement and will accept the lot if there are at most 1 defective item in the sample. If let say 2% are defective in the entire lot, then it consists of 2 defective items. What is the probability that the manager will accept the lot?

37. An airport limousine can accommodate up to four passengers on any one trip. The company will accept a maximum of six reservations for a trip and a passenger must have a reservation. From previous records, 30% of all those making reservations do not appear for the trip. Assume that passengers make reservations independently of one another. If six reservations are made, what is the probability that at least one individual with a reservation cannot be accommodated on the trip?

References

Bebbington M, Lai CD, Zitikis R (2006) Useful periods for lifetime distributions with bathtub shaped hazard rate functions. IEEE Trans Reliab 55(2):245–251

Birnbaum ZW, Esary JD, Marshall AW (1966a) Characterization of wearout of components and systems. Ann Math Stat 37(816):825

Birnbaum ZW, Esary JD (1966b) Some inequalities for reliability functions. In: Proceedings fifth Berkeley symposium on mathematical statistics and probability. University of California Press

Cordeiro GM, Silva GO, Ortega EM (2016) An extended-G geometric family. J Stat Distrib Appl 3(3)

Gurvich MR, Dibenedetto AT, Rande SV (1997) A new statistical distribution for characterizing the random strength of brittle materials. J Mater Sci 32:2559–2564

Hjorth U (1980) A reliability distribution with increasing, decreasing, constant and bathtub-shaped failure rate. Technometrics 22:99–107

Johnson NI, Kotz S (1970) Distributions in statistics: continuous univariate distributions-1. Houghton Mifflin Co, Boston

Kosambi DD (1949) Characteristic properties of series distributions. In: Proceedings of the National Academy of Sciences, India, vol 15, pp 109–113

Lai CD, Xie M, Murthy DNP (2003) Modified Weibull model. IEEE Trans Reliab 52:33–37

Marshall W, Olkin I (1997) A new method for adding a parameter to a family of distributions with application to the exponential and Weibull families. Biometrika 84:641–652

Mudholkar GS, Srivastava DK (1993) Exponential Weibull family for analyzing bathtub failure-rate data. IEEE Trans Reliab 42:299–302

Mudholkar GS, Srirastava DK, Kollia GD (1996) A generalization of the Weibull distribution with application to analysis of survival data. J Am Stat Assoc 91:1575–1583

Nadarajah S, Kotz S (2005) On some recent modifications of Weibull distribution. IEEE Trans Reliab 54(4):561–562

Nikulin M, Haghighi F (2006) A Chi-squared test for the generalized power Weibull family for the head-and-neck cancer censored data. J Math Sci 133(3):1333–1341

Pham H (2002) A Vtub-shaped hazard rate function with applications to system safety. Int J Reliab Appl 3(1):1–16

Pham H (2006) Springer handbook of engineering statistics. Springer

Pham H (2019) A distribution function and its applications in software reliability. Int J Perform Eng 15(5):1306–1313

Pham H, Lai C-D (2007) On recent generalization of the Weibull distribution. IEEE Trans Reliab 56(3):454–458

Phani KK (1987) A new modified Weibull distribution function. Commun Am Ceramic Soc 70:182–184

Slymen DJ, Lachenbruch PA (1984) Survival distributions arising from two families and generated by transformations. Commun Stat Theory Methods 13:1179–1201

Weibull W (1951) A statistical distribution function of wide applicability. J Appl Mech 18:293–297

Chapter 3
Statistical Inference

3.1 Introduction

Inference about the values of parameters involved in statistical distributions is known as estimation. The engineer might be interested in estimating the mean life of an electronic device based on the failure times of a random sample placed on lifef test. An interesting question that commonly would ask from many practitioners is: how close this estimator would be the true value of the parameter being estimated from a known distribution.

An estimator is a procedure which provides an estimate of a population parameter from a random sample. In other word, an estimator is a statistic and hence a random variable, since it depends strictly on the sample. An estimator is called a point estimator if it provides a single value as an estimate from a random sample.

In this chapter, it is assumed that the population distribution by type is known, but the distribution parameters are unknown and they have to be estimated by using collected failure data. This chapter is devoted to the theory of estimation and discusses several common estimation techniques such as maximum likelihood, method of moments, least squared, and Bayesian methods. We also discuss the confidence interval estimates, tolerance limit estimates, sequential sampling and criteria for model selection.

3.2 Statistical Inference

Statistical inference is the process of drawing conclusions about unknown characteristics of a population from which data were taken. Techniques used in this process include paramter estimation, confidence intervals, hypothesis testing, goodness of fit tests, and sequential testing. As we know, any distribution function often involves some parameters. The problem of point estimation is that of estimating the parameters of a population. For example, parameter λ from an exponential distribution; μ and

© The Author(s), under exclusive license to Springer Nature Switzerland AG 2022
H. Pham, *Statistical Reliability Engineering*, Springer Series in Reliability Engineering,
https://doi.org/10.1007/978-3-030-76904-8_3

σ^2 from a normal distribution; or n and p from the binomial. For simplicity, denote these parameters by the notation θ which may be a vector consisting of several distribution parameters. Given a real life application and the observed data, the common statistical problem will consist of how to determine the unknown distribution F or pdf f. And if F is assumed to be known, then how can one determine the unknown distribution parameters θ. If a statistic $Y = h(X_1, X_2,..., X_n)$ is taken as an estimate of the parameters θ, the first point to recognize is that in any particular sample the observed value of $h(X_1, X_2,..., X_n)$ may differ from θ. The performance of $h(X_1, X_2,..., X_n)$ as an estimate of θ is to be judged in relation to the sampling distribution of $h(X_1, X_2,..., X_n)$. For example, assume n independent samples from the exponential density $f(x; \lambda) = \lambda e^{-\lambda x}$ for $x > 0$ and $\lambda > 0$, then the joint pdf or sample density (for short) is given by

$$f(x_1, \lambda) \cdot f(x_1, \lambda) \ldots f(x_1, \lambda) = \lambda^n e^{-\lambda \sum_{i-1}^{n} x_i} \tag{3.1}$$

The problem here is to find a "good" point estimate of λ which is denoted by $\hat{\lambda}$. In other words, we shall find a function $h(X_1, X_2,...,X_n)$ such that, if $x_1, x_2, ..., x_n$ are the observed experimental values of $X_1, X_2, ..., X_n$, then the value $h(x_1, x_2, ..., x_n)$ will be a good point estimate of λ. By "good' we mean the following properties shall be implied:

- Unbiasedness
- Consistency
- Efficiency (i.e., minimum variance)
- Sufficiency
- Asymptotic efficiency.

In other words, if $\hat{\lambda}$ is a good point estimate of λ, then one can select the function $h(X_1, X_2,...,X_n)$ such that $h(X_1, X_2,...,X_n)$ is not only an unbiased estimator of λ but also the variance of $h(X_1, X_2,...,X_n)$ is a minimum. It should be noted that the estimator $\hat{\lambda}$ is a random variable where λ is a number.

The random variable $h(X_1, X_2,...,X_n)$ is called a statistic and it represents the estimator of the unknown parameters θ. We will now present the following definitions.

Definition 3.1 For a given positive integer n, the statistic $Y = h(X_1, X_2,..., X_n)$ is called an *unbiased* estimator of the parameter θ if the expectation of Y is equal to a parameter θ, that is,

$$E(Y) = \theta \tag{3.2}$$

In other words, an estimator of θ is called unbiased if its expected value is equal to the population value of the quantity it estimates.

An estimator Y is a best unbiased estimator of parameter θ if $E(Y) = \theta$ for all $\theta \in \Theta$ and $\text{var}(Y) \leq \text{var}\left(\tilde{Y}\right)$ for any other unbiased estimator $\tilde{Y}$ such that.

$E\left(\tilde{Y}\right) = \theta$. If an estimator is not unbiased we can define its *bias* by

$$b(\theta) = E(Y) - \theta.$$

It sometimes happens that $b(\theta)$ depends on the number of observations and approaches to zero as n increases. In this case, Y is said to be *asymptotically unbiased*. Note that the expectation represents a long-run average, bias is seen as an average deviation of the estimator from the true value. When considering bias, care needs to be taken with functions of parameters. In general, if $E(Y) = \theta$ and $g(\theta)$ is some function of a parameter,

$$E[g(Y)] \neq g(\theta).$$

For the exponential distribution for example, assume $X_1, X_2, \ldots, X_n$ be a random sample from the exponential distribution with pdf

$$f(x; \mu) = \frac{1}{\mu} e^{-\frac{x}{\mu}} \quad x > 0, \mu > 0$$

Let $Y = \frac{1}{n}\left(\sum_{i=1}^{n} x_i\right)$. Using the maximum likelihood estimation (MLE) method (see Sect. 3.3.2), we can show that Y is an unbiased estimator of μ. That is, $E(Y) = \mu$. Let $g(Y) = \frac{n}{\sum_{i=1}^{n} x_i}$. It should be noted that $E[g(Y)] \neq \frac{1}{\mu}$.

Definition 3.2 The statistic Y is said to be a *consistent* estimator of the parameter θ if Y converges stochastically to a parameter θ as n approaches infinity. If ϵ is an arbitrarily small positive number when Y is consistent, then.

$$\lim_{n \to \infty} P(|Y - \theta| \leq \epsilon) = 1 \tag{3.3}$$

In other words, we would like our estimators tend towards the true value when the sample size n closes to infinity or to a very large number. It can be shown that this will happen if the estimator Y is either unbiased or asymptotically unbiased and also has a variance that tends to zero as n increases.

Example 3.1 Let X be a life time with mean μ which is unknown. One can easily show that the statistic.

$$h(X_1, X_2, \ldots, X_n) = \frac{\sum_{i=1}^{n} X_i}{n}$$

is unbiased and a consistent of μ.

Definition 3.3 If the limiting distribution of.

$$\frac{\sqrt{n}(Y_n - \theta)}{a} \sim N(0, 1)$$

where Y_n is an estimator of θ based on a sample of size n, the *asymptotic efficiency* of Y_n is defined as

$$AE = \frac{1}{a^2 I(\theta)} \text{ where } I(\theta) = -E\left(\frac{\partial^2 \ln f}{\partial \theta^2}\right) \tag{3.4}$$

and f is the probability density function of the parent population. If $AE = 1$, the sequence of estimators is said to be asymptotically efficient or best asymptotically normal (BAN).

Definition 3.4 The statistic Y is said to be *sufficient* for θ if the conditional distribution of X, given $Y = y$, is independent of θ.

Definition 3.5 The statistic Y will be called the *minimum variance unbiased estimator* of the parameter θ if Y is unbiased and the variance of Y is less than or equal to the variance of every other unbiased estimator of θ. An estimator that has the property of minimum variance in large samples is said to be efficient.

In other words, among unbiased estimates of θ, the one that has smallest variance is preferred. For the variance of any unbiased estimator Y of θ, we have the lower limit

$$Var(Y) \geq \frac{1}{n \, I(\theta)} \tag{3.5}$$

where $I(\theta)$, defined by Eq. (3.4), is called the amount of information per observation (also see Sect. 3.4).

Definition 3.5 is useful in finding a lower bound on the variance of all unbiased estimators. In fact, a *minimum variance unbiased estimator* is one whose variance is least for all possible choices of unbiased estimators. Let X be a random variable of life time with mean μ which is unknown. Consider the following two statistics

$$h_1(X_1, X_2, \dots, X_n) = \frac{1}{4}X_1 + \frac{1}{2}X_2 + \frac{1}{4}X_3$$

and

$$h_2(X_1, X_2, \dots, X_n) = \frac{1}{3}(X_1 + X_2 + X_3)$$

We can easily show that both h_1 and h_2 give unbiased estimates of μ. However, based on the unbiased estimates, we cannot tell which one is better, but we can tell which estimator is better by comparing the variances of both h_1 and h_2 estimators. It is easy to see that var(h_1) > var(h_2), then h_2 is a better estimator h_1. Mainly, the smaller the variance the better!

The mean squared error is also another criterion for comparing estimators by combining the variance and the biased parts.

Definition 3.6 The *mean squared error* (MSE) is defined as the expected value of the square of the deviation of the estimate from the parameter being estimated, and is equal to the variance of the estimate plus the square of the bias. That is,

$$MSE = E\left(\hat{\theta} - \theta\right)^2 = \text{var}\left(\hat{\theta}\right)$$
$$+ E\left[E\left(\hat{\theta}\right) - \theta\right]^2 = \text{var}\left(\hat{\theta}\right) + (bias)^2.$$

Obviously, a small value of MSE is a desirable feature for an estimator. For unbiased estimators case, seeking a small MSE is identical to seeking a small variance since the second term on the right hand side is equal to zero.

Example 3.2
Assume $\hat{\theta}_1$ and $\hat{\theta}_2$ are the two estimators of parameter θ. Suppose that $E\left(\hat{\theta}_1\right) = 0.9\theta, E\left(\hat{\theta}_2\right) = \theta,$ var($\hat{\theta}_1$) $= 2$, and var($\hat{\theta}_2$) $= 3$. Which estimator would you prefer?
Solution: We have

$$MSE \text{ of } \theta_1 = \text{var}\left(\hat{\theta}\right) + (bias)^2 = 2 + 0.01\,\theta^2.$$
$$MSE \text{ of } \theta_2 = \text{var}\left(\hat{\theta}\right) + (bias)^2 = 3 + 0 = 3.$$

Thus, if $|\theta| < 10$ then $\hat{\theta}_1$ would be prefered. If $|\theta| > 10$ then $\hat{\theta}_2$ is prefered.

We will later discuss how to establish a lower bound on the variance using an inequality known as the Cramér-Rao inequality. We now discuss some basic methods of parameter estimation.

3.3 Parameter Estimation

Once a distribution function is specified with its parameters, and data have been collected, one is in a position to evaluate its goodness of fit, that is, how well it fits the observed data. The goodness of fit is assessed by finding parameter values of a model that best fits the data—a procedure called parameter estimation. There are two general methods of parameter estimation. They are the methods of moments and maximum likelihood estimate (MLE).

3.3.1   The Method of Moments

The unknown distribution parameters usually can be estimated by their respective moments, such as means, variances etc. In general, we can choose the simplest moments to equate.

The expected value of a function $g(X)$ under the distribution $F(x)$ is given by

$$E(g(X)) = \begin{cases} \int_{-\infty}^{\infty} g(x)f(x)dx & \text{if X is continuous} \\ \sum_{i=0}^{n} g(x_i)p(x_i) & \text{if X is discrete} \end{cases} \tag{3.6}$$

The kth moment of $F(x)$ is defined as follows:

$$\mu_k = E(X^k) \quad \text{for } k = 1, 2, \ldots \tag{3.7}$$

Obviously, when $k = 1$ it is the expected value of X, that is $\mu_1 = E(X)$. When X only assumes positive values and has a continuous pdf $f(x)$ and cdf $F(x)$, then

$$E(X) = \int_0^{\infty} xf(x)dx = \int_0^{\infty} \left(\int_0^x dy \right) f(x)dx$$

$$= \int_0^{\infty} \left(\int_y^{\infty} f(x)dx \right) dy = \int_0^{\infty} (1 - F(y))dy. \tag{3.8}$$

The kth central moments around the mean, μ_k^c, is defined as follows:

$$\mu_k^c = E[(X - \mu_1)^k] \text{ for } k = 1, 2, \ldots \tag{3.9}$$

The 1st central moment is 0. The second central moment is exactly the variance of $F(x)$. That is,

$$\mu_2^c = E[(X - \mu_1)^2]. \tag{3.10}$$

Example 3.3
Suppose that $f(x) = \lambda e^{-\lambda x}$ for $x \geq 0$, $\lambda > 0$. Then $F(x) = 1 - e^{-\lambda x}$ and the population mean (or the first moment) is.

$$\mu_1 = \int_0^{\infty} e^{-\lambda x}dx = \frac{1}{\lambda}$$

The sample mean is $\bar{x}$ so the estimator $\hat{\lambda}$ of λ that makes these two the same is

$$\hat{\lambda} = \frac{1}{\bar{x}} = \frac{n}{\sum_{i=1}^{n} x_i}$$

Note that when X is discrete, assuming the values $\{1,2,\ldots\}$, then we obtain

$$E(X) = 1 + \sum_{i=1}^{\infty} [1 - F(i)].$$

Example 3.4

For a normal distribution with two unknown parameters pdf, i.e. $N(\mu, \sigma^2)$.

$$f(x) = \frac{1}{\sigma\sqrt{2\pi}} e^{-\frac{1}{2}\left(\frac{x-\mu}{\sigma}\right)^2} \quad -\infty < x < \infty$$

Then the first and second moments about the mean of the sample are

$$\mu_1 = \bar{x} = \frac{\sum_{i=1}^{n} x_i}{n}$$

and

$$\mu_2^c = E[(X - \mu_1)^2] = \frac{\sum_{i=1}^{n} (x_i - \bar{x})^2}{n}.$$

For the population of corresponding moments are

$$\mu_1 = \mu \quad \text{and} \quad \mu_2^c = \sigma^2.$$

So the method of moments estimators are

$$\hat{\mu} = \bar{x} \quad \text{and} \quad \hat{\sigma}^2 = \frac{\sum_{i=1}^{n} (x_i - \bar{x})^2}{n}.$$

3.3.2 Maximum Likelihood Estimation Method

The method of maximum likelihood estimation (MLE) is one of the most useful techniques for deriving point estimators. As a lead-in to this method, a simple example will be considered. The assumption that the sample is representative of the population will be exercised both in the example and later discussions.

Example 3.5

Consider a sequence of 25 Bernoulli trials (binomial situation) where each trial results in either success or failure. From the 25 trials, 6 failures and 19 successes result. Let p be the probability of success, and $1-p$ the probability of failure. Find the estimator of p, $\hat{p}$, which maximizes that particular outcome.

Solution: The sample density function can be written as

$$g(19) = \binom{25}{19} p^{19}(1-p)^6.$$

The maximum of $g(19)$ occurs when

$$p = \hat{p} = \frac{19}{25}$$

so that

$$g\left(19|p = \frac{19}{25}\right) \geq g\left(19|p \neq \frac{19}{25}\right)$$

Now $g(19)$ is the probability or "likelihood" of 6 failures in a sequence of 25 trials. Select $p = \hat{p} = \frac{19}{25}$ as the probability or likelihood maximum value and, hence, $\hat{p}$ is referred to as the maximum likelihood estimate. The reason for maximizing $g(19)$ is that the sample contained six failures, and hence, if it is representative of the population, it is desired to find an estimate which maximizes this sample result. Just as $g(19)$ was a particular sample estimate, in general, one deals with a sample density:

$$f(x_1, x_2, \ldots, x_n) = f(x_1; \theta) f(x_2; \theta) \cdots f(x_n; \theta) \qquad (3.11)$$

where x_1, x_2, ..., x_n are random, independent observation from a population with density function $f(x)$. For the general case, it is desired to find an estimate or estimates, $\hat{\theta}_1, \hat{\theta}_2, \ldots, \hat{\theta}_m$ (if such exist) where

$$f(x_1, x_2, \ldots, x_n; \theta_1, \theta_2, \ldots, \theta_m) > f(x_1, x_2, \ldots, x_n; \theta\prime_1, \theta\prime_2, \ldots, \theta\prime_m) \qquad (3.12)$$

Notation $\theta\prime_1, \theta\prime_2, \ldots, \theta\prime_n$ refers to any other estimates different than $\hat{\theta}_1, \hat{\theta}_2, \ldots, \hat{\theta}_m$.

Once data have been collected and the likelihood function of a model given the data is determined, one is in a position to make statistical inferences about the population, that is, the probability distribution that underlies the data. We are interested in finding the parameter value that corresponds to the desired probability distribution. The principle of maximum likelihood estimation (MLE), originally developed by R. A.

Fisher in the 1920s, states that the desired probability distribution is the one that makes the observed data "most likely," which means that one must seek the value of the parameter vector that maximizes the likelihood function.

Let us now discuss the method of MLE. Consider a random sample $X_1, X_2, ..., X_n$ from a distribution having pdf $f(x; \theta)$. This distribution has a vector $\theta = (\theta_1, \theta_2, ..., \theta_m)'$ of unknown parameters associated with it, where m *is* the number of unknown parameters. Assuming that the random variables are independent, then the likelihood function, $L(X; \theta)$, is the product of the probability density function evaluated at each sample point:

$$L(X, \theta) = \prod_{i=1}^{n} f(X_i; \theta) \tag{3.13}$$

where $X = (X_1, X_2, ..., X_n)$. The maximum likelihood estimator $\hat{\theta}$ is found by maximizing $L(X; \theta)$ with respect to θ. In practice, it is often easier to maximize $\ln[L(X;\theta)]$ to find the vector of MLEs, which is valid because the logarithm function is monotonic. In other words, the maximum of $L(X; \theta)$ will occur at the same value of θ as that for $\ln[L(X; \theta)]$. The logarithm turns the multiplication of terms like $f(X_i; \theta)$ into the addition of $\ln f(X_i; \theta)$. Thus, the log likelihood function, denoted as $\ln L(\theta)$, is given by

$$\ln L(\theta) = \sum_{i=1}^{n} \ln f(X_i; \theta) \tag{3.14}$$

and is asymptotically normally distributed since it consists of the sum of n independent variables and the implication of the central limit theorem. Since $L(X; \theta)$ is a joint probability density function for $X_1, X_2, ..., X_n$, it must integrate equal to 1, that is,

$$\int_0^\infty \int_0^\infty \cdots \int_0^\infty L(X; \theta) dX = 1$$

Assuming that the likelihood is continuous, the partial derivative of the left-hand side with respect to one of the parameters, θ_i, yields

$$\frac{\partial}{\partial \theta_i} \int_0^\infty \int_0^\infty \cdots \int_0^\infty L(X; \theta) dX = \int_0^\infty \int_0^\infty \cdots \int_0^\infty \frac{\partial}{\partial \theta_i} L(X; \theta) dX$$

$$= \int_0^\infty \int_0^\infty \cdots \int_0^\infty \frac{\partial \ln L(X; \theta)}{\partial \theta_i} L(X; \theta) dX$$

$$= E\left[\frac{\partial \ln L(X;\theta)}{\partial \theta_i}\right]$$

$$= E[U_i(\theta)] \text{ for } i = 1, 2, \ldots, m \tag{3.15}$$

where $U(\theta) = (U_1(\theta), U_2(\theta), \ldots U_n(\theta))'$ is often called the score vector and the vector $U(\theta)$ has components

$$U_i(\theta) = \frac{\partial[\ln L(X;\theta)]}{\partial \theta_i} \text{ for } i = 1, 2, \ldots, m \tag{3.16}$$

which, when equated to zero and solved, yields the MLE vector $\boldsymbol{\theta}$. In other words, the MLE vector $\boldsymbol{\theta}$ satisfy

$$\frac{\partial[\ln L(X;\theta)]}{\partial \theta_i} = 0 \text{ for } i = 1, 2, \ldots, m$$

This means that if more than one parameter is to be estimated, the partial derivatives with respect to each parameter are then set equal to zero and the resulting differential equations are solved for the estimates. It is worth to note that one needs to make sure whether the solution is actually a maximum and not a minimum or a point of inflection. Often, the maximum likehood estimators are biased.

Suppose that we can obtain a non-trivial function of $X_1, X_2, \ldots, X_n$, say $h(X_1, X_2, \ldots, X_n)$, such that, when θ is replaced by $h(X_1, X_2, \ldots, X_n)$, the likelihood function L will achieve a maximum. In other words,

$$L(X, h(X)) \geq L(X, \theta)$$

for every θ. The statistic $h(X_1, X_2, \ldots, X_n)$ is called a maximum likelihood estimator of θ and will be denoted as

$$\hat{\theta} = h(x_1, x_2, \ldots, x_n)$$

The observed value of $\hat{\theta}$ is called the MLE of θ. In other words, the MLE of θ is

$$\hat{\theta} = \arg \max_{\theta \in \Theta} \ln L(\theta) \tag{3.17}$$

where Θ is the parameter space. Based on the asymptotic maximum likelihood theory, the MLE $\hat{\theta}$ is consistent and asymtotically efficient with limiting distribution

$$\sqrt{n}\left(\hat{\theta} - \theta_0\right) \rightarrow N\left(0, I^{-1}(\theta_0)\right)$$

where θ_0 is the true parameter value and I is the Fisher information matrix. The asymptotic variance matrix $I^{-1}(\theta_0)$ can be estimated consistently by an empirical variance matrix of the influence functions evaluated at $\hat{\theta}$.

In general, the mechanics for obtaining the MLE can be obtained as follows:

Step 1. Find the joint density function $L(X, \theta)$.

Step 2. Take the natural log of the join density, $\ln L(\theta)$.

Step 3. Take the partial derivatives of $\ln L(\theta)$ (or $L(\theta)$) with respect to each parameter.

Step 4. Set partial derivatives to "zero".

Step 5. Solve for parameter(s).

Example 3.6

A sample of size n is drawn, without replacement, from a population of size N composed of k individuals of type 1 and $(N-k)$ individuals of type 2. Assume that a population size N is unknown. The number X of individuals of type 1 in the sample is a hypergeometric random variable with pdf.

$$P[X = x] = \frac{\binom{k}{x}\binom{N-k}{n-x}}{\binom{N}{n}} \qquad x = 0, 1, 2, \ldots, n \qquad (3.18)$$

Obtain the MLE of N when k and n are known.

Solution: Let

$$P(x, N) = \frac{\binom{k}{x}\binom{N-k}{n-x}}{\binom{N}{n}}.$$

Then

$$\frac{P(x, N)}{P(x, N-1)} = \frac{\frac{\binom{k}{x}\binom{N-k}{n-x}}{\binom{N}{n}}}{\frac{\binom{k}{x}\binom{N-k-1}{n-x}}{\binom{N-1}{n}}}.$$

After simplifications, we obtain

$$\frac{P(x, N)}{P(x, N-1)} = \frac{N^2 - kN - nN + kn}{N^2 - kN - nN + xN}.$$

Note that $P(x, N)$ is greater than, equal to, or less than $P(x, N-1)$ according to kn is greater than, equal to, or less than xN, or equivalently, as N is less than, equal to, or greater than kn/x. We can now consider the following two cases.

Case 1: when $\frac{kn}{x}$ is not an integer.

The sequence $\{P(x, N), N = 1, 2, \ldots\}$ is increasing when $N < \frac{kn}{x}$ and is decreasing when $N > \frac{kn}{x}$. Thus, the maximum value of $P(x, N)$ occurs when $N = \lfloor \frac{kn}{x} \rfloor$ which is the largest integer less than $\frac{kn}{x}$.

Case 2: when $\frac{kn}{x}$ is an integer.

In this case the maximum value of $P(x, N)$ occurs when both $P(x,N)$ and $P(x, N-1)$ are equal. This implies that

$$\frac{kn}{x} - 1 = \left\lfloor \frac{kn}{x} \right\rfloor$$

as an estimate of the population. Therefore, the MLE of N is $\hat{N} = \lfloor \frac{kn}{x} \rfloor$ where $\lfloor y \rfloor$ denotes the greatest integer less than y.

Example 3.7

Let $X_1, X_2, \ldots, X_n$ be a random sample from the exponential distribu- tion with pdf.

$$f(x; \lambda) = \lambda e^{-\lambda x} \quad x > 0, \lambda > 0$$

The joint pdf of $X_1, X_2, \ldots, X_n$, is given by

$$L(X, \lambda) = \lambda^n e^{-\lambda \sum_{i=1}^{n} x_i}$$

and

$$\ln L(\lambda) = n \ln \lambda - \lambda \sum_{i=1}^{n} x_i$$

The function $\ln L$ can be maximized by setting the first derivative of $\ln L$, with respect to λ, equal to zero and solving the resulting equation for λ. Therefore,

$$\frac{\partial \ln L}{\partial \lambda} = \frac{n}{\lambda} - \sum_{i=1}^{n} x_i = 0$$

This implies that

$$\hat{\lambda} = \frac{n}{\sum_{i=1}^{n} x_i}.$$

The observed value of $\hat{\lambda}$ is the MLE of λ.

Example 3.8

Let $X_1, X_2, \ldots, X_n$, denote a random sample from the normal distribution $N(\mu, \sigma^2)$ with unknown mean μ and known σ^2 where pdf is.

$$f(x; \mu) = \frac{1}{\sigma\sqrt{2\pi}}e^{-\frac{(x-\mu)^2}{2\sigma^2}} \qquad -\infty < x < \infty, \ \mu \in (-\infty, \infty). \qquad (3.19)$$

Find the maximum likelihood estimator of μ.
Solution: The likelihood function of a sample of size n is

$$L\left(\underset{\sim}{x}, \mu\right) = \prod_{i=1}^{n} f(x_i; \mu) = \prod_{i=1}^{n} \frac{1}{\left(\sigma\sqrt{2\pi}\right)}e^{-\frac{(x_i-\mu)^2}{2\sigma^2}}$$

$$= \frac{1}{\left(\sigma\sqrt{2\pi}\right)^n}e^{-\frac{1}{2\sigma^2}\sum_{i=1}^{n}(x_i-\mu)^2}.$$

The log of likelihood function is

$$\ln L = -n\ln(\sqrt{2\pi}\sigma) - \frac{1}{2\sigma^2}\sum_{i=1}^{n}(x_i - \mu)^2.$$

Thus we have

$$\frac{\partial \ln L}{\partial \mu} = \frac{1}{\sigma^2}\sum_{i=1}^{n}(x_i - \mu) = 0.$$

Solving the above equation, we obtain

$$\hat{\mu} = \frac{\sum_{i=1}^{n} x_i}{n} = \bar{x}.$$

Thus, the MLE of μ is $\bar{x}$.

Example 3.9

In an exponential censored case, the non-conditional joint pdf that r items have failed is given by.

$$f(x_1, x_2, \ldots, x_r) = \lambda^r e^{-\lambda\sum_{i=1}^{r} x_i} \quad (r \text{ failed items}) \qquad (3.20)$$

and the probability distribution that $(n-r)$ items will survive is

$$P(X_{r+1} > t_1, X_{r+2} > t_2, \ldots, X_n > t_{n-r}) = e^{-\lambda \sum_{j=1}^{n-r} t_j}$$

Thus, the joint density function is

$$L(X, \lambda) = f(x_1, x_2, \ldots, x_r) \, P(X_{r+1} > t_1, \ldots, X_n > t_{n-r})$$

$$= \frac{n!}{(n-r)!} \lambda^r e^{-\lambda (\sum_{i=1}^{r} x_i + \sum_{j=1}^{n-r} t_j)}$$

Let

$$T = \sum_{i=1}^{r} x_i + \sum_{j=1}^{n-r} t_j \tag{3.21}$$

then

$$\ln L = \ln \left(\frac{n!}{(n-r)!} \right) + r \ln \lambda - \lambda T$$

and

$$\frac{\partial \ln L}{\partial \lambda} = \frac{r}{\lambda} - T = 0.$$

Hence,

$$\hat{\lambda} = \frac{r}{T}. \tag{3.22}$$

Note that with the exponential, regardless of the censoring type or lack of censoring, the MLE of λ is the number of failures divided by the total operating time.

Example 3.10

Let $X_1, X_2, \ldots, X_n$ represent a random sample from the distribution with pdf.

$$f(x; \theta) = e^{-(x-\theta)} \quad \text{for } \theta \leq x \leq \infty \text{ and } -\infty < \theta < \infty. \tag{3.23}$$

The likelihood function is given by

$$L(\theta; X) = \prod_{i=1}^{n} f(x_i; \theta) \text{ for } \theta \leq x_i \leq \infty \text{ all } i$$

$$= \prod_{i=1}^{n} e^{-(x_i-\theta)} = e^{-\sum_{i=1}^{n} x_i + n\theta}.$$

For fixed values of $x_1, x_2, \ldots, x_n$, we wish to find that value of θ which maximizes $L(\theta; X)$. Here we cannot use the techniques of calculus to maximize $L(\theta; X)$. Note that $L(\theta; X)$ is largest when θ is as large as possible. However, the largest value of θ is equal to the smallest value of X_i in the sample. Thus, $\hat{\theta} = \min\{X_i\}\ 1 \le i \le n$.

Example 3.11

Let $X_1, X_2, \ldots, X_n$, denote a random sample from the normal distribution $N(\mu, \sigma^2)$. Then the likelihood function is given by.

$$L(X, \mu, \sigma^2) = \left(\frac{1}{2\pi}\right)^{\frac{n}{2}} \frac{1}{\sigma^n} e^{-\frac{1}{2\sigma^2} \sum_{i=1}^{n} (x_i - \mu)^2}$$

and

$$\ln L = -\frac{n}{2} \log(2\pi) - \frac{n}{2} \log \sigma^2 - \frac{1}{2\sigma^2} \sum_{i=1}^{n} (x_i - \mu)^2.$$

Thus we have

$$\frac{\partial \ln L}{\partial \mu} = \frac{1}{\sigma^2} \sum_{i=1}^{n} (x_i - \mu) = 0$$

$$\frac{\partial \ln L}{\partial \sigma^2} = -\frac{n}{2\sigma^2} - \frac{1}{2\sigma^4} \sum_{i=1}^{n} (x_i - \mu)^2 = 0.$$

Solving the two equations simultaneously, we obtain

$$\hat{\mu} = \frac{\sum_{i=1}^{n} x_i}{n} \quad \text{and} \quad \hat{\sigma}^2 = \frac{1}{n} \sum_{i=1}^{n} (x_i - \bar{x})^2. \tag{3.24}$$

Note that the MLEs, if they exist, are both sufficient and efficient estimates. They also have an additional property called invariance, i.e., for an MLE of θ, then $\mu\ (\theta)$ is the MLE of $\mu\ (\theta)$. However, they are not necessarily unbiased, i.e., $E(\hat{\theta}) = \theta$. The point in fact is σ^2:

$$E(\hat{\sigma}^2) = \left(\frac{n-1}{n}\right)\sigma^2 \ne \sigma^2$$

Therefore, for small n, σ^2 is usually adjusted for its bias and the best estimate of σ^2 is

$$\hat{\sigma}^2 = \left(\frac{1}{n-1}\right) \sum_{i=1}^{n} (x_i - \bar{x})^2$$

Sometimes it is difficult, if not impossible, to obtain maximum likelihood estimators in a closed form, and therefore numerical methods must be used to maximize the likelihood function. For illustration see the following example.

Example 3.12
Suppose that $X_1, X_2, \ldots, X_n$ is a random sample from the Weibull distribution with pdf.

$$f(x, \alpha, \lambda) = \alpha \lambda x^{\alpha-1} e^{-\lambda x^{\alpha}} \tag{3.25}$$

The likelihood function is

$$L(X, \alpha, \lambda) = \alpha^n \lambda^n \left(\prod_{i=1}^{n} x_i^{\alpha-1}\right) e^{-\lambda \sum_{i=1}^{n} x_i^{\alpha}}$$

Then

$$\ln L = n \log \alpha + n \log \lambda + (\alpha - 1) \sum_{i=1}^{n} \log x_i - \lambda \sum_{i=1}^{n} x_i^{\alpha}$$

$$\frac{\partial \ln L}{\partial \alpha} = \frac{n}{\alpha} + \sum_{i=1}^{n} \log x_i - \lambda \sum_{i=1}^{n} x_i^{\alpha} \log x_i = 0$$

$$\frac{\partial \ln L}{\partial \lambda} = \frac{n}{\lambda} - \sum_{i=1}^{n} x_i^{\alpha} = 0$$

As noted, solutions of the above two equations for α and λ are extremely difficult and require either graphical or numerical methods.

Example 3.13
Let $X_1, X_2, \ldots, X_n$ be a random sample from the gamma distribution with pdf.

$$f(x, \lambda, \alpha) = \frac{\lambda^{(\alpha+1)} x^{\alpha} e^{-\lambda x}}{(\alpha!)^n} \tag{3.26}$$

then the likelihood function and log of the likelihood function, respectively, are

$$L(X, \lambda, \alpha) = \frac{\lambda^{n(\alpha+1)} \prod_{i=1}^{n} x_i^{\alpha} e^{-\lambda \sum_{i=1}^{n} x_i}}{(\alpha!)^n}$$

$$\ln L = n(\alpha + 1) \log \lambda + \alpha \sum_{i=1}^{n} \log x_i - \lambda \sum_{i=1}^{n} x_i - n \log(\alpha!).$$

Taking the partial derivatives, we obtain

$$\frac{\partial \ln L}{\partial \alpha} = n \log \lambda + \sum_{i=1}^{n} \log x_i - n \frac{\partial}{\partial \alpha}[\log \alpha!] = 0$$

$$\frac{\partial \ln L}{\partial \lambda} = \frac{n(\alpha + 1)}{\lambda} - \sum_{i=1}^{n} x_i = 0 \tag{3.27}$$

The solutions of the two equations at Eq. (3.27) for α and λ are extremely difficult and require either graphical or numerical methods.

Example 3.14

Let $t_1, t_2, \ldots, t_n$ be failure times of a random variable having the loglog distribution, also known as Pham distribution (Pham 2002), with two parameters a and α as follows (see also Eq. (2.82), Chap. 2):

$$f(t) = \alpha \ln(a) \, t^{\alpha-1} \, a^{t^{\alpha}} \, e^{1-a^{t^{\alpha}}} \quad \text{for } t > 0, \alpha > 0, a > 1 \tag{3.28}$$

From Chap. 2, Eq. (2.83), the Pham cdf is given by:

$$F(t) = 1 - e^{1-a^{t^{\alpha}}}$$

We now estimate the values of a and α using the MLE method. From Eq. (3.28), the likelihood function is

$$L(a, \alpha) = \prod_{i=1}^{n} \alpha \ln a \cdot t_i^{\alpha-1} e^{1-a^{t_i^{\alpha}}} a^{t_i^{\alpha}}$$

$$= \alpha^n (\ln a)^n \left(\prod_{i=1}^{n} t_i \right)^{\alpha-1} a^{\sum_{i=1}^{n} t_i^{\alpha}} e^{n - \sum_{i=1}^{n} a^{t_i^{\alpha}}}$$

The log likelihood function is

$$\log L(a, \alpha) = n \log \alpha + n \ln(\ln a) + (\alpha - 1) \left(\sum_{i=1}^{n} \ln t_i \right)$$

$$+ \ln a \cdot \sum_{i=1}^{n} t_i^{\alpha} + n - \sum_{i=1}^{n} a^{t_i^{\alpha}} \tag{3.29}$$

The first derivatives of the log likelihood function with respect to a and α are, respectively,

$$\frac{\partial}{\partial a} \log L(a, \alpha) = \frac{n}{a \ln a} + \frac{1}{a} \cdot \sum_{i=1}^{n} t_i^{\alpha} - \sum_{i=1}^{n} t_i^{\alpha} a^{t_i^{\alpha}-1} \tag{3.30}$$

and

$$\frac{\partial}{\partial \alpha} \log L(a, \alpha) = \frac{n}{\alpha} + \sum_{i=1}^{n} \ln t_i + \ln a \sum_{i=1}^{n} \ln t_i \, t_i^{\alpha}$$

$$- \sum_{i=1}^{n} t_i^{\alpha} a^{t_i^{\alpha}} \ln a \, \ln t_i \tag{3.31}$$

Setting eqs. (3.30) and (3.31) equal to zero, we can obtain the MLE of a and α by solving the following simultaneous equations:

$$\frac{n}{\ln a} + \sum_{i=1}^{n} t_i^{\alpha} - \sum_{i=1}^{n} t_i^{\alpha} a^{t_i^{\alpha}} = 0$$

$$\frac{n}{\alpha} + \sum_{i=1}^{n} \ln t_i + \ln a \cdot \sum_{i=1}^{n} \ln t_i \cdot t_i^{\alpha} \left(1 - a^{t_i^{\alpha}}\right) = 0$$

After rearrangements, we obtain

$$\ln a \sum_{i=1}^{n} t_i^{\alpha} \left(a^{t_i^{\alpha}} - 1\right) = n$$

$$\ln a \cdot \sum_{i=1}^{n} \ln t_i \cdot t_i^{\alpha} \cdot \left(a^{t_i^{\alpha}} - 1\right) - \frac{n}{\alpha} = \sum_{i=1}^{n} \ln t_i$$

Example 3.15

Let $t_1, t_2, \ldots, t_n$ be failure times of a random variable having the vtub-shaped failure rate function. The probability density function $f(t)$ of its vtub-shaped failure rate is given by (see Eq. (1) in Chap. 2):

$$f(t) = \left(at \ln(bt) + \frac{a}{b}\right) e^{-\{at[\frac{t}{2} \ln(bt) - \frac{t}{4} + \frac{1}{b}]\}} \quad \text{for } t > 0, a > 0, b > 0$$

We now estimate the two unknown parameters a and b using the MLE method. From the pdf above, the likelihood function is given by

$$L(a, b) = \prod_{i=1}^{n} \left(at_i \ln(bt_i) + \frac{a}{b}\right) e^{-a \sum_{i=1}^{n} t_i \left[\frac{t_i}{2} \ln(bt_i) - \frac{t_i}{4} + \frac{1}{b}\right]}.$$

The log likelihood function is

$$\ln L(a, b) = \sum_{i=1}^{n} \ln\left(at_i \ln(bt_i) + \frac{a}{b}\right)$$

$$- a \sum_{i=1}^{n} t_i \left[\frac{t_i}{2} \ln(bt_i) - \frac{t_i}{4} + \frac{1}{b}\right].$$

By taking the first derivatives of the log likelihood function above with respect to a and b and equate them to zero, and after some algebra, the MLE of a and b can be easily obtained by solving the following equations:

$$\begin{cases} a = \frac{2b \sum_{i=1}^{n} \frac{bx_i - 1}{(bx_i \ln(bx_i) + 1)}}{\sum_{i=1}^{n} x_i(bx_i - 2)} \\ \sum_{i=1}^{n} \frac{bx_i \ln(bx_i) + 1}{abx_i \ln(bx_i) + a} = \sum_{i=1}^{n} x_i\left[\frac{x_i}{2} \ln(bx_i) - \frac{x_i}{4} + \frac{1}{b}\right]. \end{cases}$$

Example 3.16

Suppose that $X_1, X_2, \ldots, X_n$ are independent random variable, each with the uniform distribution on $[a - d, a + d]$ where a is known and d is positive and unknown. Find the maximum likelihood estimator of d.

Solution: For $i = 1,2,\ldots, n$, the pdf of X_i is given by

$$f(d, x_i) = \begin{cases} \frac{1}{2d} & \text{if } a - d \leq x_i \leq a + d \\ 0 & \text{otherwise.} \end{cases} \qquad (3.32)$$

The likelihood function is

$$L(d) = \begin{cases} \left(\frac{1}{2d}\right)^n & \text{if } a - d \leq x_1, x_2, \ldots, x_n \leq a + d \\ 0 & \text{otherwise.} \end{cases}$$

To maximize L(c,d) is the same as to minimize $(2d)^n$. In other words, L(d) will be maximized by the smallest possible d with L(d) > 0. This implies for all $i = 1,2,\ldots,n$ $d \geq -x_i + a$ and $d \geq x_i - a$. Thus, $\hat{d} = \max_{1 \leq i \leq n} \{|x_i - a|\}$.

3.4 Invariance and Lower Bound on the Variance

In this section we discuss some properties of MLEs and how to establish a lower bound on the variance using an inequality known as the Cramér-Rao inequality.

Theorem 3.1 (Invariance Principle)
   If $\hat{\theta}$ is the MLE of parameter θ then $g(\hat{\theta})$ is the MLE of parameter $g(\theta)$.

In other words, if the likelihood $L(X, \theta)$ *is maximized at the point.*
$\hat{\theta} = h(x_1, x_2, \ldots, x_n)$ *then the function* $L(X, g(\theta))$ *is maximized at the point*

$$g\left(\hat{\theta}\right) = g(h(x_1, x_2, \ldots, x_n)).$$

Example 3.17

If $\hat{\theta}$ is the MLE of the variance σ^2 then $\sqrt{\hat{\theta}}$ is the MLE of the standard deviation σ.

Theorem 3.2 (Likelihood Principle) *Consider two sets of data,* x *and* y, *obtained from the same population, although possibly according to different sampling plans. If the ratio of their likelihoods,* $\frac{L_1(x,\theta)}{L_2(y,\theta)}$, *does not depend on* θ, *then both data sets provide the same information about the parameter* θ *and consequently should lead to the same conclusion about* θ.

Example 3.18

Assume that we want to estimate θ, the probability of success of Bernoulli trials. One experimental design consists of fixing n, the number of observations, and recording the number of successes. If we observe x successes, the likelihood is.

$$L_1(x, \theta) = \binom{n}{x} \theta^x (1 - \theta)^{n-x}.$$

Suppose that one decide to fix x and take observations until x successes are recorded. The probability that the observations will end on the nth trial is given by negative binomial distribution, and the likelihood is

$$L_2(n, \theta) = \binom{n-1}{x-1} \theta^x (1 - \theta)^{n-x}.$$

Note that in the first case x was random and n fixed; in the second it is the other way around. Thus, the ratio of these two likelihoods is given by

$$\frac{L_1(x, \theta)}{L_2(n, \theta)} = \frac{\binom{n}{x}}{\binom{n-1}{x-1}},$$

which does not depend on the parameter θ. The MLE of θ is $\hat{\theta} = \frac{x}{n}$ in either case. Hence the additional information that in the second case the last experiment led to success does not affect our estimate of the parameter θ.

Theorem 3.3 (Cramér-Rao inequality) *Let $X_1, X_2, ..., X_n$ denote a random sample from a distribution with pdf $f(x; \theta)$ for $\theta_1 < \theta < \theta_2$, where θ_1 and θ_2 are known. Let $Y = h(X_1, X_2, ..., X_n)$ be an unbiased estimator of θ. The lower bound inequality on the variance of Y, Var(Y), is given by.*

$$Var(Y) \geq \frac{1}{nE\left\{\left[\frac{\partial \ln f(x;\theta)}{\partial\theta}\right]^2\right\}} = -\frac{1}{nE\left(\frac{\partial^2 \ln f(x;\theta)}{\partial\theta^2}\right)} \tag{3.33}$$

or

$$Var(Y) \geq \frac{1}{n\,I(\theta)} \tag{3.34}$$

where

$$I(\theta) = E\left\{\left[\frac{\partial \ln f(x; \theta)}{\partial\theta}\right]^2\right\}$$

also known as Fisher's Information (see Problem 15).

Theorem 3.4 *An estimator $\hat{\theta}$ is said to be asymptotically efficient if $\sqrt{n}\hat{\theta}$ has a variance that approaches the Cramér-Rao lower bound for large n, that is,*

$$\lim_{n\to\infty} Var(\sqrt{n}\hat{\theta}) = -\frac{1}{nE\left(\frac{\partial^2 \ln f(x;\theta)}{\partial\theta^2}\right)}. \tag{3.35}$$

Example 3.19

Let $X_1, X_2, ..., X_n$ denote a random sample from the Bernoulli distribution with a pdf.

$$f(x) = p^x(1 - p)^{1-x} \quad x = 0, 1 \quad 0 \leq p \leq 1$$

where p is the probability of success in each trial.

Let Y represents the number of successes in n independent trials, that is, $Y = \sum_{i=1}^{n} X_i$ then we can easily show that $W = \frac{Y}{n}$ is an unbiased estimator of p. We now determine if W is the minimum variance unbiased estimator of p. The variance of W is given by

$$V(W) = \frac{1}{n^2}V(Y) = \frac{1}{n^2}np(1 - p) = \frac{p(1 - p)}{n}.$$

To obtain the lower bound of the Cramer-Rao inequality, we need to compute the following

$$\ln f = x \ln p + (1 - x)\ln(1 - p)$$

$$\frac{\partial}{\partial p} \ln f = \frac{x}{p} - \frac{(1-x)}{(1-p)}$$

$$\frac{\partial^2}{\partial p^2} \ln f = -\frac{x}{p^2} - \frac{(1-x)}{(1-p)^2}.$$

Then

$$E\left\{\frac{\partial^2}{\partial p^2} \ln f\right\} = -\frac{p}{p^2} - \frac{(1-p)}{(1-p)^2} = -\frac{1}{p(1-p)}.$$

Thus,

$$Var(W) \geq -\frac{1}{nE\left(\frac{\partial^2 \ln f(x;\theta)}{\partial \theta^2}\right)}$$

This means that W is the

$$= -\frac{1}{n\left(-\frac{1}{p(1-p)}\right)} = \frac{p(1-p)}{n} = Var(W).$$

minimum variance unbiased estimator of p.

Example 3.20

Let $T_1, T_2, \ldots, T_n$ denote a random sample from an exponential distribution with the mean μ and its pdf is given by.

$$f(t) = \frac{1}{\mu} e^{-\frac{t}{\mu}} \quad t \geq 0,\ \mu > 0.$$

Note that, from example 3.5, the MLE of μ is $\hat{\mu} = \frac{\sum_{i=1}^{n} T_i}{n}$. Define $Y = \frac{\sum_{i=1}^{n} T_i}{n}$. Is Y the minimum variance unbiased estimator of μ?

The variance of Y is given by

$$Var(Y) = Var\left(\frac{\sum_{i=1}^{n} T_i}{n}\right) = \frac{1}{n^2} \sum_{i=1}^{n} Var(T_i) = \frac{1}{n^2}(n\mu^2) = \frac{\mu^2}{n}.$$

We now compute the lower bound using the Cramer-Rao inequality.

$$\ln f = -\ln \mu - \frac{t}{\mu}$$

$$\frac{\partial \ln f}{\partial \mu} = -\frac{1}{\mu} + \frac{t}{\mu^2}$$

$$\frac{\partial^2 \ln f}{\partial \mu^2} = \frac{1}{\mu^2} - \frac{2t}{\mu^3}.$$

Then

$$E\left\{\frac{\partial^2 \ln f}{\partial \mu^2}\right\} = \frac{1}{\mu^2} - \frac{2E(T)}{\mu^3} = \frac{1}{\mu^2} - \frac{2\mu}{\mu^3} = -\frac{1}{\mu^2}.$$

Thus,

$$Var(Y) \geq -\frac{1}{nE\left(\frac{\partial^2 \ln f(x;\theta)}{\partial \theta^2}\right)}$$

$$= -\frac{1}{n\left(-\frac{1}{\mu^2}\right)} = \frac{\mu^2}{n} = Var(Y).$$

This shows that Y is the minimum variance unbiased estimator of μ.

3.5 Maximum Likelihood Estimation with Censored Data

Censored data arises when an individual's life length is known to occur only in a certain period of time. In other words, a censored observation contains only partial information about the random variable of interest. In this section, we consider two types of censoring. The first type is called Type-I censoring where the event is observed only if it occurs prior to some pre-specified time. The second type is Type-II censoring in which the study continues until the failure of the first r units (or components), where r is some predetermined integer ($r < n$).

Examples of Type-II censoring are often used in testing of equipment life. Here items are put on test at the same time, and the test is terminated when r of the n items have failed and without replacement. Such an experiment may, however, save time and resources because it could take a very long time for all items to fail. Both Type-I and Type-II censoring arise in many reliability applications.

For example, there is a batch of transistors or tubes; we put them all on test at $t = 0$, and record their times to failure. Some transistors may take a long time to burn out, and we will not want to wait that long to end the experiment. Therefore, we might stop the experiment at a pre-specified time t_c, in which case we have Type-I censoring, or we might not know beforehand what value of the fixed censoring time is good so we decide to wait until a pre-specified number of units have failed, r, of all the transistors has burned out, in which case we have Type-II censoring.

Censoring times may vary from individual to individual or from application to application. We now discuss a generalized censoring times case, call a multiple-censored data.

3.5.1 Parameter Estimate with Multiple-Censored Data

A sample of n units are drawn at random and put on test with a pdf f and cdf F. The likelihood function for the multiple-censored data is given by

$$L = f(t_{1,f}, \ldots, t_{r,f}, t_{1,s}, \ldots, t_{m,s}) = \frac{n!}{(n-r)!} \prod_{i=1}^{r} f(t_{i,f}) \prod_{j=1}^{m} [1 - F(t_{j,s})] \quad (3.36)$$

where $f(.)$ is the density function and $F(.)$ is the distribution function. There are r failures at times $t_{1,f}, \ldots, t_{r,f}$ and m units (here, $m = n\,s- r$) with censoring times $t_{1,s}, \ldots, t_{m,s}$. Note that this includes Type-I censoring by simply setting $t_{i,f} = t_{i,n}$ and $t_{j,s} = t_0$ in the likelihood function in Eq. (3.36). Also, the likelihood function for Type-II censoring is similar to Type-I censoring except $t_{j,s} = t_r$ in Eq. (3.36). In other words, the likelihood function for the first r observations from a sample size n drawn from the model in both Type-I and Type-II censoring is given by

$$L = f(t_{1,n}, \ldots, t_{r,n}) = \frac{n!}{(n-r)!} \prod_{i=1}^{r} f(t_{i,n}) \, [1 - F(t_*)]^{n-r} \quad (3.37)$$

where $t_* = t_0$, the time of cessation of the test for Type-I censoring and $t_* = t_r$, the time of the rth failure for Type-II censoring.

Example 3.21
Consider a two-parameter probability density distribution with multiple-censored data and distribution function with failure rate bathtub shape, as given by (Chen 2000):

$$f(t) = \lambda \beta t^{\beta-1} \exp\left[t^{\beta} + \lambda(1 - e^{t^{\beta}})\right], \qquad t, \lambda, \beta > 0 \quad (3.38)$$

and

$$F(t) = 1 - \exp\left[\lambda(1 - e^{t^{\beta}})\right], \qquad t, \lambda, \beta > 0 \quad (3.39)$$

respectively. Substituting the functions $f(t)$ and $F(t)$ in Eqs. (3.38) and (3.39) into Eq. (3.37), we obtain the logarithm of the likelihood function:

$$\ln L = \ln \frac{n!}{(n-r)!} + r \ln \lambda + r \ln \beta + \sum_{i=1}^{r} (\beta - 1) \ln t_i$$

$$+ (m+r)\lambda + \sum_{i=1}^{r} t_i^{\beta} - \left(\sum_{i=1}^{r} \lambda e^{t_i^{\beta}} + \sum_{j=1}^{m} \lambda e^{t_j^{\beta}}\right). \quad (3.40)$$

The function $\ln L$ can be maximized by setting the partial derivative of $\ln L$ with respect to λ and β, equal to zero and solving the resulting equations simul-taneously for λ and β. Therefore, we obtain

$$\frac{\partial \ln L}{\partial \lambda} = \frac{r}{\lambda} + (m+r) - \sum_{i=1}^{r} e^{t_i^{\beta}} - \sum_{j=1}^{m} e^{t_j^{\beta}} \equiv 0$$

$$\frac{\partial \ln L}{\partial \beta} = \frac{r}{\beta} + \sum_{i=1}^{r} \ln t_i + \sum_{i=1}^{r} t_i^{\beta} \ln t_i$$

$$- \lambda \left[\sum_{i=1}^{r} e^{t_i^{\beta}} t_i^{\beta} \ln t_i + \sum_{j=1}^{m} e^{t_j^{\beta}} t_j^{\beta} \ln t_j \right] \equiv 0.$$

This implies that

$$\hat{\lambda} = \frac{r}{\left(\sum_{i=1}^{r} e^{t_i^{\hat{\beta}}} + \sum_{j=1}^{m} e^{t_j^{\hat{\beta}}} \right) - m - r} \tag{3.41}$$

and $\hat{\beta}$ is the solution of

$$\frac{r}{\hat{\beta}} + \sum_{i=1}^{r} \ln t_i + \sum_{i=1}^{r} t_i^{\hat{\beta}} \ln t_i$$

$$= \frac{r}{\left(\sum_{i=1}^{r} e^{t_i^{\hat{\beta}}} + \sum_{j=1}^{m} e^{t_j^{\hat{\beta}}} \right) - m - r} \left[\sum_{i=1}^{r} e^{t_i^{\hat{\beta}}} t_i^{\hat{\beta}} \ln t_i + \sum_{j=1}^{m} e^{t_j^{\hat{\beta}}} t_j^{\hat{\beta}} \ln t_j \right] \tag{3.42}$$

We now discuss two special cases as follows.

Case I: Type-I or Type-II Censoring Data.

From Eq. (3.37), the likelihood function for the first r observations from a sample size n drawn from the model in both Type-I and Type-II censoring is

$$L = f(t_{1,n}, \ldots, t_{r,n}) = \frac{n!}{(n-r)!} \prod_{i=1}^{r} f(t_{i,n}) \left[1 - F(t_*) \right]^{n-r}$$

where $t_* = t_0$, the time of cessation of the test for Type-I censoring and $t_* = t_r$, the time of the rth failure for Type-II censoring Eqs. (3.41) and (3.42) become

$$\hat{\lambda} = \frac{r}{\sum_{i=1}^{r} e^{t_i^{\hat{\beta}}} + (n-r)e^{t_*^{\hat{\beta}}} - n}. \tag{3.43}$$

$$\frac{r}{\hat{\beta}} + \sum_{i=1}^{r} \ln t_i + \sum_{i=1}^{r} t_i^{\hat{\beta}} \ln t_i$$

$$= \frac{r}{\sum_{i=1}^{r} e^{t_i^{\hat{\beta}}} + (n-r)e^{t_*^{\hat{\beta}}} - n} \left[\sum_{i=1}^{r} e^{t_i^{\hat{\beta}}} t_i^{\hat{\beta}} \ln t_i + \sum_{j=1}^{m} e^{t_j^{\hat{\beta}}} t_j^{\hat{\beta}} \ln t_j \right]. \tag{3.44}$$

Case II: Complete Censored Data.

Simply replace r with n in Eqs. (3.41) and (3.42) and ignore the t_j portions. The maximum likelihood equations for the λ and β are given by

$$\hat{\lambda} = \frac{n}{\sum_{i=1}^{n} e^{t_i^{\hat{\beta}}} - n} \tag{3.45}$$

$$\frac{n}{\hat{\beta}} + \sum_{i=1}^{n} \ln t_i + \sum_{i=1}^{n} t_i^{\hat{\beta}} \ln t_i = \frac{n}{\sum_{i=1}^{n} e^{t_i^{\hat{\beta}}} - n} \times \sum_{i=1}^{n} e^{t_i^{\hat{\beta}}} t_i^{\hat{\beta}} \ln t_i. \tag{3.46}$$

3.5.2 Confidence Intervals of Estimates

The asymptotic variance-covariance matrix of the parameters (λ and β) is obtain- ed by inverting the Fisher information matrix

$$I_{ij} = E\left[-\frac{\partial^2 L}{\partial \theta_i \partial \theta_j}\right], \quad i, j = 1, 2 \tag{3.47}$$

where $\theta_1, \theta_2 = \lambda$ or β (Nelson et al.1992). This leads to

$$\begin{bmatrix} Var(\hat{\lambda}) & Cov(\hat{\lambda}, \hat{\beta}) \\ Cov(\hat{\lambda}, \hat{\beta}) & Var(\hat{\beta}) \end{bmatrix}$$

$$= \begin{bmatrix} E\left(-\frac{\partial^2 \ln L}{\partial^2 \lambda}|_{\hat{\lambda},\hat{\beta}}\right) & E\left(-\frac{\partial^2 \ln L}{\partial\lambda\partial\beta}|_{\hat{\lambda},\hat{\beta}}\right) \\ E\left(-\frac{\partial^2 \ln L}{\partial\beta\partial\lambda}|_{\hat{\lambda},\hat{\beta}}\right) & E\left(-\frac{\partial^2 \ln L}{\partial^2 \beta}|_{\hat{\lambda},\hat{\beta}}\right) \end{bmatrix}^{-1} \tag{3.48}$$

We can obtain an approximate $(1-\alpha)100\%$ confidence intervals on parameter λ and β based on the asymptotic normality of the MLEs (Nelson et al. 1992) as follows:

$$\hat{\lambda} \pm Z_{\alpha/2}\sqrt{Var(\hat{\lambda})} \quad \text{and} \quad \hat{\beta} \pm Z_{\alpha/2}\sqrt{Var(\hat{\beta})} \tag{3.49}$$

where $Z_{\alpha/2}$ is upper percentile of standard normal distribution.

Equation Sect. 1.

3.5.3 Applications

Consider a helicopter main rotor blade part code xxx-015-001-107 based on the system database collected from October 1995 to September 1999 (Pham 2002). The data set is shown in Table 3.1. In this application, we consider several distribution functions including Weibull, lognormal, normal, and loglog distribution functions. From Example 3.14, the Pham pdf (see Eq. 3.28) with parameters a and α is.

$$f(t) = \alpha \cdot \ln a \cdot t^{\alpha-1} \cdot a^{t^{\alpha}} \cdot e^{1-a^{t^{\alpha}}} \text{ for } t > 0, \alpha > 0, a > 0$$

and its corresponding log likelihood function (see Eq. 3.30) is

$$\log L(a, \alpha) = n \log \alpha + n \ln(\ln a) + (\alpha - 1) \left(\sum_{i=1}^{n} \ln t_i \right)$$

$$+ \ln a \cdot \sum_{i=1}^{n} t_i^{\alpha} + n - \sum_{i=1}^{n} a^{t_i^{\alpha}}$$

We next determine the confidence intervals for parameter estimates a and α. For the log-likelihood function given in Eq. (3.30), we can obtain the Fisher information matrix H as $H = \begin{bmatrix} h_{11} & h_{12} \\ h_{21} & h_{22} \end{bmatrix}$ where $h_{11} = E\left[-\frac{\partial^2 \log L}{\partial a^2} \right]$

Table 3.1 Main rotor blade data (hour)

1634.3	2094.3	3318.2
1100.5	2166.2	2317.3
1100.5	2956.2	1081.3
819.9	795.5	1953.5
1398.3	795.5	2418.5
1181	204.5	1485.1
128.7	204.5	2663.7
1193.6	1723.2	1778.3
254.1	403.2	1778.3
3078.5	2898.5	2943.6
3078.5	2869.1	2260
3078.5	26.5	2299.2
26.5	26.5	1655
26.5	3180.6	1683.1
3265.9	644.1	1683.1
254.1	1898.5	2751.4
2888.3	3318.2	
2080.2	1940.1	

$$h_{12} = h_{21} = E\left[-\frac{\partial^2 \log L}{\partial a \, \partial \alpha}\right] \quad \text{and} \quad h_{22} = E\left[-\frac{\partial^2 \log L}{\partial \alpha^2}\right]$$

The variance matrix, V, can be obtained as follows:

$$V = [H]^{-1} = \begin{bmatrix} v_{11} & v_{12} \\ v_{21} & v_{22} \end{bmatrix} \qquad (3.50)$$

The variances of a and α are

$$Var(a) = v_{11} \quad Var(\alpha) = v_{22}$$

One can approximately obtain the $100(1-\beta)\%$ confidence intervals for a and α based on the normal distribution as $[\hat{a} - z_{\frac{\beta}{2}}\sqrt{v_{11}}, \hat{a} + z_{\frac{\beta}{2}}\sqrt{v_{11}}]$ and $[\hat{\alpha} - z_{\frac{\beta}{2}}\sqrt{v_{22}}, \hat{\alpha} + z_{\frac{\beta}{2}}\sqrt{v_{22}}]$, respectively, where v_{ij} is given in Eq. (3.50) and z_β is $(1-\beta/2)100\%$ of the standard normal distribution. After we obtain $\hat{a}$ and $\hat{\alpha}$, the MLE of reliability function can be computed as

$$\hat{R}(t) = e^{1-\hat{a}t^{\hat{\alpha}}}$$

Let us define a partial derivative vector for reliability $R(t)$ as

$$v[R(t)] = \left[\frac{\partial R(t)}{\partial a} \quad \frac{\partial R(t)}{\partial \alpha}\right]$$

then the variance of $R(t)$ can be obtained as follows:

$$Var[R(t)] = v[R(t)] \cdot V \cdot (v[R(t)])^T \qquad (3.51)$$

where V is given in Eq. (3.50).

One can approximately obtain the $(1-\beta)100\%$ confidence interval for $R(t)$ as

$$\left[\hat{R}(t) - z\beta\sqrt{Var[R(t)]}, \ \hat{R}(t) + z\beta\sqrt{Var[R(t)]}, \right].$$

The MLE parameter estimations of Pham distribution using the data set in Table 3.1 are given as follows:

$$\hat{\alpha} = 1.1075 \quad Var[\hat{\alpha}] = 0.0162$$
$$95\% \text{ CI for } \hat{\alpha} : [0.8577, 1.3573]$$
$$\hat{a} = 1.0002 \quad Var[\hat{a}] = 2.782e^{-8}$$
$$95\% \text{CI for a: } [0.9998, 1.0005]$$
$$\text{MTTF} = 1608.324 \quad \text{MRL}(t = \text{MTTF}) = 950.475$$

Figure 3.1 shows the loglog reliability and its 95% confidence interval of main rotor blade, respectively. Figure 3.2 shows the reliability comparisons between the normal, the lognormal, Weibull, and the loglog distributions for the main rotor blade data set.

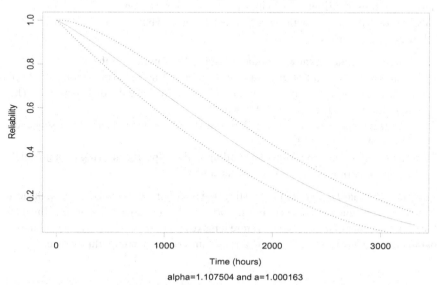

Fig. 3.1 Estimated reliability and its confidence interval for a main rotor blade data set

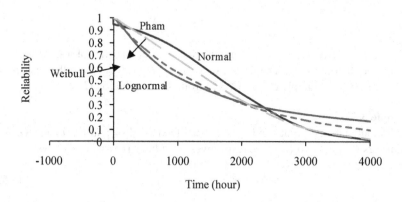

Fig. 3.2 Reliability comparisons for a main rotor blade data set

3.6　Statistical Change-Point Estimation Methods

The change-point problem has been widely studied in reliability applications such as biological sciences, survival analysis, and environmental statistics.

Assume there is a sequence of random variables $X_1, X_2, \ldots, X_n$, that represent the inter-failure times and exists an index change-point τ, such that $X_1, X_2, \ldots, X_\tau$ have a common distribution F with density function $f(t)$ and $X_{\tau+1}, X_{\tau+2}, \ldots, X_n$ have the distribution G with density function $g(t)$, where $F \neq G$. Consider the following assumptions:

1.　There is a finite unknown number of units, N, to put under the test.
2.　At the beginning, all of the units have the same lifetime distribution F. After τ failures are observed, the remaining $(N-\tau)$ items have the distribution G. The change-point τ is assumed unknown.
3.　The sequence $\{X_1, X_2, \ldots, X_\tau\}$ is statistically independent of the sequence $\{X_{\tau+1}, X_{\tau+2}, \ldots, X_n\}$.
4.　The lifetime test is performed according to the Type-II censoring plan in which the number of failures, n, is pre-determined.

Note that in hardware reliability testing, the total number of units to put on the test N can be determined in advance. But in software, the parameter N can be defined as the initial number of faults, and therefore it makes more sense for it to be an unknown parameter. Let $T_1, T_2, \ldots, T_n$ be the arrival times of sequential failures. Then

$$T_1 = X_1$$
$$T_2 = X_1 + X_2$$
$$\vdots$$
$$T_n = X_1 + X_2 + \ldots X_n \qquad (3.52)$$

The failure times $T_1, T_2, \ldots, T_\tau$ are the first τ order statistics of a sample of size N from the distribution F. The failure times $T_{\tau+1}, T_{\tau+2}, \ldots, T_n$ are the first $(n-\tau)$ order statistics of a sample of size $(N-\tau)$ from the distribution G.

Example 3.22
The Weibull change-point model of given life time distributions F and G with parameters (λ_1, β_1) and (λ_2, β_2), respectively, can be expressed as follows:

$$F(t) = 1 - \exp\left(-\lambda_1 t^{\beta_1}\right) \qquad (3.53)$$

$$G(t) = 1 - \exp\left(-\lambda_2 t^{\beta_2}\right). \qquad (3.54)$$

Assume that the distributions belong to parametric families $\{F(t\,|\,\theta_1)\,, \theta_1 \in \Theta_1\}$ and $\{G(t\,|\,\theta_2)\,, \theta_2 \in \Theta_2\}$. Assume $T_1, T_2, \ldots, T_\tau$ are the first τ order statistics of a sample with size N from the distribution $\{F(t\,|\,\theta_1)\,, \theta_1 \in \Theta_1\}$ and $T_{\tau+1}, T_{\tau+2}, \ldots, T_n$

are the first $(n-\tau)$ order statistics of a sample of size $(N-\tau)$ from the distribution $\{G(t|\theta_2), \theta_2 \in \Theta_2\}$ where N is unknown. The log likelihood function can be expressed as follows (Zhao 2003):

$$
\begin{aligned}
L(\tau, & N, \theta_1, \theta_2 | T_1, T_2, \ldots, T_n) \\
&= \sum_{i=1}^{n} (N - i + 1) + \sum_{i=1}^{\tau} f(T_i | \theta_1) \\
&\quad + \sum_{i=\tau+1}^{n} g(T_i | \theta_2) + (N - \tau) \log(1 - F(T_\tau | \theta_1)) \\
&\quad + (N - n) \log(1 - G(T_n | \theta_2)).
\end{aligned}
\tag{3.55}
$$

If the parameter N is known where hardware reliability is commonly considered, then the likelihood function is given by

$$
\begin{aligned}
L(\tau, & \theta_1, \theta_2 | T_1, T_2, \ldots, T_n) \\
&= \sum_{i=1}^{\tau} f(T_i | \theta_1) + \sum_{i=\tau+1}^{n} g(T_i | \theta_2) \\
&\quad + (N - \tau) \log(1 - F(T_\tau | \theta_1)) + (N - n) \log(1 - G(T_n | \theta_2)).
\end{aligned}
$$

The MLE of the change-point value $\hat{\tau}$ and $(\hat{N}, \hat{\theta}_1, \hat{\theta}_2)$ can be obtained by taking partial derivatives of the log likelihood function in Eq. (3.55) with respect to the unknown parameters that maximizes the function. It should be noted that there is no closed form for $\hat{\tau}$ but it can be obtained by calculating the log likelihood for each possible value of τ, $1 \leq \tau \leq (n - 1)$, and selecting as $\hat{\tau}$ the value that maximizes the log-likelihood function.

3.6.1 Application: A Software Model with a Change Point

In this application we examine the case where the sample size N is unknown. Consider a software reliability model developed by Jelinski and Moranda (1972), often called the Jelinski-Moranda model. The assumptions of the model are as follows:

1. There are N initial faults in the program.
2. A detected fault is removed instantaneously and no new fault is introduced.
3. Each failure caused by a fault occurs independently and randomly in time according to an exponential distribution.
4. The functions F and G are exponential distributions with failure rate parameters λ_1 and λ_2, respectively.

Based on the assumptions, the inter-failure times $X_1, X_2, \ldots, X_n$ are independently exponentially distributed. Specifically,$X_i = T_i - T_{i-1}, i = 1, 2, \ldots \tau$, are exponen-tially distributed with parameter $\lambda_1(N - i + 1)$ where λ_1 is the initial fault detection rate of the first τ failures and $X_j = T_j - T_{j-1}$, $j = \tau + 1, \tau + 2, \ldots n$, are exponentially distributed with parameter $\lambda_2(N - \tau - j + 1)$ where λ_2 is the fault detection rate of the first $(n - \tau)$ failures. If $\lambda_1 = \lambda_2$ it means that each fault removal is the same and the change-point model becomes the Jelinski-Moranda software reliability model (Jelinski and Moranda 1972).

The MLEs of the parameters $(\tau, N, \lambda_1, \lambda_2)$ can be obtained by solving the following equations simultaneously:

$$\hat{\lambda}_1 = \frac{\tau}{\sum_{i=1}^{\tau} \left(\hat{N} - i + 1\right)x_i} \tag{3.56}$$

$$\hat{\lambda}_2 = \frac{(n - \tau)}{\sum_{i=\tau+1}^{n} \left(\hat{N} - i + 1\right)x_i} \tag{3.57}$$

$$\sum_{i=1}^{n} \frac{1}{(\hat{N} - i + 1)} = \hat{\lambda}_1 \sum_{i=1}^{\tau} x_i + \hat{\lambda}_2 \sum_{i=\tau+1}^{n} x_i. \tag{3.58}$$

To illustrate the model, we use the data set as in Table 3.2 to obtain the unknown parameters $(\tau, N, \lambda_1, \lambda_2)$ using Eqs. (3.56)–(3.58). The data in Table 3.2 (Musa et al. 1987) shows the successive inter-failure times for a real-time command and control system. The table reads from left to right in rows, and the recorded times are execution

Table 3.2 Successive inter-failure times (in seconds) for a real-time command system

3	30	113	81	115	9	2	91	112	15
138	50	77	24	108	88	670	120	26	114
325	55	242	68	422	180	10	1146	600	15
36	4	0	8	227	65	176	58	457	300
97	263	452	255	197	193	6	79	816	1351
148	21	233	134	357	193	236	31	369	748
0	232	330	365	1222	543	10	16	529	379
44	129	810	290	300	529	281	160	828	1011
445	296	1755	1064	1783	860	983	707	33	868
724	2323	2930	1461	843	12	261	1800	865	1435
30	143	108	0	3110	1247	943	700	875	245
729	1897	447	386	446	122	990	948	1082	22
75	482	5509	100	10	1071	371	790	6150	3321
1045	648	5485	1160	1864	4116				

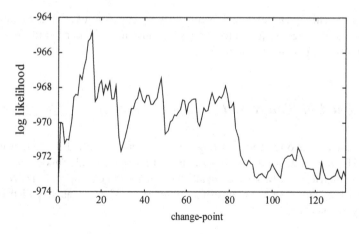

Fig. 3.3 The log-likelihood function versus the number of failures

times, in seconds. There are 136 failures in total. Figure 3.3 plots the log-likelihood function *vs* number of failures. The MLEs of the parameters $(\tau, N, \lambda_1, \lambda_2)$ with one change-point are given by

$$\hat{\tau} = 16, \hat{N} = 145, \hat{\lambda}_1 = 1.1 \times 10^{-4}, \hat{\lambda}_2 = 0.31 \times 10^{-4}.$$

If we do not consider a change-point in the model, the MLEs of the parameters N and λ can be given as

$$\hat{N} = 142, \hat{\lambda} = 0.35 \times 10^{-4}.$$

From Fig. 3.3, we can observe that it is worth considering the change-points in the reliability functions.

3.7 Goodness of Fit Tests

The problem at hand is to compare some observed sample distribution with a theoretical distribution. In fact, a practitioner often wonders how to test some hypothesis about the distribution of a population. If the test is concerned with the agreement between the distribution of a set of observed sample values and a theoretical distribution, we call it a "test of goodness of fit".

The basic question in validating distribution models is whether the shape of the fitted model corresponds to that of the data. To do that, we may just simply make a direct comparison of the observed data with what we expect to see from the fitted distribution model. In this section, two common tests of goodness of fit that will be discussed are the Chi-square test, χ^2, and the Kolmogorov-Smirnov (KS) test.

Note that the chi-square test requires all cell frequencies to exceed 5, whereas this restriction is no longer required for KS test because there is no need to classify the observations in carrying out the KS test.

3.7.1   The Chi-Square Test

The chi-square test often requires large samples and is applied by comparing the observed frequency distribution of the sample to the expected value under the assumption of the distribution. More specifically, consider a large sample of size N. Let $a_0 < a_1 < \ldots < a_k$ be the upper points of k subintervals of the frequency distribution. Basically, the statistic

$$\chi^2 = \sum_{i=1}^{k} \left(\frac{x_i - \mu_i}{\sigma_i} \right)^2 \tag{3.59}$$

has a chi-square (χ^2) distribution with k degrees of freedom where μ_i and σ_i. are the mean and the standard deviation from the normal distribution in the ith subinterval, $i = 1, 2, \ldots, k$.

Let f_i and E_i, for $i = 1, 2, \ldots, k$, be the observed frequency and the expected frequency in the ith subinterval, respectively. The expected frequency E_i in the ith subinterval is.
$E_i = N (F_i - F_{i-1})$.
The chi-square test statistic is defined as

$$S = \sum_{i=1}^{k} \frac{(f_i - E_i)^2}{E_i}. \tag{3.60}$$

It should be noted that

$$\sum_{i=1}^{k} f_i = \sum_{i=1}^{k} E_i = N$$

where N is the sample size. Therefore, the chi-square statistic S can be rewritten as

$$S = \sum_{i=1}^{k} \frac{f_i^2}{E_i} - N. \tag{3.61}$$

The value of S is approximately distributed of $\chi^2_{1-\alpha, k-1}$. Thus, if $S \geq \chi^2_{1-\alpha, k-1}$. then the distribution $F(x)$ does not fit the observed data.

If the values of the parameters (some unknown parameters) of the distribution have to be estimated from the sample, then we need to reduce the number of degrees of freedom of χ^2 by the number of estimated parameters. In other words, the degrees of freedom are:

(the number of frequencies − 1 − number of parameters estimated with the same data).

In this case, the distribution $F(x)$ does not fit the observed data if $S \geq \chi^2_{1-\alpha,k-1-r}$ where r is the number of estimated parameters.

It should be noted that in order to get a good approximation the sample size N needs to be large and values of E_i should not be small, not much less than 5. The steps of chi-squared goodness of fit test are as follows:

1. Divide the sample data into the mutually exclusive cells (normally 8–12) such that the range of the random variable is covered.
2. Determine the frequency, f_i, of sample observations in each cell.
3. Determine the theoretical frequency, E_i, for each cell (area under density function between cell boundaries X_n—total sample size). Note that the theoretical frequency for each cell should be greater than 1. To carry out this step, it normally requires estimates of the population parameters which can be obtained from the sample data.
4. Form the statistic

$$S = \sum_{i=1}^{k} \frac{(f_i - E_i)^2}{E_i}$$

where $E_i = N (F_i - F_{i-1})$.

5. From the χ^2 tables, choose a value of χ^2 with the desired significance level and with degrees of freedom ($=k-r-1$), where r is the number of population parameters estimated.
6. Reject the hypothesis that the sample distribution is the same as theoretical distribution if

$$S \geq \chi^2_{1-\alpha,\, k-r-1}$$

where α is called the significance level.

Example 3.23
Given the data in Table 3.3, can the data be represented by the exponential distribution with a significance level of α?

Solution: From the above calculation, $\hat{\lambda} = 0.00263$, $R_i = e^{-\lambda t_i}$ and $F_i = 1 - R_i$. Given that a value of significance level α is 0.1, from Eq. (3.60) we obtain

Table 3.3 Sample
observations in each cell
boundary ($N = 60$)

Cell boundaries	f_i	$Q_i = Fi * 60$	$E_i = N(F_i - F_{i-1}) = Q_i - Q_{i-1}$
0–100	10	13.8755	13.8755
100–200	9	24.5422	10.6667
200–300	8	32.7421	8.1999
300–400	8	39.0457	6.3036
400–500	7	43.8915	4.8458
500–600	6	47.6168	3.7253
600–700	4	50.4805	2.8637
700–800	4	52.6819	2.2010
800–900	2	54.3743	1.6924
900–1,000	1	55.6753	1.3010
>1,000	1	60.0000	4.3247

$$S = \sum_{i=1}^{11} \frac{(f_i - E_i)^2}{E_i} = 8.7332$$

Here $k = 11$, $r = 1$. Then $k-r-1 = 11-1-1 = 9$. From Table A.3 in Appendix A, the value of χ^2 with nine degrees of freedom is 14.68, that is,

$$\chi^2_{9df}(0.90) = 14.68$$

Since $S = 8.7332 < 14.68$, we would not reject the hypothesis of exponential with $\lambda = 0.00263$.

Example 3.24

The number of defective soldering points found on a circuit board in a given day is given in Table 3.4. Now the question here is: Can the data be represented by the Poisson distribution with a significance level of α?

Solution: From Table 3.4, one can estimate the rate: $\hat{\lambda} = 56/33 = 1.7$. The expected frequency E_i is given by:

Table 3.4 Sample
observations in each cell
boundary ($N = 33$)

Number of defective	Observed frequency f_i	Expected frequency E_i	Total number of defectives
0	6	6.04	0
1	8	10.26	8
2	12	8.71	24
3	4	4.92	12
4 or more	3	3.07	12
Total	33		56

$$E_i = NP(D = i) = N\frac{e^{-\lambda}\lambda^i}{i!}.$$

Note that.

$P(D = d) = \frac{e^{-\lambda}\lambda^d}{d!}$ for d = 0, 1, ..., 4

Given that a value of significance level α is 0.1, from Eq. (3.60) we obtain

$$S = \sum_{i=0}^{4} \frac{(f_i - E_i)^2}{E_i} = 1.912$$

From Table A.3 in Appendix A, the value of χ^2 with three degrees of freedom $(k-1-1 = 3)$ is

$$\chi^2_{3df}(0.90) = 6.25$$

Since $S = 1.912 < 6.25$, we would not reject the hypothesis of Poisson model.

Example 3.25

Consider a sample of 100 failure times of electric generators (in hours). We now determine whether the distribution of failure times is normal. Table 3.5 gives the observed frequencies and expected frequencies over 8 intervals. Given that the estimated values of the mean μ and standard deviation σ are $\mu = 0.122$ and $\sigma = 0.011$.

we obtain

$$S = \sum_{i=1}^{8} \frac{(f_i - E_i)^2}{E_i} = 5.404$$

From Table A.3 in Appendix A, the value of χ^2 with 5 degrees of freedom (i.e., $k-1-r = 8-1-2 = 5$) is

Table 3.5 Sample observations of 100 failure times ($N = 100, k = 8$)

Intervals	Observed frequency f_i	Expected frequency E_i
325–400	7	6.10
400–475	9	7.70
475–550	17	12.60
550–625	12	16.80
625–700	18	18.10
700–725	11	15.90
725–800	12	11.40
>800	14	11.40
	N = 100	

$$\chi^2_{5df}(0.90) = 9.24$$

Since $S = 5.404 < 9.24$, we would not reject the hypothesis of normality.

3.7.2 Kolmogorov–Smirnov (KS) Test

Both the χ^2 and KS tests are non-parameters. However, the χ^2 assumes large sample normality of the observed frequency about its mean while the KS test only assumes a continuous distribution. Suppose the hypothesis is that the sample comes from a specified population distribution with c.d.f. $F_0(x)$. The test statistic compares the observed (empirical) distribution of the sample $F_n(x)$, to $F_0(x)$ and considers the maximum absolute deviation between the two cdfs ($F_n(x)$ and $F_0(x)$). Let $X_{(1)} \leq X_{(2)} \leq X_{(3)} \leq \ldots \leq X_{(n)}$ denote the ordered sample values. Define the observed distribution function, $F_n(x)$, as follows:

$$F_n(x) = \begin{cases} 0 \text{ for } x \leq x_{(1)} \\ \dfrac{i}{n} \text{ for } x_{(i)} < x \leq x_{(i+1)} \\ 1 \text{ for } x > x_{(n)} \end{cases} \tag{3.62}$$

Assume the testing hypothesis

$$H_0 : F(x) = F_0(x)$$

where $F_0(x)$ *is* a given continuous distribution and $F(x)$ is an unknown distribution. The KS test statistic can be computed as follows:

$$D_n = \max_{-\infty < x < \infty} |F_n(x) - F_0(x)|. \tag{3.63}$$

Since $F_0(x)$ is a continuous increasing function, we can evaluate $|F_n(x) - F_0(x)|$ for each n. Table A.4 in Appendix A (Smirnov 1948) gives certain critical values $d_{n,\alpha}$ of the distribution of D_n, which is the maximum absolute difference between sample and population cumulative distributions, for various sample sizes n and a is the level of significance. For example, the critical value $d_{n,\alpha}$ for $n = 10$ at a 0.05 level of significance is 0.409. This means that in 5 percent of random samples of size 10, the maximum absolute deviation between the sample observed cumulative distribution and the population cumulative distribution will be at least 0.409.

If $D_n \leq d_{n,\alpha}$ then we would not reject the hypothesis distribution H_0. If $D_n > d_{n,\alpha}$, then we reject the hypothetical distribution. The value $d_{n,\alpha}$ can be found in Table A.4 in Appendix A. This test implies that

$$P\{D_n > d_{n,\alpha}\} = \alpha.$$

Substituting for D_n and rearranging gives a $100(1-\alpha)\%$ confidence level for $F(x)$ as follows:

$$\hat{F}(x) - d_{n,\alpha} \le F(x) \le \hat{F}(x) + d_{n,\alpha}$$

or, equivalently, that

$$P\{\hat{F}(x) - d_{n,\alpha} \le F(x) \le \hat{F}(x) + d_{n,\alpha}\} = 1 - \alpha.$$

Example 3.26

(*Continued from Example* 3.25) Can the data given in Table 3.4 be represented by the Poisson distribution with a significance level of α using KS test?

To obtain the calculation in Table 3.6:

$F_n(x)$: is the cumulative observed frequency/33.

$F_0(x)$: is the estimated from cumulative $P(d)$ (see table below).

where

$P(d) = \frac{e^{-\lambda} \lambda^d}{d!}$ for d = 0, 1, ..., 4

and $\lambda = 1.7$. Thus,

Table 3.6 KS test calculations ($N = 33$)

Number of defective	Observed frequency	Cum. observed frequency	$F_n(x)$	$F_0(x)$	$|F_n(x) - F_0(x)|$
0	6	6	0.182	0.183	0.001
1	8	14	0.424	0.494	0.07
2	12	26	0.788	0.758	0.03
3	4	30	0.909	0.907	0.002
4 or more	3	33	1.000	1.000	0.000

Table 3.7 KS test

Number of defective	Observed frequency f_i	$P(d)$	Cumulative $P(d)$
0	6	0.183	0.183
1	8	0.311	0.494
2	12	0.264	0.758
3	4	0.149	0.907
4 or more	3	0.093	1.000
Total	33	1.000	

$$D_n = \max_{-\infty < x < \infty} |F_n(x) - F_0(x)| = 0.070$$

From Table A.4 in Appendix A, here $N = 33$ and $\alpha = 10\%$ then we obtain $d_{n,\alpha} = 0.21$. Since $0.070 < 0.21$ $(D_n \leq d_{n,\alpha})$ therefore, the hypothesis of the Poisson model is accepted.

Example 3.27
Determine whether the following failure data (in days) of a system:
   10.50, 1.75, 6.10, 1.30, 15.00, 8.20, 0.50, 20.50, 11.05, 4.60.

be represented as a sample from an exponential population distribution with constant rate $\lambda = 0.20$ failures per day at the $(\alpha =)$ 5% level of the significance using KS test?

Solution: Under the hypothesis that failure times are exponential distribution, so the theoretical pdf and cdf are given by:

$$f(x) = 0.2e^{-0.2x} \text{ for } x > 0$$

and

$$F(x) = 1 - e^{-0.2x}$$

respectively. From Table 3.8, the maximum difference D_n is 0.2061. From Table A.4 in Appendix A, here $n = 10$ and $\alpha = 5\%$ then we obtain the critical value $d_{n,\alpha} = 0.409$. Since $D_n \leq d_{n,\alpha}$ therefore, the null hypothesis, that failure times are exponentially distributed with constant rate $\lambda = 0.20$, cannot be rejected at the 5% level of significance.

Table 3.8 The observed and theoretical cdf values

Failure time x	$F_n(x)$	$F_0(x)$	$|F_n(x) - F_0(x)|$
0.50	0.10	0.0952	0.0048
1.30	0.20	0.2289	0.0289
1.75	0.30	0.2953	0.0047
4.60	0.40	0.6015	0.2015
6.10	0.50	0.7048	0.2048
8.20	0.60	0.8061	0.2061
10.50	0.70	0.8776	0.1776
11.05	0.80	0.8903	0.0903
15.00	0.90	0.9503	0.0503
20.50	1.00	0.9835	0.0165

3.8 Test for Independence

The n elements of a sample may be classified according to two different criteria. It is of interest to know whether the two methods of classification are statistically independent. Assume that the first method of classification has r levels and the second method of classification has c levels. Let O_{ij} be the observed frequency for level i of the first classification method and level j of the second classification method as shown in Table 3.9 for $i = 1,2,\ldots,r$ and $j = 1,2,\ldots,c$.

We are interested in testing the hypothesis that the row and column methods of classification are independent. If we reject this hypothesis, we conclude there is some interaction between the two criteria of classification. The test statistic

$$\lambda^2_{vahe} = \sum_{i=1}^{r} \sum_{j=1}^{c} \frac{\left(O_{ij} - E_{ij}\right)^2}{E_{ij}} \sim \lambda^2_{(r-1)(c-1)} \tag{3.64}$$

and we would reject the hypothesis of independence if $\lambda^2_{value} > \lambda^2_{\alpha,(r-1)(c-1)}$ where E_{ij}, the expected number in each cell, is

$$E_{ij} = \frac{1}{n} \left(\sum_{i=1}^{r} O_{ij} \right) \left(\sum_{j=1}^{c} O_{ij} \right).$$

For example,

Table 3.9 Observed frequency for level i and level j of classification

	1	2	3 ...	c	Total
1	O_{11}	O_{12}	O_{13}	O_{1c}	$\sum_{j=1}^{c} O_{1j}$
2	O_{21}	O_{22}	O_{23}	O_{2c}	$\sum_{j=1}^{c} O_{2j}$
3	O_{31}	O_{32}	O_{33}	O_{3c}	$\sum_{j=1}^{c} O_{3j}$
...					
r	O_{r1}	O_{r2}	O_{r3}	O_{rc}	$\sum_{j=1}^{c} O_{rj}$
Total	$\sum_{i=1}^{r} O_{i1}$	$\sum_{i=1}^{r} O_{i2}$	$\sum_{i=1}^{r} O_{i3}$	$\sum_{i=1}^{r} O_{ic}$	n

Table 3.10 A random sample of 395 people

VarA/VarB	Observed	Variable B				Total
		15–20	21–30	31–40	41–65	
	Yes	60	54	46	41	201
Variable A	No	40	44	53	57	194
	Total	100	98	99	98	395

$$E_{21} = \frac{1}{n}\left(\sum_{i=1}^{r} O_{i1}\right)\left(\sum_{j=1}^{c} O_{2j}\right)$$

$$E_{32} = \frac{1}{n}\left(\sum_{i=1}^{r} O_{i2}\right)\left(\sum_{j=1}^{c} O_{3j}\right).$$

Example 3.28

A random sample of 395 people were surveyed. Each person was asked their interest in riding a bicycle (Variable A) and their age (Variable B). The data that resulted from the survey is summarized in Table 3.10.

Is there evidence to conclude, at the 0.05 level, that the desire to ride a bicycle depends on age? In other words, is age independent of the desire to ride a bicycle?

Solution: Here $r = 2$ and $c = 4$. The expected number in each cell can be obtained as shown in Table 3.11:

$$E_{11} = \frac{1}{395}\left(\sum_{i=1}^{2} O_{i1}\right)\left(\sum_{j=1}^{4} O_{1j}\right) = \frac{1}{395}(100)(201) = 50.8861$$

$$E_{12} = \frac{1}{395}\left(\sum_{i=1}^{2} O_{i2}\right)\left(\sum_{j=1}^{4} O_{1j}\right) = \frac{1}{395}(98)(201) = 49.8684$$

$$E_{13} = \frac{1}{395}(99)(201) = 50.3772$$

$$\ldots$$

$$E_{23} = \frac{1}{395}(99)(194) = 48.6228$$

$$E_{24} = \frac{1}{395}(98)(194) = 48.1316$$

The test statistic value

$$\lambda_{value}^2 = \sum_{i=1}^{2}\sum_{j=1}^{4}\frac{\left(O_{ij} - E_{ij}\right)^2}{E_{ij}}$$

Table 3.11 The expected values

		Variable B				
VarA/VarB	Expected	15–20	21–30	31–40	41–65	Total
	Yes	50.8861	49.8684	50.3772	49.8684	201
Variable A	No	49.1139	48.1316	48.6228	48.1316	194
	Total	100	98	99	98	395

$$= \frac{(60 - 50.8861)^2}{50.8861} + \frac{(54 - 49.8684)^2}{49.8684} + \cdots + \frac{(57 - 48.1316)^2}{48.1316}$$
$$= 8.0062$$

Here $\alpha = 0.05$. So $\chi^2_{0.05,(2-1)(4-1)} = \chi^2_{0.05,3} = 7.815$. Since $\chi^2_{\text{value}} = 8.0062 > 7.815$ we, therefore, reject the null hypothesis. That is, there is sufficient evidence, at the 0.05 level, to conclude that the desire to ride a bicycle depends on age.

3.9 Statistical Trend Tests of Homogeneous Poisson Process

Let us assume that repairable units observe from time $t = 0$, with successive failure times denoted by t_1, t_2, ... An equivalent representation of the failure process is in terms of the counting process $\{N(t), t \geq 0\}$, where $N(t)$ equals the number of failures in $(0, t]$. Repairable units are those units which can be restored to fully satisfactory performance by a method other than replacement of entire system (Ascher and Feingold 1984). It is assumed that simultaneous failures are not possible and the repair times are negligible compared to times between failures. In other word, it is assumed that repair times equal 0. The observed failure times during the interval $(0,T]$ for a specific unit are denoted by t_1, t_2, $...t_{N(T)}$.

Trend analysis is a common statistical method used to investigate the changes in a system device or unit over time. Based on the historical failure data of a unit or a group of similar units, one can test to determine whether the recurrent failures exhibit an increasing or decreasing trend subject to statistical tests. In other words, there are several methods to test whether the observed trend based on the failure data is statistically significant or there is no trend. This implies that, for the case there is no trend, the failure rate is constant or the failure process is homogeneous Poisson process. A number of tests such as Laplace Test and Military Handbook Test have been developed for testing the following null hypotheses (see Ascher and Feingold 1984):

H_0: *No trend* (Failure times observed are independently and identically distributed (iid), homogeneous Poisson process).

Against the alternative hypothesis.

H_1: *Monotonic trend*, i.e. the process is nonhomogeneous Poisson process (*NHPP*).

By rejection of null hypothesis, (case H_1) failure rate is not constant and the cumulative no of fault vs. time have concave or convex shape, respectively. Let t_i denote the running time of repairable item at the ith failure, $i = 1,...,n$ and $N(t_i)$ be the total number of failures observed till time t_n (failure truncated case) or the observation is terminated at time T censored time for time truncated case which is discusses in the following subsections. Following we discuss the Laplace trend test and Military Handbook test. These tests can be used to detect existence of trends in the data set of successive failure times.

3.9.1 Laplace Trend Test

The Laplace trend test can be used to detect existence of trends in the failure data of successive event times. Depend on the data type whether is time truncated or failure truncated, we can calculate Laplace trend index as follows:

Time Truncated Test: Suppose that observation is terminated at time T and $N(t_n)$ failures are recorded at time $t_1 < t_2 < ... < t_n < T$. The test statistic U of Laplace trend test in this case is given by

$$U = \sqrt{12.N(t_n)}\left(\frac{\sum_1^n t_i}{T.N(t_n)} - 0.5\right) \qquad (3.65)$$

Failure Truncated Test: In the cases where the observation terminates at a failure event, say t_n, instead of T, the test statistics for failure truncated data is given by

$$U = \sqrt{12\,N(t_{n-1})}\left(\frac{\sum_1^{n-1} t_i}{t_n\,N(t_{n-1})} - 0.5\right). \qquad (3.66)$$

In both test cases, the test statistic U is approximately standard normally $N(0, 1)$ distributed when the null hypothesis H_0 is true.

At significance level α of 5% for example, the lower and upper bound of confidence interval are -1.96 and $+ 1.96$, respectively. If U is less than lower bound or greater than upper bound then there is a trend in the data. In these cases, we can reject the null hypothesis of trend free at 5% and accept the alternative hypothesis. In the first case, there is reliability growth (U is less than lower bound) and in second case (U is greater than upper bound) there is reliability deterioration. Hence, in these cases, one need to be some probability distributions that can fit to the failure data set, other than exponential distribution.

In general, at significance level α, the rejection criteria of the null hypothesis is given by.

$$\text{Reject } H_0 : U < -z_{\alpha/2} \quad \text{or} \quad U > z_{\alpha/2} \qquad (3.67)$$

where z_α is $(1 - \alpha)$th quantile of the standard normal distribution.

3.9.2 Military Handbook Test

This test is based on the assumption that a failure intensity of $r(t) = \lambda \beta t^{\beta-1}$ is appropriate (MIL-HDBK-189 1981). When $\beta = 1$, the failure intensity reduces to $r(t) = \lambda$, which means the failure process follows a HPP. Hence the test determines whether an estimate of β is significantly different from 1 or not. The null hypothesis test in this case is $\beta = 1$ i.e. no trend in data (HPP) vs. alternative hypothesis $\beta \neq 1$, there is trend in the failure history (NHPP):

Time Truncated Test: During a test period T suppose that n failure are recorded at time $t_1 < t_2 < ... < t_n < T$. The mathematical test in this case is given by:

$$S = 2 \sum_{i=1}^{n} \ln \frac{T}{t_i}. \tag{3.68}$$

According to (MIL-HDBK-189, 1981), S has chi-square distribution with $2n$ degree of freedom. In this case, the rejection criteria is given by:

$$\text{Reject } H_0 \text{ when } S < \chi^2_{2n,\, 1-\alpha/2} \quad \text{or} \quad S > \chi^2_{2n,\, \alpha/2}. \tag{3.69}$$

In other words, the hypothesis (H_0) of trend free is rejected for the values beyond the interval boundaries. Low values of S correspond to deteriorating while large values of S correspond to improving.

Failure Truncated Test: If the data are failure truncated at t_n instead of time truncated at T, S has chi-square distribution with $2(n-1)$ degree of freedom when the null hypothesis H_0 is true. The test statistic S of Military handbook test in this case is given by:

$$S = 2 \sum_{i=1}^{n-1} \ln \left(\frac{t_n}{t_i} \right). \tag{3.70}$$

Here, the rejection criterion is given by:

$$\text{Reject } H_0 \text{ when } S < \chi^2_{2(n-1),\, 1-\alpha/2} \quad \text{or} \quad S > \chi^2_{2(n-1),\, \alpha/2}. \tag{3.71}$$

The hypothesis of HPP (H_0) is rejected for the values beyond the following interval:

$$\left[\chi^2_{2(n-1),1-\alpha/2},\, \chi^2_{2(n-1),\alpha/2} \right].$$

3.10   Least Squared Estimation

A problem of curve fitting, which is unrelated to normal regression theory and MLE estimates of coefficients but uses identical formulas, is called the method of least squares. This method is based on minimizing the sum of the squared distance from the best fit line and the actual data points. It just so happens that finding the MLEs for the coefficients of the regression line also involves these sums of squared distances.

Normal Linear Regression

Regression considers the distributions of one variable when another is held fixed at each of several levels. In the bivariate normal case, consider the distribution of X as a function of given values of Z where $X = \alpha + \beta Z$. Consider a sample of n observations (x_i, z_i), we can obtain the likelihood and its natural log for the normal distribution as follows:

$$f(x_1, x_2, \ldots, x_n) = \frac{1}{2\pi^{\frac{n}{2}}} \left(\frac{1}{\sigma^2}\right)^{\frac{n}{2}} e^{-\frac{1}{2\sigma^2} \sum_{i=1}^{n} (x_i - \alpha - \beta z_i)^2}$$

$$\ln L = -\frac{n}{2} \log 2\pi - \frac{n}{2} \log \sigma^2 - \frac{1}{2\sigma^2} \sum_{i=1}^{n} (x_i - \alpha - \beta z_i)^2. \qquad (3.72)$$

Taking the partial derivatives of $\ln L$ with respect to α and β, we have

$$\frac{\partial \ln L}{\partial \alpha} = \sum_{i=1}^{n} (x_i - \alpha - \beta z_i)^2 = 0$$

$$\frac{\partial \ln L}{\partial \beta} = \sum_{i=1}^{n} z_i (x_i - \alpha - \beta z_i) = 0.$$

The solution of the simultaneous equations is

$$\hat{\alpha} = \bar{X} - \beta \bar{Z}$$

$$\hat{\beta} = \frac{\sum_{i=1}^{n} (X_i - \bar{X})(Z_i - \bar{Z})}{\sum_{i=1}^{n} (Z_i - \bar{Z})^2}. \qquad (3.73)$$

Least Squared Straight Line Fit

Assume there is a linear relationship between X and $E(Y/x)$, that is, $E(Y/x) = a + bx$. Given a set of data, we want to estimate the coefficients a and b that minimize the sum of the squares. Suppose the desired polynomial, $p(x)$, is written as

$$\sum_{i=0}^{m} a_i x^i$$

where $a_0, a_1, \ldots, a_m$ are to be determined. The method of least squares chooses as the solutions those coefficients minimizing the sum of the squares of the vertical distances from the data points to the presumed polynomial. This means that the polynomial termed "best" is the one whose coefficients minimize the function L, where

$$L = \sum_{i=1}^{n} [y_i - p(x_i)]^2.$$

Here, we will treat only the linear case, where $X = \alpha + \beta Z$. The procedure for higher-order polynomials is identical, although the computations become much more tedious. Assume a straight line of the form $X = \alpha + \beta Z$. For each observation (x_i, z_i): $X_i = \alpha + \beta Z_i$, let

$$Q = \sum_{i=1}^{n} (x_i - \alpha - \beta z_i)^2.$$

We wish to find α and β estimates such as to minimize Q. Taking the partial differentials, we obtain

$$\frac{\partial Q}{\partial \alpha} = -2 \sum_{i=1}^{n} (x_i - \alpha - \beta z_i) = 0$$

$$\frac{\partial Q}{\partial \beta} = -2 \sum_{i=1}^{n} z_i (x_i - \alpha - \beta z_i) = 0.$$

Note that the above are the same as the MLE equations for normal linear regression. Therefore, we obtain the following results:

$$\hat{\alpha} = \bar{x} - \beta \bar{z}$$

$$\hat{\beta} = \frac{\sum_{i=1}^{n} (x_i - \bar{x})(z_i - \bar{z})}{\sum_{i=1}^{n} (z_i - \bar{z})^2}. \tag{3.74}$$

The above gives an example of least squares applied to a linear case. It follows the same pattern for higher-order curves with solutions of 3, 4, and so on, from the linear systems of equations.

3.11 Interval Estimation

A point estimate is sometimes inadequate in providing an estimate of an unknown parameter since it rarely coincides with the true value of the parameter. An alternative way is to obtain a confidence interval estimation of the form $[\theta_L, \theta_U]$ where θ_L is the lower bound and θ_U is the upper bound. A confidence interval (CI) is an interval estimate of a population parameter that also specifies the likelihood that the interval contains the true population parameter. This probability is called the level of confidence, denoted by $(1-\alpha)$, and is usually expressed as a percentage.

Point estimates can become more useful if some measure of their error can be developed, i.e., some sort of tolerance on their high and low values could be developed. Thus, if an interval estimator is $[\theta_L, \theta_U]$ with a given probability $(1-\alpha)$, then θ_L and θ_U will be called $100(1-\alpha)\%$ confidence limits for the given parameter θ and the interval between them is a $100(1-\alpha)\%$ confidence interval and $(1-\alpha)$ is also called the confidence coefficient.

3.11.1 Confidence Intervals for the Normal Parameters

The one-dimensional normal distribution has two parameters: mean μ and variance σ^2. The simultaneous employment of both parameters in a confidence statement concerning percentages of the population will be discussed in the next section on tolerance limits. Hence, individual confidence statements about μ and σ^2 will be discussed here.

Confidence Limits for the Mean μ with Known σ^2.

It is easy to show that the statistic

$$Z = \frac{\bar{X} - \mu}{\sigma/\sqrt{n}}$$

is a standard normal distribution where

$$\bar{X} = \frac{1}{n} \sum_{i=1}^{n} X_i$$

Hence, a $100(1 - \alpha)\%$ confidence interval for the mean μ is given by

$$P\left[\bar{X} - Z_{\frac{\alpha}{2}} \frac{\sigma}{\sqrt{n}} < \mu < \bar{X} + Z_{\frac{\alpha}{2}} \frac{\sigma}{\sqrt{n}} \right] = 1 - \alpha \qquad (3.75)$$

In other words,

$$\mu_L = \bar{X} - Z_{\frac{\alpha}{2}} \frac{\sigma}{\sqrt{n}} \quad \text{and} \quad \mu_U = \bar{X} + Z_{\frac{\alpha}{2}} \frac{\sigma}{\sqrt{n}}$$

For the normal distribution, an empirical rule relates the standard deviation to the proportion of the observed values of the variable in a data set that lie in an interval around the sample mean can be approximately expressed as follows:

- 68% of the values lie within: $\bar{x} \pm s_x$
- 95% of the values lie within: $\bar{x} \pm 2s_x$
- 99.7% of the values lie within: $\bar{x} \pm 3s_x$

Example 3.29
Draw a sample of size 4 from a normal distribution with known variance $= 9$, say $x_1 = 2, x_2 = 3, x_3 = 5, x_4 = 2$. Determine the location of the true mean (μ). The sample mean can be calculated as.

$$\bar{x} = \frac{\sum_{i=1}^{n} x_i}{n} = \frac{2+3+5+2}{4} = 3$$

Assuming that $\alpha = 0.05$ and from the standard normal distribution (Table A.1 in Appendix A), we obtain

$$P\left[3 - 1.96\frac{3}{\sqrt{4}} < \mu < 3 + 1.96\frac{3}{\sqrt{4}}\right] = 0.95$$

$$P[0.06 < \mu < 5.94] = 0.95$$

This example shows that there is a 95% probability that the true mean is somewhere between 0.06 and 5.94. Now, μ is a fixed parameter and does not vary, so how do we interpret the probability? If the samples of size 4 are repeatedly drawn, a different set of limits would be constructed each time. With this as the case, the interval becomes the random variable and the interpretation is that, for 95% of the time, the interval so constructed will contain the true (fixed) parameter.

Confidence Limits for the Mean μ with Unknown σ^2.
Let

$$S = \sqrt{\frac{1}{n-1}\sum_{i=1}^{n}(X_i - \bar{X})^2} \tag{3.76}$$

It can be shown that the statistic

$$T = \frac{\bar{X} - \mu}{\frac{S}{\sqrt{n}}}$$

has a t distribution with $(n-1)$ degrees of freedom (see Table A.2 in Appendix A). Thus, for a given sample mean and sample standard deviation, we obtain

$$P\left[|T| < t_{\frac{\alpha}{2},n-1}\right] = 1 - \alpha$$

Hence, a $100(1 - \alpha)\%$ confidence interval for the mean μ is given by

$$P\left[\bar{X} - t_{\frac{\alpha}{2},n-1}\frac{S}{\sqrt{n}} < \mu < \bar{X} + t_{\frac{\alpha}{2},n-1}\frac{S}{\sqrt{n}}\right] = 1 - \alpha \qquad (3.77)$$

Example 3.30

A problem on the variability of a new product was encountered. An experiment was run using a sample of size $n = 25$; the sample mean was found to be $\bar{X} = 50$ and the variance $\sigma^2 = 16$. From Table A.2 in Appendix A, $t_{\frac{\alpha}{2},n-1} = t_{.975,24} = 2.064$. A 95% confidence limit for μ is given by.

$$P\left[50 - 2.064\sqrt{\frac{16}{25}} < \mu < 50 + 2.064\sqrt{\frac{16}{25}}\right] = 0.95$$

$$P[48.349 < \mu < 51.651] = 0.95$$

Note that, for one-sided limits, choose t_α, or $t_{1-\alpha}$.

Example 3.31

Consider a normal distribution with unknown mean μ and unknown standard deviation σ. Suppose we draw a random sample of size n $= 16$ from this population, and the sample values are:

16.16	9.33	12.96	11.49
12.31	8.93	6.02	10.66
7.75	15.55	3.58	11.34
11.38	6.53	9.75	9.47

Compute the confidence interval for the mean μ and at confidence level $1 - \alpha = 0.95$.

Solution: A C.I. for μ at confidence level 0.95 when σ is unknown is obtained from the corresponding t-test. The C.I is

$$\bar{X} - t_{1-\frac{\alpha}{2},n-1}\frac{S}{\sqrt{n}} \le \mu \le \bar{X} + t_{1-\frac{\alpha}{2},n-1}\frac{S}{\sqrt{n}}$$

The sample mean and sample variance, respectively, are:

$$\bar{X} = 10.20 \; and \; S^2 = 10.977$$

The sample standard deviation is: $S = 3.313$. Also, $t_{1-\frac{\alpha}{2},n-1} = t_{0.975,15} = 2.131$.

The 95% C.I. for μ is

$$10.20 - 2.131\frac{3.313}{\sqrt{16}} \le \mu \le 10.20 + 2.131\frac{3.313}{\sqrt{16}}$$

or

$$8.435 \le \mu \le 11.965$$

Confidence Limits on σ^2.

Note that $n\frac{\widehat{\sigma^2}}{\sigma^2}$ has a χ^2 distribution with $(n-1)$ degrees of freedom. Correcting for the bias in $\hat{\sigma}^2$, then $(n-1)\hat{\sigma}^2/\sigma^2$ has this same distribution. Hence,

$$P\left[\chi^2_{\frac{\alpha}{2},n-1} < \frac{(n-1)S^2}{\sigma^2} < \chi^2_{1-\frac{\alpha}{2},n-1}\right] = 1 - \alpha$$

or

$$P\left[\frac{\sum(x_i - \bar{x})^2}{\chi^2_{1-\frac{\alpha}{2},n-1}} < \sigma^2 < \frac{\sum(x_i - \bar{x})^2}{\chi^2_{\frac{\alpha}{2},n-1}}\right] = 1 - \alpha. \tag{3.78}$$

Similarly, for one-sided limits, choose $\chi^2(\alpha)$ or $\chi^2(1-\alpha)$.

3.11.2 Confidence Intervals for the Exponential Parameters

The pdf and cdf of the exponential distribution are given as

$$f(x) = \lambda e^{-\lambda x} \quad x > 0, \lambda > 0$$

and

$$F(x) = 1 - e^{-\lambda x}$$

respectively. From Eq. (3.22), it was shown that the distribution of a function of the estimate

$$\hat{\lambda} = \frac{r}{\sum_{i=1}^{r} x_i + (n-r)x_r} \tag{3.79}$$

derived from a test of n identical components with common exponential failure density (failure rate λ), whose testing was stopped after the rth failure, was chi-squared (X^2), i.e.,

$$2r\frac{\lambda}{\hat{\lambda}} = 2\lambda T \quad (\chi^2 \text{distribution with } 2r \text{ degrees of freedom})$$

where T is the total accrued time on all units. Knowing the distribution of $2\lambda T$ allows us to obtain the confidence limits on the parameter as follows:

$$P\left[\chi^2_{1-\frac{\alpha}{2},2r} < 2\lambda T < \chi^2_{\frac{\alpha}{2},2r}\right] = 1 - \alpha$$

or, equivalently, that

$$P\left[\frac{\chi^2_{1-\frac{\alpha}{2},2r}}{2T} < \lambda < \frac{\chi^2_{\frac{\alpha}{2},2r}}{2T}\right] = 1 - \alpha$$

This means that in $(1-\alpha)\%$ of samples with a given size n, the random interval

$$\left(\frac{\chi^2_{1-\frac{\alpha}{2},2r}}{2T}, \frac{\chi^2_{\frac{\alpha}{2},2r}}{2T}\right) \tag{3.80}$$

will contain the population of constant failure rate. In terms of $\theta = 1/\lambda$ or the mean time between failure (MTBF), the above confidence limits change to

$$P\left[\frac{2T}{\chi^2_{\frac{\alpha}{2},2r}} < \theta < \frac{2T}{\chi^2_{1-\frac{\alpha}{2},2r}}\right] = 1 - \alpha.$$

If testing is stopped at a fixed time rather than a fixed number of failures, the number of degrees of freedom in the lower limit increases by two. Table 3.12 shows

Table 3.12 Confidence limits for θ	Confidence limits	Fixed number of failures	Fixed time
	One-sided lower limit	$\dfrac{2T}{\chi^2_{\alpha,2r}}$	$\dfrac{2T}{\chi^2_{\alpha,2r+2}}$
	One-sided upper limit	$\dfrac{2T}{\chi^2_{1-\alpha,2r}}$	$\dfrac{2T}{\chi^2_{1-\alpha,2r}}$
	Two-sided limits	$\dfrac{2T}{\chi^2_{\alpha/2,2r}}, \dfrac{2T}{\chi^2_{1-\alpha/2,2r}}$	$\dfrac{2T}{\chi^2_{\alpha/2,2r+2}}, \dfrac{2T}{\chi^2_{1-\alpha/2,2r}}$

the confidence limits for θ, the mean of an exponential density.

Example 3.32
(*Two-sided*) From the goodness of fit example, $T = 22{,}850$, testing stopped after $r = 60$ failures. We can obtain $\hat{\lambda} = 0.00263$ and $\hat{\theta} = 380.833$. Assu-ming that $\alpha = 0.1$, then, from the above formula, we obtain.

$$P\left[\frac{2T}{\chi^2_{0.05,120}} < \theta < \frac{2T}{\chi^2_{0.95,120}}\right] = 0.9$$

$$P\left[\frac{45{,}700}{146.568} < \theta < \frac{45{,}700}{95.703}\right] = 0.9$$

$$P[311.80 < \theta < 477.52] = 0.9$$

Example 3.33
(*One-sided lower*) Assuming that testing stopped after 1,000 h with four failures, then.

$$P\left[\frac{2T}{\chi^2_{0.10,10}} < \theta\right] = 0.9$$

$$P\left[\frac{2{,}000}{15.987} < \theta\right] = 0.9$$

$$P[125.1 < \theta] = 0.9$$

3.11.3 Confidence Intervals for the Binomial Parameters

Consider a sequence of n Bernoulli trials with k successes and $(n - k)$ failures. We now determine one-sided upper and lower and two-sided limits on the parameter p, the probability of success. For the lower limit, the binomial sum is set up such that the chance probability of k or more successes with a true p as low as p_L is only $\alpha/2$. This means the probability of k or more successes with a true p higher than p_L is $\left(1 - \frac{\alpha}{2}\right)$:

$$\sum_{i=k}^{n} \binom{n}{i} p_L^i (1 - p_L)^{n-i} = \frac{\alpha}{2}$$

Similarly, the binomial sum for the upper limit is

$$\sum_{i=k}^{n} \binom{n}{i} p_U^i (1 - p_U)^{n-i} = 1 - \frac{\alpha}{2}$$

or, equivalently, that

$$\sum_{i=0}^{k-1} \binom{n}{i} p_U^i (1 - p_U)^{n-i} = \frac{\alpha}{2}$$

Solving for p_L and p_U in the above equations,

$$P[p_L < p < p_U] = 1 - \alpha$$

For the case of one-sided limits, merely change $\alpha/2$ to α.

Example 3.34

Given $n = 100$ with 25 successes, and 75 failures, an 80% two-sided confidence limits on p can be obtained as follows:

$$\sum_{i=25}^{100} \binom{100}{i} p_L^i (1 - p_L)^{100-i} = 0.10$$

$$\sum_{i=0}^{24} \binom{100}{i} p_U^i (1 - p_U)^{100-i} = 0.10$$

Solving the above two equations simultaneously, we obtain

$$p_L \approx 0.194 \text{ and } p_U \approx 0.313$$
$$P[0.194 < p < 0.313] = 0.80$$

Example 3.35

Continuing with Example 3.34, find an 80% one-sided confidence limit on p.

Solution: We now can set the top equation to 0.20 and solve for p_L. It is easy to obtain $p_L = 0.211$ and $P[p > 0.211] = 0.80$. Let us define $\bar{p} = k/n$, the number of successes divided by the number of trials. For large values of n and if $np > 5$ and $n(1 - p) > 5$, and from the central limit theorem (Feller 1966, 1968), the statistic

$$Z = \frac{(\bar{p} - p)}{\sqrt{\frac{\bar{p}(1-\bar{p})}{n}}} \tag{3.81}$$

approximates to the standard normal distribution. Hence,

$$P[-z_{\frac{\alpha}{2}} < Z < z_{\frac{\alpha}{2}}] = 1 - \alpha$$

Then

$$P\left[\bar{p} - z_{\frac{\alpha}{2}}\sqrt{\frac{\bar{p}(1-\bar{p})}{n}} < p < \bar{p} + z_{\frac{\alpha}{2}}\sqrt{\frac{\bar{p}(1-\bar{p})}{n}}\right] = 1 - \alpha \qquad (3.82)$$

Example 3.36
Given $n = 900$, $k = 180$, and $\alpha = 0.05$. Then we obtain $p = 180/900 = 0.2$ and

$$P\left[0.2 - 1.96\sqrt{\tfrac{0.2(0.8)}{900}} < p < 0.2 + 1.96\sqrt{\tfrac{0.2(0.8)}{900}}\right] = 0.95$$
$$P[0.174 < p < 0.226] = 0.95$$

3.11.4 Confidence Intervals for the Poisson Parameters

Limits for the Poisson parameters are completely analogous to the binomial except that the sample space is infinite instead of finite. The lower and upper limits can be solved simultaneously in the following equations:

$$\sum_{i=k}^{\infty} \frac{\lambda_L^i e^{-\lambda_L}}{i!} = \frac{\alpha}{2} \quad \text{and} \quad \sum_{i=k}^{\infty} \frac{\lambda_U^i e^{-\lambda_U}}{i!} = 1 - \frac{\alpha}{2} \qquad (3.83)$$

or, equivalently

$$\sum_{i=k}^{\infty} \frac{\lambda_L^i e^{-\lambda_L}}{i!} = \frac{\alpha}{2} \quad \text{and} \quad \sum_{i=0}^{k-1} \frac{\lambda_U^i e^{-\lambda_U}}{i!} = \frac{\alpha}{2}.$$

Example 3.37
One thousand article lots are inspected resulting in an average of 10 defects per lot. Find 90% limits on the average number of defects per 1000 article lots. Assume $\alpha = 0.1$,

$$\sum_{i=10}^{\infty} \frac{\lambda_L^i e^{-\lambda_L}}{i!} = 0.05 \quad \text{and} \quad \sum_{i=0}^{9} \frac{\lambda_U^i e^{-\lambda_U}}{i!} = 0.05.$$

Solving the above two equations simulteneously for λ_L and λ_U, we obtain.
$P[5.45 < \lambda < 16.95] = 0.90$.
The one-sided limits are constructed similarly to the case for binomial limits.

3.12 Non-Parametric Tolerance Limits

Non-parametric tolerance limits are based on the smallest and largest observation in the sample, designated as X_S and X_L, respectively. Due to their non-parametric nature, these limits are quite insensitive and to gain precision proportional to the parametric methods requires much larger samples. An interesting question here is to determine the sample size required to include at least $100(1-\alpha)\%$ of the population between X_S and X_L with given probability γ.

For two-sided tolerance limits, if $(1-\alpha)$ is the minimum proportion of the population contained between the largest observation X_L and smallest observation X_S with confidence $(1-\gamma)$, then it can be shown that

$$n(1-\alpha)^{n-1} - (n-1)(1-\alpha)^n = \gamma. \tag{3.84}$$

Therefore, the number of observations required is given by

$$n = \left\lfloor \frac{(2-\alpha)}{4\alpha} \chi_{1-\gamma,4}^2 + \frac{1}{2} \right\rfloor + 1 \tag{3.85}$$

where a value of $\chi_{1-\gamma,4}^2$ is given in Table A.3 of Appendix A.

Example 3.38
Determine the tolerance limits which include at least 90% of the population with probability 0.95. Here,
$\alpha = 0.1$, $\gamma = 0.95$ and $\chi^2_{0.05,4} = 9.488$.
and therefore, a sample of size

$$n = \left\lfloor \frac{(2-0.1)}{4(0.1)} (9.488) + \frac{1}{2} \right\rfloor + 1 = 46$$

is required. For a one-sided tolerance limit, the number of observations required is given by

$$n = \left\lfloor \frac{\log(1-\gamma)}{\log(1-\alpha)} \right\rfloor + 1. \tag{3.86}$$

Example 3.39
As in Example 3.38, we wish to find a lower tolerance limit, that is, the number of observations required so that the probability is 0.95 that at least 90% of the population will exceed X_S is given by.

$$n = \left\lfloor \frac{\log(1-0.95)}{\log(1-0.1)} \right\rfloor + 1 = 30.$$

One can easily generate a table containing the sample size required to include a given percentage of the population between X_S and X_L with given confidence, or sample size required to include a given percentage of the population above or below X_S or X_L, respectively.

3.13   Sequential Sampling

A sequential sampling scheme is one in which items are drawn one at a time and the results at any stage determine if sampling or testing should stop. Thus, any sampling procedure for which the number of observations is a random variable can be regarded as sequential sampling. Sequential tests derive their name from the fact that the sample size is not determined in advance, but allowed to "float" with a decision (accept, reject, or continue test) after each trial or data point.

In general, let us consider the hypothesis

$$H_0 : f(x) = f_0(x) \quad \text{vs} \quad H_1 : f(x) = f_1(x)$$

For an observation test, say X_1, if $X_1 \leq A$, then we will accept the testing hypothesis ($H_0 : f(x) = f_0(x)$); if $X_1 \geq A$, then we will reject H_0 and accept $H_1 : f(x) = f_1(x)$. Otherwise, we will continue to perform at least one more test. The interval $X_1 \leq A$ is called the acceptance region. The interval $X_1 \geq A$ is called the rejection or critical region (see Fig. 3.4).

A "good" test is one that makes the α and β errors as small as possible where

$$P\{\text{Type I error}\} = P\{\text{Reject H}_0 \mid \text{H}_0 \text{ is True}\} = \alpha$$
$$P\{\text{Type II error}\} = P\{\text{Accept H}_0 \mid \text{H}_0 \text{ is False}\} = \beta$$

Type I error, also known as a "*false positive*", the error of rejecting a null hypothesis H_0 when it is actually true. In other words, this is the error of accepting an alternative hypothesis H_1 (the real hypothesis of interest) when the results can be attributed to chance. So the probability of making a type I error in a test with rejection region R is $P(R \mid H_0$ is true).

Type II error, also known as a "*false negative*", the error of not rejecting a null hypothesis when the alternative hypothesis is the true state of nature. In other words, this is the error of failing to accept an alternative hypothesis when you don't have

Fig. 3.4 A sequential sampling scheme

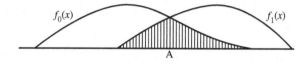

adequate power. So the probability of making a type II error in a test with rejection region R is 1-P(R | H_1 is true). The power of the test can be P(R | H_1 is true).

However, there is not much freedom to do this without increasing the sample size. The common procedure is to fix the β error and then choose a critical region to minimize the error or maximize the "power" (power = $1-\beta$) of the test, or to choose the critical region so as to equalize the α and β errors to reasonable levels.

A criterion, similar to the MLE, for constructing tests is called the "sequential probability ratio", which is the ratio of the sample densities under H_1 over H_0. The sequential probability ratio is given by

$$\lambda_n = \frac{\prod_{i=1}^n f_1(x_i)}{\prod_{i=1}^n f_0(x_i)} > k \tag{3.87}$$

Here, $x_1, x_2, ..., x_n$ are n independent random observations and k is chosen to give the desired error.

Recall from the MLE discussion in Sect. 3.3 that $f_1(x_1), f_1(x_2),..., f_1(x_n)$ are maximized under H_1 when the parameter(s), e.g., $\theta = \theta_1$ and, similarly, $f_0(x_1)$, $f_0(x_2),..., f_0(x_n)$ are maximized when $\theta = \theta_0$. Thus, the ratio will become large if the sample favors H_1 and will become small if the sample favors H_0. Therefore, the test will be called a sequential probability ratio test if we.

1. Stop sampling and reject H_0 as soon as $\lambda_n \geq A$.
2. Stop sampling and accept H_0 as soon as $\lambda_n \leq B$.
3. Continue sampling as long as $B < \lambda_n < A$, where $A > B$.

For example, the test will continue iff

$$\frac{\beta}{1 - \alpha} < \lambda_n < \frac{1 - \beta}{\alpha}$$

The choice of A and B with the above test, suggested by Wald (1947), can be determined as follows:

$$B = \frac{\beta}{1 - \alpha} \text{ and } A = \frac{1 - \beta}{\alpha}$$

The basis for α and β are therefore

$$P[\lambda_n \geq A | H_0] = \alpha$$
$$P[\lambda_n \leq B | H_1] = \beta$$

3.13.1 *Exponential Distribution Case*

Let

$$V(t) = \sum_{i=1}^{r} X_i + \sum_{j=1}^{n-r} t_j$$

where X_i are the times to failure and t_j are the times to test termination without failure. Thus, $V(t)$ is merely the total operating time accrued on both successful and unsuccessful units where the total number of units is n. The hypothesis to be tested is.

$H_0: \theta = \theta_0$ vs $H_1: \theta = \theta_1$.

For the failed items,

$$g(x_1, x_2, \ldots, x_r) = \left(\frac{1}{\theta}\right)^r e^{-\frac{\sum_{i=1}^{r} x_i}{\theta}}$$

For the non-failed items,

$$P(X_{r+1} > t_1, X_{r+2} > t_2, \ldots, X_n > t_{n-r}) = e^{-\frac{\sum_{j=1}^{n-r} t_j}{\theta}}$$

The joint density for the first r failures among n items, or likelihood function, is

$$f(x_1, x_2, \ldots, x_r, t_{r+1}, \ldots, t_n) = \frac{n!}{(n-r)!}\left(\frac{1}{\theta}\right)^r e^{-\frac{1}{\theta}\left(\sum_{i=1}^{r} x_i + \sum_{j=1}^{n-r} t_j\right)}$$

$$= \frac{n!}{(n-r)!}\left(\frac{1}{\theta}\right)^r e^{-\frac{V(t)}{\theta}}$$

and the sequential probability ratio is given by

$$\lambda_n = \frac{\prod_{i=1}^{n} f_1(x_i, \theta_1)}{\prod_{i=1}^{n} f_0(x_i, \theta_0)} = \frac{\frac{n!}{(n-r)!}\left(\frac{1}{\theta_1}\right)^r e^{-\frac{V(t)}{\theta_1}}}{\frac{n!}{(n-r)!}\left(\frac{1}{\theta_0}\right)^r e^{-\frac{V(t)}{\theta_0}}}$$

$$= \left(\frac{\theta_0}{\theta_1}\right)^r e^{-V(t)\left[\frac{1}{\theta_1} - \frac{1}{\theta_0}\right]}$$

Now, it has been shown that for sequential tests, the reject and accept limits, A and B, can be equated to simple functions of α and β. Thus, we obtain the following test procedures:

Continue test: $\frac{\beta}{1-\alpha} \equiv B < \lambda_n < A \equiv \frac{1-\beta}{\alpha}$.

Reject H_0: $\lambda_n > A \equiv \frac{1-\beta}{\alpha}$.

Accept H_0: $\lambda_n < B \equiv \frac{\beta}{1-\alpha}$.

Working with the continue test inequality, we now have

$$\frac{\beta}{1-\alpha} < \left(\frac{\theta_0}{\theta_1}\right)^r e^{-V(t)\left[\frac{1}{\theta_1}-\frac{1}{\theta_0}\right]} < \frac{1-\beta}{\alpha}$$

Taking natural logs of the above inequality, we obtain

$$\ln\left(\frac{\beta}{1-\alpha}\right) < r\ln\left(\frac{\theta_0}{\theta_1}\right) - V(t)\left[\frac{1}{\theta_1}-\frac{1}{\theta_0}\right] < \ln\left(\frac{1-\beta}{\alpha}\right)$$

The above inequality is linear in $V(t)$ and r, and therefore the rejection line $V(t)$, say $V_r(t)$, can be obtained by setting

$$r\ln\left(\frac{\theta_0}{\theta_1}\right) - V_r(t)\left[\frac{1}{\theta_1}-\frac{1}{\theta_0}\right] = \ln\left(\frac{1-\beta}{\alpha}\right)$$

or, equivalently,

$$V_r(t) = \frac{r\ln\left(\frac{\theta_0}{\theta_1}\right)}{\left[\frac{1}{\theta_1}-\frac{1}{\theta_0}\right]} - \frac{\ln\left(\frac{1-\beta}{\alpha}\right)}{\left[\frac{1}{\theta_1}-\frac{1}{\theta_0}\right]}$$

Similarly, the acceptance line $V(t)$), say $V_a(t)$, (see Fig. 3.5) can be obtained by setting

$$r\log\left(\frac{\theta_0}{\theta_1}\right) - V_a(t)\left[\frac{1}{\theta_1}-\frac{1}{\theta_0}\right] = \log\left(\frac{\beta}{1-\alpha}\right)$$

This implies that

Fig 3.5 Test procedure

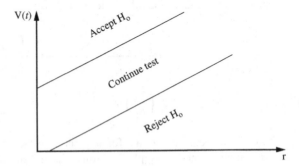

$$V_a(t) = \frac{r \ln\left(\frac{\theta_0}{\theta_1}\right)}{\left[\frac{1}{\theta_1} - \frac{1}{\theta_0}\right]} - \frac{\ln\left(\frac{\beta}{1-\alpha}\right)}{\left[\frac{1}{\theta_1} - \frac{1}{\theta_0}\right]}$$

Thus, continue to perform the test when

$$-h_1 + rs < V(t) < h_0 + rs$$

where

$$h_0 = -\frac{\ln\left(\frac{\beta}{1-\alpha}\right)}{\left[\frac{1}{\theta_1} - \frac{1}{\theta_0}\right]}, \, s = \frac{\ln\left(\frac{\theta_0}{\theta_1}\right)}{\left[\frac{1}{\theta_1} - \frac{1}{\theta_0}\right]}, \, h_1 = \frac{\ln\left(\frac{1-\beta}{\alpha}\right)}{\left[\frac{1}{\theta_1} - \frac{1}{\theta_0}\right]}, \text{ and}$$

$$V(t) = \sum_{i=1}^{r} X_i + \sum_{j=1}^{n-r} t_j$$

Example 3.40
Given that $H_0: \theta = 500$ vs $H_1: \theta = 250$ and $\alpha = \beta = 0.1$. The acceptance and rejection lines are given by

$$V_a(t) = 346.6r + 1098.6$$

and

$$V_r(t) = 346.6r - 1098.6$$

respectively. Both are linear functions in terms of r, the number of first r failures in the test. For an exponential distribution

θ	$P(A)$
0	0
θ_1	β
$\log\left(\frac{\theta_0}{\theta_1}\right) \frac{\log\left(\frac{1-\beta}{\alpha}\right)}{\left[\frac{1}{\theta_1} - \frac{1}{\theta_0}\right]}$	$\frac{1-\beta}{\log\left(\frac{1-\beta}{\alpha}\right) - \log\left(\frac{\beta}{1-\alpha}\right)}$
θ_1	$1-\alpha$
∞	1

From the information given in the above example, we can obtain

$$
\begin{array}{cc}
\theta & P(A) \\
0 & 0 \\
250 & 0.10 \\
346.5 & 0.5 \\
500 & 0.90 \\
\infty & 1.0
\end{array}
$$

Since there is no pre-assigned termination to a regular sequential test, it is customary to draw a curve called the "average sample number" (ASN). This curve shows the expected sample size as a function of the true parameter value. It is known that the test will be terminated with a finite observation. It should be noted that "on average", the sequential tests utilizes significantly smaller samples than fixed sample plans:

$$
\begin{array}{cc}
\theta & E(r) = ASN \\[2mm]
0 & 0 \\[2mm]
\theta_1 & \dfrac{\theta_0\beta \log\left(\frac{\beta}{1-\alpha}\right)+(1-\beta)\log\left(\frac{1-\beta}{\alpha}\right)}{\left[\log\left(\frac{\theta_0}{\theta_1}\right)-\left(\frac{\theta_0-\theta_1}{\theta_1}\right)\right]\theta_1} \\[4mm]
\dfrac{\log\left(\frac{\theta_0}{\theta_1}\right)}{\left[\frac{1}{\theta_1}-\frac{1}{\theta_0}\right]} & \dfrac{\log\left(\frac{1-\beta}{\alpha}\right)\log\left(\frac{\beta}{1-\alpha}\right)}{\left[\log\left(\frac{\theta_0}{\theta_1}\right)\right]^2} \\[4mm]
\theta_0 & \dfrac{\left[(1-\alpha)\log\left(\frac{\beta}{1-\alpha}\right)+\alpha\log\left(\frac{1-\beta}{\alpha}\right)\right]\theta_1}{\left[\log\left(\frac{\theta_0}{\theta_1}\right)-\left(\frac{\theta_0-\theta_1}{\theta_1}\right)\right]\theta_0} \\[4mm]
\infty & 0
\end{array}
$$

An approximate formula for $E(t)$, the expected time to reach decision, is

$$
E(t) \equiv \theta \log\left(\frac{n}{n-E(r)}\right)
$$

where n is the total number of units on test (assuming no replacement of failed units). If replacements are made, then

$$
E(t) = \frac{\theta}{n}E(r)
$$

Occasionally, it is desired to "truncate" a sequential plan such that, if no decision is made before a certain point, testing is stopped and a decision is made on the basis of data acquired up to that point (Pham 2000, page 251). There are a number of rules and theories on optimum truncation. In the reliability community, a $V(t)$ truncation point at $10\theta_0$ is often used to determine the $V(t)$ and r lines for truncation and the corresponding exact $\alpha = \beta$ errors (these will in general be larger for truncated tests than for the non-truncated). An approximate method draws the $V(t)$ truncation line to the center of the continue test band and constructs the r truncation line perpendicular to that point.

3.13.1.1 Bernoulli Distribution Case

$$H_0: p = p_0 \text{ vs } H_1: p = p_1.$$

where p denotes the proportion defective units in the population, and α and β are pre-determined. Let us define the following random variable X:

$$X = \begin{cases} 0 & \text{if good} \\ 1 & \text{if defective} \end{cases}$$

where p denotes the proportion defective items in the population. The Bernoulli distribution is given by

$$P(x) = p^x (1-p)^{1-x} \quad \text{for } x = 0, 1$$

The sequential probability ratio is given by

$$\lambda_n = \frac{\prod_{i=1}^{n} P(x_i, p_1)}{\prod_{i=1}^{n} P(x_i, p_0)} = \frac{\prod_{i=1}^{n} p_1^{x_i}(1-p_1)^{1-x_i}}{\prod_{i=1}^{n} p_0^{x_i}(1-p_0)^{1-x_i}} = \left(\frac{p_1}{p_0}\right)^{\sum_{i=1}^{n} x_i} \left(\frac{1-p_1}{1-p_0}\right)^{n-\sum_{i=1}^{n} x_i}$$
$$= \left(\frac{p_1}{p_0}\right)^{d_n} \left(\frac{1-p_1}{1-p_0}\right)^{n-d_n}$$

where $d_n = \sum_{i=1}^{n} x_i$ is the cumulative number of defective units and n is the cumulative sample size. As we know, sampling process will continue as long as

$$\frac{\beta}{1-\alpha} < \lambda_n < \frac{1-\beta}{\alpha}$$

Taking the natural log, we have

$$\ln\left(\frac{\beta}{1-\alpha}\right) < \ln(\lambda_n) < \ln\left(\frac{1-\beta}{\alpha}\right)$$

or equivalently that,

$$\ln\left(\frac{\beta}{1-\alpha}\right) < d_n \ln\left(\frac{p_1}{p_0}\right) + (n-d_n)\ln\left(\frac{1-p_1}{1-p_0}\right) < \ln\left(\frac{1-\beta}{\alpha}\right)$$

After simplifications, we can obtain

$$s_b n + b_1 < d_n < s_b n + b_0 \tag{3.88}$$

where
$$s_b = -\frac{\log\left(\frac{1-p_1}{1-p_0}\right)}{\log\left(\frac{p_1(1-p_0)}{p_0(1-p_1)}\right)}, \quad b_1 = \frac{\log\left(\frac{\beta}{1-\alpha}\right)}{\log\left(\frac{p_1(1-p_0)}{p_0(1-p_1)}\right)}, \quad b_0 = \frac{\log\left(\frac{1-\beta}{\alpha}\right)}{\log\left(\frac{p_1(1-p_0)}{p_0(1-p_1)}\right)}. \text{ Thus the acceptance line}$$
and rejection line are, respectively, as follows:

$$\text{Acceptance line} = s_b n + b_1 \tag{3.89}$$

and

$$\text{Rejection line} = s_b n + b_0. \tag{3.90}$$

Similar tests can be constructed for other distribution parameters following the same general scheme.

Example 3.41
The following data were drawn one observation at a time in the order records:
g b g g b g b g b g g b g b g.

where b denotes a defective item and g denotes a good item. The experiment was performed to test the following hypothesis:
H_0: $p = p_0 = 0.10$ versus H_1: $p = p_1 = 0.2$

where p denotes the proportion defective items in the population. It is desired to reject H_0 when it is true with probability 0.05 and to accept H_0 when H_1 is true with probability 0.20. Using the sequential testing plan, we wish to determine whether we would accept or reject a lot on the basis of the observations above.

Given $p_0 = 0.10$, $p_1 = 0.20$, $\alpha = 0.05$, $\beta = 0.20$. From Eq. (3.88) we have

$$s_b = -\frac{\ln\left(\frac{1-p_1}{1-p_0}\right)}{\ln\left(\frac{p_1(1-p_0)}{p_0(1-p_1)}\right)} = -\frac{\ln\left(\frac{1-0.20}{1-0.10}\right)}{\ln\left(\frac{0.2(1-0.10)}{0.10(1-0.20)}\right)} = 0.1452$$

$$b_1 = \frac{\ln\left(\frac{\beta}{1-\alpha}\right)}{\ln\left(\frac{p_1(1-p_0)}{p_0(1-p_1)}\right)} = \frac{\ln\left(\frac{0.2}{1-0.05}\right)}{\ln\left(\frac{0.2(1-0.1)}{0.1(1-0.2)}\right)} = -1.9214,$$

$$b_0 = \frac{\ln\left(\frac{1-\beta}{\alpha}\right)}{\ln\left(\frac{p_1(1-p_0)}{p_0(1-p_1)}\right)} = \frac{\ln\left(\frac{1-0.2}{0.05}\right)}{\ln\left(\frac{0.2(1-0.1)}{0.1(1-0.2)}\right)} = 3.4190,$$

Thus,

$$s_b n + b_1 < d_n < s_b n + b_0$$

Table 3.13 The sampling procedure

n	x_n	d_n	Acceptance #	Rejection #
1	0	0	–	–
2	1	1	–	–
3	0	1	–	–
4	0	1	–	–
5	1	2	–	5
6	0	2	–	5
7	1	3	–	5
8	0	3	–	5
9	1	4	–	5
10	0	4	–	5
11	0	4	–	6
12	1	5	–	6
13	0	5	–	6
14	1	6	0	6
15	0	6	0	6

$$0.1452n - 1.9214 < d_n < 0.1452n + 3.4190$$

The acceptance line is: $0.1452n - 1.9214$.

The rejection line is: $0.1452n + 3.4190$.

From Table 3.13, the cumulative number of defective units $d_{14} = 6$ falls above the rejection line when $n = 14$. Therefore, the sampling is stopped at the 14th observation and that the lot would be rejected. That is, reject the hypothesis H_0: p = 0.10.

3.14 Bayesian Methods

The Bayesian approach to statistical inference is based on a theorem first presented by the Reverend Thomas Bayes. To demonstrate the approach, let X have a pdf $f(x)$, which is dependent on θ. In the traditional statistical inference approach, θ is an unknown parameter, and hence, is a constant. We now describe our prior belief in the value of θ by a pdf $h(\theta)$. This amounts to quantitatively assessing subjective judgment and should not be confused with the so-called objective probability assessment derived from the long-term frequency approach. Thus, θ will now essentially be treated as a random variable θ with pdf $h(\theta)$.

Consider a random sample $X_1, X_2, \ldots, X_n$ from $f(x)$ and define a statistic Y as a function of this random sample. Then there exists a conditional pdf $g(y \mid \theta)$ of Y for a given θ. The joint pdf for y and θ is

$$f(\theta, y) = h(\theta)g(y|\theta)$$

If θ is continuous, then

$$f_1(y) = \int_\theta h(\theta)g(y|\theta)d\theta$$

is the marginal pdf for the statistic y. Given the information y, the conditional pdf for θ is

$$k(\theta|y) = \frac{h(\theta)g(y|\theta)}{f_1(y)} \quad \text{for } f_1(y) > 0$$
$$= \frac{h(\theta)g(y|\theta)}{\int_\theta h(\theta)g(y|\theta)d\theta}$$

If θ is discrete, then

$$f_1(y) = \sum_k P(\theta_k)P(y|\theta_k)$$

and

$$P(\theta_i|y_i) = \frac{P(\theta_k)P(y_i|\theta_i)}{\sum_k P(\theta_k)P(y_j|\theta_k)}$$

where $P(\theta_j)$ is a prior probability of event θ_i and $P(\theta_j|y_j)$ is a posterior probability of event y_j given θ_i. This is simply a form of Bayes' theorem. Here, $h(\theta)$ is the prior pdf that expresses our belief in the value of θ before the data $(Y = y)$ became available. Then $k(\theta \mid y)$ *is* the posterior pdf of given the data $(Y = y)$.

Note that the change in the shape of the prior pdf $h(\theta)$ to the posterior pdf $k(\theta \mid y)$ due to the information is a result of the product of $g(y \mid \theta)$ and $h(\theta)$ because $f_1(y)$ is simply a normalization constant for a fixed y The idea in reliability is to take "prior" data and combine it with current data to gain a better estimate or confidence interval or test than would be possible with either singularly. As more current data is acquired, the prior data is "washed out" (Pham 2000).

Case 1: Binomial Confidence Limits—Uniform Prior. Results from ten missile tests are used to form a one-sided binomial confidence interval of the form

$$P[R \geq R_L] = 1 - \alpha$$

From subsection 3.11.3, we have

Table 3.14 Lower limits as a function of the number of missile test successes

K	RL	Exact level
10	0.79	0.905
9	0.66	0.904
8	0.55	0.900
7	0.45	0.898
6	0.35	0.905

$$\sum_{i=k}^{10} \binom{10}{i} R_L^i (1 - R_L)^{10-i} = \alpha$$

Choosing $\alpha = 0.1$, lower limits as a function of the number of missile test successes are shown in Table 3.14. Assume from previous experience that it is known that the true reliability of the missile is somewhere between 0.8 and 1.0 and furthermore that the distribution through this range is uniform. The prior density on R is then.

$g(R) = 5 \ 0.8 < R < 1.0$

From the current tests, results are k successes out of ten missile tests, so for the event A that contained k successes:

$$P(A|R) = \binom{10}{k} R^k (1 - R)^{10-k}$$

Applying Bayes' theorem, we obtain

$$g(R|A) = \frac{g(R)P(A|R)}{\int_R g(R)P(A|R)dR}$$

$$= \frac{5\binom{10}{k} R^k (1 - R)^{10-k}}{\int_{0.8}^{1.0} 5\binom{10}{k} R^k (1 - R)^{10-k}dR}$$

For the case of $k = 10$,

$$g(R|A) = \frac{R^{10}}{\int_{0.8}^{1.0} R^{10}dR}$$

$$\frac{11 \ R^{10}}{0.914} = 12.035 \ R^{10}$$

To obtain confidence limits incorporating the "new" or current data,

Table 3.15 Comparison
between limits applying the
Bayesian method and those
that do not

K	R_L (uniform [0.8,1] prior)	R_L (no prior)	Exact level
10	0.855	0.79	0.905
9	0.822	0.66	0.904
8	0.812	0.55	0.900
7	0.807	0.45	0.898
6	0.805	0.35	0.905

$$\int_{R_L}^{1.0} g(A|R)dR = 0.9$$

$$\int_{R_L}^{1.0} 12.035 R^{10} dR = 0.9$$

After simplifications, we have

$$R_L^{11} = 0.177, \quad R_L = 0.855.$$

Limits for the 10/10, 9/10, 8/10, 7/10, and 6/10 cases employing the Bayesian method are given in Table 3.15 along with a comparison with the previously calculated limits not employing the prior assumption. Note that the lower limit of 0.8 on the prior cannot be washed out.

Case 2: Binomial Confidence Limits—Beta Prior. The prior density of the beta function is

$$g(R) = \frac{(\alpha + \beta + 1)!}{\alpha! \, \beta!} R^{\alpha} (1 - R)^{\beta}$$

The conditional binomial density function is

$$P(A|R) = \binom{10}{i} R^i (1 - R)^{10-i}$$

Then we have

$$g(R|A) = \frac{\frac{(\alpha+\beta+1)!}{\alpha! \, \beta!} R^{\alpha} (1 - R)^{\beta} \binom{10}{k} R^k (1 - R)^{10-k}}{\frac{(\alpha+\beta+1)!}{\alpha! \, \beta!} \int_0^1 R^{\alpha} (1 - R)^{\beta} \binom{10}{k} R^k (1 - R)^{10-k} dR}.$$

After simplifications, we obtain

$$g(R|A) = \frac{R^{\alpha+k}(1-R)^{\beta+10-k}}{\int_0^1 R^{\alpha+k}(1-R)^{\beta+10-k}dR}.$$

Multiplying and dividing by

$$\frac{(\alpha+\beta+11)!}{(\alpha+\beta)!(\beta+10-k)!}$$

puts the denominator in the form of a beta function with integration over the entire range, and hence, equal to 1. Thus,

$$g(R|A) = \binom{\alpha+\beta+10}{\alpha+k} R^{\alpha+k}(1-R)^{\beta+10-k}$$

which again is a beta density function with parameters

$$(\alpha+k) = \alpha' \text{ and } (\beta+10-k) = \beta'$$

Integration over $g(R|A)$ from R_L to 1.0 with an integral set to $1-\alpha$ and a solution of R_L will produce $100(1-\alpha)\%$ lower confidence bounds on R, that is,

$$\int_{R_L}^{1.0} g(R|A)dR = 1-\alpha.$$

Case 3: Exponential Confidence Limits—Gamma Prior. For this situation, assume interest is in an upper limit on the exponential parameter X. The desired statement is of the form

$$p[\lambda < \lambda_U] = 1-\alpha$$

If 1000 h of test time was accrued with one failure, a 90% upper confidence limit on λ would be

$$p[\lambda < 0.0039] = 0.9$$

From a study of prior data on the device, assume that λ has a gamma prior density of the form

$$g(\lambda) = \frac{\lambda^{n-1} e^{-\frac{\lambda}{\beta}}}{(n-1)! \beta^n}.$$

With an exponential failure time assumption, the current data in terms of hours of test and failures can be expressed as a Poisson, thus,

$$p(A|\lambda) = \frac{(\lambda T)^n \, e^{-\lambda T}}{r!}$$

where $n =$ number of failures, $T =$ test time, and $A =$ event which is r failures in T hours of test. Applying Bayes' results, we have

$$g(\lambda|A) = \frac{\frac{\lambda^{n-1} e^{-\frac{\lambda}{\beta}}}{(n-1)! \, \beta^n} \frac{(\lambda T)^r e^{-\lambda T}}{r!}}{\int_{\lambda=0}^{\infty} \frac{\lambda^{n-1} e^{-\frac{\lambda}{\beta}}}{(n-1)! \, \beta^n} \frac{(\lambda T)^r e^{-\lambda T}}{r!} d\lambda}$$

$$= \frac{\lambda^{n+r-1} e^{-\lambda\left(\frac{1}{\beta}+T\right)}}{\int_0^{\infty} \lambda^{n+r-1} e^{-\lambda\left(\frac{1}{\beta}+T\right)} d\lambda}.$$

Note that

$$\int_0^{\infty} \lambda^{n+r-1} e^{-\lambda\left(\frac{1}{\beta}+T\right)} d\lambda = \frac{(n+r-1)!}{\left(\frac{1}{\beta}+T\right)^{n+r}}$$

Hence,

$$g(\lambda|A) = \frac{\lambda^{n+r-1} e^{-\lambda\left(\frac{1}{\beta}+T\right)} \left(\frac{1}{\beta}+T\right)^{n+r}}{(n+r-1)!}$$

Thus, $g(\lambda|A)$ is also a gamma density with parameters $(n+r-1)$ and $\frac{1}{\left(\frac{1}{\beta}+T\right)}$. This density can be transformed to the χ^2 density with $2(n+r)$ degree of freedom by the following change of variable. Let

$$\lambda' = 2\lambda \left(\frac{1}{\beta}+T\right)$$

then

$$d\lambda = \frac{1}{2} \left(\frac{1}{\frac{1}{\beta}+T}\right) d\lambda'$$

We have

$$h(\lambda'|A) = \frac{(\lambda')^{\frac{2(n+r)}{2}-1} e^{-\frac{\lambda'}{2}}}{\left[\frac{2(n+r)}{2} - 1\right]! \, 2^{\frac{2(n+r)}{2}}}$$

To obtain a $100(1-\alpha)\%$ upper confidence limit on λ, solve for λ' in the integral

$$\int_0^{\lambda'} h(s|A)ds = 1 - \alpha$$

and convert λ' to λ via the above transformation.

Example 3.42

Given a gamma prior with $n = 2$ and $\beta = 0.0001$ and current data as before (i.e., 1000 h of test with one failure), the posterior density becomes.

$$g(\lambda|A) = \frac{\lambda^2 \, e^{-\lambda(11,000)}(11,000)^2}{2}$$

converting to χ^2 via the transformation $\lambda' = 2\lambda$ (11,000) and

$$h(\lambda'|A) = \frac{(\lambda')^{\frac{6}{2}-1}e^{-\frac{\lambda'}{2}}}{[\frac{6}{2} - 1]! \, 2^{\frac{6}{2}}}$$

which is χ^2 with six degrees of freedom. Choosing $\alpha = 0.1$ then

$$\chi^2_{6,1-\alpha} = \chi^2_{6,0.9} = 10.6$$

and

$$p[\lambda' < 10.6] = 0.9$$

But $\lambda' = 2\lambda(11,000)$, hence,

$$p\left[\lambda' = 2\lambda(11,000) < 10.6\right] = 0.9$$

or

$$p[\lambda < 0.0005] = 0.9$$

The latter limit conforms to 0.0039 derived without the use of a prior density, i.e., an approximate eight fold improvement. The examples above involved the development of tighter confidence limits where a prior density of the parameter could be utilized.

In general, for legitimate applications and where prior data are available, employment of Bayesian methods can reduce cost or give results with less risk for the same dollar value (Pham 2000).

3.15   Statistical Model Selection

In this section we discuss some common criteria that can be used for distribution and model selections (Pham 2014, 2019,2020). Let y_i be the observed data value and $\hat{y}_i$ is the fitted value from the fit for $i = 1,2 ,\ldots,n$; and n and k are the number of observations and number of estimated parameters, respectively.

Define the sum of squared error (SSE) and total sum of squares (SST) as follows:

$$SSE = \sum_{i=1}^{n} \left(y_i - \hat{y}_i\right)^2 \tag{3.91}$$

and

$$SST = \sum_{i=1}^{n} (y_i - \bar{y})^2$$

where

$$\bar{y} = \frac{\sum_{i=1}^{n} y_i}{n}.$$

Mean Squared Error (MSE). The mean squared error (MSE) measures the total deviation of the response values from the fitted response values and is defined as:

$$MSE = \frac{\sum_{i=1}^{n} \left(y_i - \hat{y}_i\right)^2}{n - k} = \frac{SSE}{n - k} \tag{3.92}$$

where y_i is the observed data value; $\hat{y}_i$ is the fitted value from the fit for $i = 1,2 ,\ldots,n$; and n and k are the number of observations and number of estimated parameters, respectively. SSE is the sum of squared error. Note that MSE considers the penalty term with respect to the degrees of freedom when there are many parameters and consequently assigns a larger penalty to a model with more parameters. The smaller the value MSE value, the better the model fit.

Root Mean Squared Error (RMSE). The RMSE is the square root of the variance of the residuals or the MSE and is given as

$$RMSE = \sqrt{MSE} = \sqrt{\frac{SSE}{n - k}} \tag{3.93}$$

The RMSE indicates the absolute fit of the model to the data–how close the observed data points are to the model's predicted values. Lower values of RMSE indicate better fit.

Coefficient of Determinations (R^2). R^2 is the square of the correlation between the response values and the predicted response values. It measures the amount of variation accounted for the fitted model. The R^2 is defined as

$$R^2 = 1 - \frac{\text{SSE}}{\text{SST}} \tag{3.94}$$

R^2 assumes every independent variable in the model explains the variation in the dependent variable. It gives the percentage of explained variation as if all independent variables in the model affect the dependent variable. It is always increase with model size. The larger R^2, the better is the model's performance.

Adjusted R^2. Adjusted R^2 takes into account the number of estimated parameters in the model and is defined as:

$$R^2_{adj} = 1 - \left(\frac{n-1}{n-k}\right)(1 - R^2) \tag{3.95}$$

where n and k are the number of observations and number of estimated parameters, respectively. The adjusted R^2 gives the percentage of variation explained by only those independent variables that actually affect the dependent variable. The larger adjusted R^2, the better is the model's goodness-of-fit.

The predictive-ratio risk (PRR). PRR measures the total deviation of the response values from the fit to the response values against the fitted values, and is defined as (Pham 2006; Pham and Deng 2003):

$$PRR = \sum_{i=1}^{n} \left(\frac{y_i - \hat{y}_i}{\hat{y}_i}\right)^2 \tag{3.96}$$

The predictive-power (PP). PP measures the total deviation of the response values from the fit to the response values against the response values, and is defined as follows:

$$PP = \sum_{i=1}^{n} \left(\frac{y_i - \hat{y}_i}{y_i}\right)^2 \tag{3.97}$$

For all these three criteria—MSE, PRR, and PP—the smaller the value, the better the model fits.

Normalized-Rank Euclidean Distance criterion (RED). In general, let s denotes the number of models with d criteria, C_{ij1} represents the ranking based on specified criterion of model i with respect to (w.r.t.) criteria j, and C_{ij2} the criteria value of model i w.r.t. criteria j where $i = 1, 2, \ldots, s$ and $j = 1, 2, \ldots, d$. The normalized-rank Euclidean distance value, D_i, for $i = 1, 2, \ldots, s$, measures the distance of the two-dimensional normalized criteria from the origin for ith model using Euclidean distance function (Pham 2019). The RED criteria function is defined as follows:

$$D_i = \sum_{j=1}^{d} \left\{ \left(\sqrt{\left[\sum_{k=1}^{2} \left(\frac{C_{ijk}}{\sum_{i=1}^{s} C_{ijk}} \right)^2 \right]} \right) w_j \right\} \qquad (3.98)$$

where s = total number of models.

d = total number of criteria.

w_j = the weight of the jth criteria for $j = 1,2,...,d$

$$k = \begin{cases} 1 \text{ represent criteria j value} \\ 2 \text{ represent criteria j ranking.} \end{cases}$$

Thus, the smaller the RED value, D_i, it represents the better rank as compare to higher RED value.

Akaike Information Criterion (AIC). The AIC is one of the most common criteria has been used in past years to choose the best model from a set of candidate models. It is defined as (Akaike 1973):

$$AIC = -2 \log(L) + 2k \qquad (3.99)$$

where L is the maximum value of the likelihood function for the model and k is the number of estimated parameters in the model. To select the best model, practitioners should choose the model that minimizes AIC. In other words, the lower value of AIC indicates better goodness-of-fit. By adding more parameters in the model, it improves the goodness of the fit but also increases the penalty imposed by adding more parameters.

Bayesian Information Criterion (BIC). The BIC was introduced by Schwarz (1978) also known as Schwarz criterion, and is defined as

$$BIC = -2 \log(L) + k \log(n) \qquad (3.100)$$

As can be seen from Eqs. (3.5) and (3.6), the difference between these two criteria, BIC and AIC, is only in the second term which depends on the sample size n that show how strongly they impacts the penalty of the number of parameters in the model. With AIC the penalty is $2\ k$ whereas with BIC the penalty is $k.\log(n)$ that penalizes large models. As n increases, the BIC tends to favor simpler models than the AIC as k smaller. This implies when $n > 8$, $k\cdot\log(n)$ exceeds $2\ k$.

Second Order Information Criterion (AICc). When the sample size is small, there is likely that AIC will select models that include many parameters. The second order information criterion, often called AICc, takes into account sample size by increasing the relative penalty for model complexity with small data sets. It is defined as:

$$AIC_c = -2 \log(L) + 2k + \frac{2k(k+1)}{n-k-1} \qquad (3.101)$$

where n denotes the sample size and k denotes the number of estimated parameters. As n gets larger, AICc converges to AIC so there's really no harm in always using AICc regardless of sample size.

Pham's information criterion (PIC). In general, the adjusted R^2 attaches a small penalty for adding more variables in the model. The difference between the adjusted R^2 and R^2 is usually slightly small unless there are too many unknown coefficients in the model to be estimated from too small a sample in the presence of too much noise. Pham (2019) presents a criterion, called Pham's information criterion (PIC), by taking into account a larger the penalty when adding too many coefficients in the model when there is too small a sample, is as follows:

$$\text{PIC} = \text{SSE} + k\left(\frac{n-1}{n-k}\right) \tag{3.102}$$

where n is the number of observations in the model, k is the number of estimated parameters or $(k-1)$ explanatory variables in the model, and SSE is the sum of squared error as given in Eq. (3.91).

Pham's criterion (PC). Pham (2020) recently introduces a criterion, called Pham's criterion (PC), that measures the tradeoff between the uncertainty in the model and the number of parameters in the model by slightly increasing the penalty when each time adding parameters in the model when there is too small a sample. The criteria is as follows (Pham 2020a):

$$PC = \left(\frac{n-k}{2}\right)\log\left(\frac{SSE}{n}\right) + k\left(\frac{n-1}{n-k}\right)$$
$$\text{where } SSE = \sum_{i=1}^{n}\left(y_i - \hat{y}_i\right)^2 \tag{3.103}$$

Table 3.16 presents a summary of criteria for model selection.

3.15.1 Applications

In this section we demonstrate a couple of real applications such as advertising budget products and heart blood pressure health using multiple regression models to illustrate the model selection.

Application 1: Advertising Budget Analysis.

In this application, we use the advertising budget data set [Advertising] to illustrate the model selection criteria where the sales for a particular product is a dependent variable of multiple regression and the three different media channels such as TV, Radio, and Newspaper are independent variables as shown in Table 3.17. The advertising dataset consists of the sales of a product in 200 different markets (200 rows),

Table. 3.16 A summary of some criteria model selection (Pham 2020a)

No	Criteria	Formula	Brief description
1	SSE	$$SSE = \sum_{i=1}^{n} (y_i - \hat{y}_i)^2$$	Measures the total deviations between the estimated values and the actual data
2	MSE	$$MSE = \frac{\sum_{i=1}^{n} (y_i - \hat{y}_i)^2}{n-k}$$	Measures the difference between the estimated values and the actual data
3	RMSE	$$RMSE = \sqrt{\frac{\sum_{i=1}^{n} (y_i - \hat{y}_i)^2}{n-k}}$$	The square root of the MSE
4	R^2	$$R^2 = 1 - \frac{\sum_{i=1}^{n} (y_i - \hat{y}_i)^2}{\sum_{i=1}^{n} (y_i - \bar{y})^2}$$	Measures the amount of variation accounted for the fitted model
5	Adj R^2	$$R^2_{adj} = 1 - \left(\frac{n-1}{n-k}\right)\left(1 - R^2\right)$$	Take into account a small penalty for adding more variables in the model
6	AIC	$$AIC = -2\log(L) + 2k$$	Measure the goodness of the fit considering the penalty of adding more parameters
7	BIC	$$BIC = -2\log(L) + k\log(n)$$	Same as AIC but the penalty term will also depend on the sample size
8	AICc	$$AIC_c = -2\log(L) + 2k + \frac{2k(k+1)}{n-k-1}$$	AICc takes into account sample size by increasing the relative penalty for model complexity with small data sets
9	PIC	$$PIC = SSE + k\left(\frac{n-1}{n-k}\right)$$ where $$SSE = \sum_{i=1}^{n} (y_i - \hat{y}_i)^2$$	Take into account a larger the penalty when there is too small a sample but too many parameters in the model
10	PRR	$$PRR = \sum_{i=1}^{n} \left(\frac{\hat{m}(t_i) - y_i}{\hat{m}(t_i)}\right)^2$$	Measures the distance of model estimates from the actual data against the model estimate
11	PP	$$PP = \sum_{i=1}^{n} \left(\frac{\hat{m}(t_i) - y_i}{y_i}\right)^2$$	Measures the distance of model estimates from the actual data against the actual data

(continued)

Table. 3.16 (continued)

No	Criteria	Formula	Brief description
12	PC	$$PC = \left(\frac{n-k}{2}\right) \log\left(\frac{SSE}{n}\right) + k\left(\frac{n-1}{n-k}\right)$$ where $SSE = \sum_{i=1}^{n} (y_i - \hat{y}_i)^2$	Increase slightly the penalty each time adding parameters in the model when there is too small a sample

Table 3.17 Advertising Budget Data in 200 different markets

TV	Radio	Newspaper	Sales
230.1	37.8	69.2	22.1
44.5	39.3	45.1	10.4
17.2	45.9	69.3	9.3
151.5	41.3	58.5	18.5
180.8	10.8	58.4	12.9
8.7	48.9	75	7.2
57.5	32.8	23.5	11.8
120.2	19.6	11.6	13.2
8.6	2.1	1	4.8

together with advertising budgets for the product in each of those markets for three different media channels: TV, Radio and Newspaper. The sales are in thousands of units and the budget is in thousands of dollars. Table 3.17 shows just the first few rows of the advertising budget data set.

Pham (2019) discusses the results of the linear regression model using this advertising data. Figures 3.6 and 3.7 present the data plot and the correction coefficients between the pairs of variables of the advertising budget data, respectively. It shows that the pair of Sales and TV advertising have the highest positive correlation. From Table 3.18, the values of R^2 with all three variables and just two variables (TV and Radio advertisings) in the model are the same. This implies that we can select the model with two variables (TV and Radio) in the regression (Pham 2019). Based on the PIC criterion, the model with the two advertising media channels (TV and Radio) is the best model from a set of seven candidate models as shown in Table 3.18.

Application 2: Heart Blood Pressure Health Analysis.

The heart blood pressure health data (Pham 2019) consists of the heart rate (pulse), systolic blood pressure and diastolic blood pressure in 86 days with 2 data points measured each day (172 rows) is used in this application. Blood pressure (BP) is one of the main risk factors for cardiovascular diseases. BP is the force of blood pushing against your artery walls as it goes through your body [Blood pressure]. The Systolic BP is the pressure when the heart beats—while the heart muscle is contracting (squeezing) and pumping oxygen-rich blood into the blood vessels. Diastolic BP is

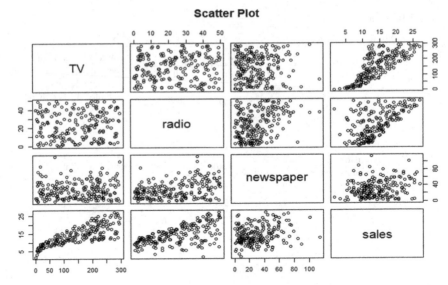

Fig. 3.6 The scatter plot of the advertising data with four variables

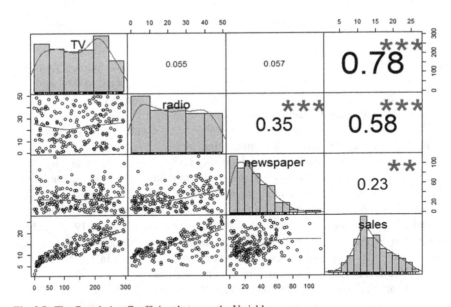

Fig. 3.7 The Correlation Coefficient between the Variables

the pressure on the blood vessels when the heart muscle relaxes. The diastolic pressure is always lower than the systolic pressure [Systolic]. The Pulse or Heart rate measures the heart rate by counting the number of beats per minute (BPM). The first few rows of the data set are shown in Table 3.19. In the table, the first row of

Table 3.18 Criteria values of independent variables (TV, Radio, Newspaper) of regression models (X_1, X_2, and X_3 be denoted as the TV, radio and newspaper, respectively)

Criteria	X_1, X_2, X_3	X_1, X_2	X_1, X_3	X_2, X_3	X_1	X_2	X_3
MSE	2.8409	2.8270	9.7389	18.349	10.619	18.275	25.933
AIC	782.36	780.39	1027.8	1154.5	1044.1	1152.7	1222.7
AICc	782.49	780.46	1027.84	1154.53	1044.15	1152.74	1222.73
BIC	795.55	790.29	1037.7	1164.4	1050.7	1159.3	1229.3
RMSE	1.6855	1.6814	3.1207	4.2836	3.2587	4.2750	5.0925
R^2	0.8972	0.8972	0.6458	0.3327	0.6119	0.3320	0.0521
Adjusted R^2	0.8956	0.8962	0.6422	0.3259	0.6099	0.3287	0.0473
PIC	5.7467	**4.7118**	6.1512	7.3141	5.2688	6.2850	7.1026

Table 3.19 Sample heart blood pressure health data set of an individual in 86 day interval (Pham 2019)

Day	Time	Systolic	Diastolic	Pulse
5	0	154	99	71
	1	144	94	75
6	0	139	93	73
6	1	128	76	85
7	0	129	73	78
7	1	125	65	74
1	0	129	80	70
1	1	130	83	72
2	0	144	83	74
2	1	124	87	84
3	0	120	77	73
3	1	124	70	80

the data set can be read as follows: on a Thursday morning, the high blood, low blood and heart rate measurements were 154, 99, and 71, respectively. Similarly, on a Thursday afternoon (i.e., the second row of the data set in Table 3.19), the high blood, low blood and heart rate measurements were 144, 94, and 75, respectively.From Fig. 3.8, the systolic BP and diastolic BP have the highest correlation. The model with only Systolic blood pressure variable seems to be the best model from the set of seven candidate models based on PIC and BIC criteria as shown in Table 3.20.

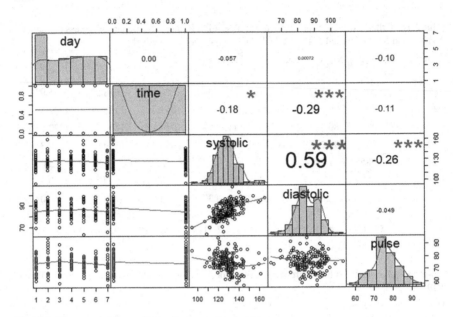

Fig. 3.8 The correlation coefficient between the variables

Table 3.20 Criteria values of variables (day, systolic, diastolic) of regression models (X_1, X_2, and X_3 be denoted as the day, systolic, diastolic, respectively)

Criteria	X_1, X_2, X_3	X_1, X_2	X_1, X_3	X_2, X_3	X_1	X_2	X_3
MSE	43.1175	43.7381	47.0101	43.5859	46.8450	44.1352	47.2311
AIC	1141.463	1142.942	1155.351	1142.342	1153.76	1143.511	1155.172
BIC	1154.053	1152.384	1164.793	1151.784	1160.055	1149.806	1161.467
RMSE	6.5664	6.6135	6.8564	6.6020	6.8443	6.6434	6.8725
R^2	0.09997	0.0816	0.01287	0.08477	0.0105	0.0678	0.00236
Adj R^2	0.08389	0.0707	0.00119	0.07394	0.00469	0.0623	−0.00351
PIC	10.6378	9.6490	9.8919	9.6375	8.8561	**8.6552**	8.8843

3.16   Problems

1. Let $X_1, X_2, \ldots, X_n$ represent a random sample from the Poisson distribution having pdf

    $f(x; \lambda) = \frac{e^{-\lambda}\lambda^x}{x!}$ for x = 0, 1, 2,... and $\lambda \geq 0$.

    Find the maximum likelihood estimator $\hat{\lambda}$ of λ.

2. Let $X_1, X_2, \ldots, X_n$ be a random sample from the distribution with a discrete pdf

    $$P(x) = p^x(1-p)^{1-x} \quad x = 0, 1 \text{ and } 0 < p < 1$$

where p is the parameter to be estimated. Find the maximum likelihood estimator $\hat{p}$ of p.

3. Assume that $X_1, X_2, \ldots, X_n$ represent a random sample from the Pareto distribution, that is,

$$F(x; \lambda, \theta) = 1 - \left(\frac{\lambda}{x}\right)^{\theta} \quad \text{for } x \geq \lambda, \; \lambda > 0, \; \theta > 0$$

This distribution is commonly used as a model to study incomes. Find the maximum likelihood estimators of λ and θ.

4. Let $Y_1 < Y_2 < \ldots < Y_n$ be the order statistics of a random sample $X_1, X_2, \ldots, X_n$ from the distribution with pdf

$$f(x; \theta) = 1 \quad \text{if } \theta - \frac{1}{2} \leq x \leq \theta + \frac{1}{2}, \; -\infty < \theta < \infty$$

Show that any statistic $h(X_1, X_2, \ldots, X_n)$ such that

$$Y_n - \frac{1}{2} \leq h(X_1, X_2, \ldots, X_n) \leq Y_1 + \frac{1}{2}$$

is a maximum likelihood estimator of θ. What can you say about the following functions?

$$\text{(a)} \quad \frac{(4Y_1 + 2Y_n + 1)}{6}$$

$$\text{(b)} \quad \frac{(Y_1 + Y_n)}{2}$$

$$\text{(c)} \quad \frac{(2Y_1 + 4Y_n - 1)}{6}$$

5. The lifetime of transistors is assumed to have an exponential distribution with pdf

$$f(t; \theta) = \frac{1}{\theta} e^{-\frac{t}{\theta}} \quad \text{for } t \geq 0, \; \theta > 0$$

A random sample of size n is observed. Determine the following:

(a) The maximum likelihood estimator of θ.
(b) The MLE of the transistor reliability function, $\hat{R}(t)$, of

$$R(t) = e^{-\frac{t}{\theta}}$$

6. Suppose that $X_1, X_2,..., X_n$ are independent random variable, each with the uniform distribution on $[c - d, c + d]$ where c is unknown and d is known $(-\infty < c < \infty, d > 0)$. Find the maximum likelihood estimator of c.

7. Suppose that $X_1, X_2,..., X_n$ are independent random variable, each with the uniform distribution on $[c - d, c + d]$ where c and d are both unknown $(-\infty < c < \infty, d > 0)$.

   (a) Find the maximum likelihood estimators of c and d.
   (b) Given the following failure time data: 20, 23, 25, 26, 28, 29, 31, 33, 34, and 35 days. Assuming that the data follow a uniform distribution on $[c - d, c + d]$, use the maximum likelihood method to obtain the unknown parameters, c and d, of the uniform distribution.

8. Suppose that $X_1, X_2, ..., X_n$ are independent random variable, each with the uniform distribution on $[-d, d]$ where d is positive and unknown. Find the maximum likelihood estimator of d.

9. Suppose on five consecutive days in a given month the number of customers who enter services at a printing shop were 48, 60, 78, 56, and 73 Test the null hypothesis that the expected numbers of customers per day were the same on the five days at the 5% level of significance using the Chi-square test.

10. Suppose that a die is rolled 120 times and the number of times each face comes up is recorded. The following results are obtained:

Face	1	2	3	4	5	6
n_i	15	21	20	15	26	23

Test whether the die is fair at the 5% level of significance using the Chi-square test.

11. Suppose that a die is rolled 250 times and the number of times each face comes up is recorded. The following results are obtained:

Face	1	2	3	4	5	6
n_i	45	37	60	55	29	24

Test whether the die is fair at the 5% level of significance using the Chi-square test.

12. The proportion of components produced by a manufacturing process from last year is as follows: 3% scrapped, 6% reworked, and 91% acceptable. This year, inspection of 500 units showed that 20 units must be scrapped and 25 units can be reworked. Can we say that the results this year are consistent with the last year data at the 5% level of significance using the Chi-square test.

13. An engineer obtained the following data that shows the cycle time in hours for the assembly of a certain electronic product:

Cycle time (hours)	Frequency
2.45	4
2.55	6
2.58	15
2.64	8
2.73	2
2.80	29
2.85	7
2.92	13
2.98	19
3.20	12

The engineer concludes that these data might represent like a sample from a normal population at the 5% level of significance. What is your opinion. Is the engineer correct? (Hint: using the KS test).

14. 27 units were placed on life test and the test was run until all units failed. The observed failure times $t_1, t_2, \ldots, t_{27}$ were give below:

4.6	12.7	20.5
5.4	13.2	20.9
5.8	13.5	21.5
6.7	13.9	22.7
7.3	14.7	23.6
7.9	17.5	24.9
8.7	17.6	25.4
9.9	19.3	25.9
12.5	20.3	35.3

Test the hypothesis that the underlying distribution of life is exponential at the 5% level of significance using the Chi-square test.

15. Show that

$$I(\theta) \equiv E\left\{\left[\frac{\partial \ln f(x;\theta)}{\partial \theta}\right]^2\right\} = -E\left(\frac{\partial^2 \ln f(x;\theta)}{\partial \theta^2}\right)$$

16. Suppose that we have k disjoint events $A_1, A_2, \ldots, A_k$ such that the probability of A_i is p_i for $i = 1,2,\ldots, k$ and $\sum_{i=1}^{k} p_i = 1$. Let's assume that among n independent trials there are $X_1, X_2, \ldots, X_k$ outcomes associated with $A_1, A_2, \ldots, A_k$, respectively. The joint probability that $X_1 = x_1, X_2 = x_2, \ldots, X_k = x_k$ is given as follows by the likelihood function

$$L(x, p) = \frac{n!}{x_1! x_2! \dots x_k!} p_1^{x_1} p_1^{x_1} \dots p_k^{x_k}$$

where $\sum_{i=1}^{k} x_i = n$. This is infact known as the multinomial distribution. Find the MLE of $p_1, p_2, \dots, p_k$ by maximizing the likelihood function above.

17. Consider the binomial distribution with unknown parameter p given by

$$P(X = k) = \binom{n}{k} p^k (1 - p)^{n-k} \quad k = 1, 2, \dots, n; \; 0 \le p \le 1$$

where n = number of trials; k = number of successes; p = single trial probability of success. Find the maximum likelihood estimator of p.

18. (a) Find the maximum likelihood estimator of θ if $t_1, t_2, \dots, t_n$ are independent observations from a population with the following probability density function

$$f(t; \theta) = \frac{\theta}{t^{\theta+1}} \quad \text{for } t \ge 1, \; \theta > 0$$

   (b) Given the following failure time data: 200, 225, 228, 245, 250, 286, 290 h. Assuming that the data follow the above pdf, obtain the MLE of θ.

19. (a) Find the maximum likelihood estimator of λ if the failure times $t_1, t_2, \dots, t_n$ are independent observations from a population with the following probability density function

$$f(t) = (1 + \lambda) t^{\lambda} \quad \text{for } 0 < t < 1.$$

   (b) The failure times are: 0.2, 0.3, 0.35, 0.45, 0.5, 0.6, 0.7, 0.75, 0.8, and 0.95 h. Obtain the MLE of λ.

20. Suppose that $X_1, X_2, \dots, X_n$ are independent random variables, each with the following probability density function:

$$f(x, \alpha, \beta) = \begin{cases} 0 & \text{if } x < \alpha \\ \frac{1}{\beta} e^{-\frac{(x-\alpha)}{\beta}} & \text{if } x \ge \alpha \end{cases}$$

where $-\infty < \alpha < \infty$ and $0 < \beta < \infty$ are both unknown.

(a) Find the maximum likelihood estimators (MLE) of α and β, say $\hat{\alpha}$ and $\hat{\beta}$, respectively.

(b) Find $E(\hat{\alpha})$ and $E(\hat{\beta})$.

(c) If $n = 7$ and $x_1 = 5.3$, $x_2 = 3.2$, $x_3 = 2.4$, $x_4 = 3.8$, $x_5 = 4.2$, $x_6 = 3.4$, $x_7 = 2.9$, find the MLE of α and β

21. Suppose on five consecutive days in a given month the number of customers who enter services at a printing shop were 48, 60, 78, 56, and 73
Test the null hypothesis that the expected numbers of customers per day were the same on the five days at the 5% level of significance using the Chi-square test.

22. Suppose that X is a discrete random variable with the following probability mass function:

X	$P(X)$
0	$\frac{2\theta}{3}$
1	$\frac{\theta}{3}$
2	$\frac{2(1-\theta)}{3}$
3	$\frac{(1-\theta)}{3}$

where $0 \leq \theta \leq 1$ is a parameter. The following 10 independent observations were taken from such a distribution:

3, 0, 2, 1, 3, 2, 1, 0, 2, 1.

What is the maximum likelihood estimate of θ?

23. Let X denote the proportion of allotted time that a randomly selected engineer spends working on a certain project, and suppose that the probability density function of X is

$$f(x) = \begin{cases} (\theta + 1)x^\theta & \text{for } 0 \leq x \leq 1 \\ 0 & \text{otherwise} \end{cases}.$$

where $\theta > -1$.
A random sample of 10 engineers yielded data

$$x_1 = 0.92, \; x_2 = 0.79, \; x_3 = 0.90, \; x_4 = 0.65, \; x_5 = 0.86$$
$$x_6 = 0.47, \; x_7 = 0.73, \; x_8 = 0.97, \; x_9 = 0.94, \; x_{10} = 0.77$$

(a) Use the method of moments to obtain an estimator of θ, and then compute the estimate for this data.
(b) Obtain the maximum likelihood estimator of θ, and then compute the estimate for the given data.

24. Let X be uniformly distributed on the interval $[0, \theta]$ and the probability density function of X is

$$f(x) = \begin{cases} \dfrac{1}{\theta} & \text{for } 0 \leq x \leq \theta \\ 0 & \text{otherwise.} \end{cases}$$

where $\theta > 0$.

A random sample of 5 inspectors yielded data

$x_1 = 10.52$, $x_2 = 12.37$, $x_3 = 8.90$, $x_4 = 14.35$, $x_5 = 9.66$

Obtain an estimator of θ using (a) the method of moments and (b) the method of maximum likelihood, and then compute the estimate for this data.

25. Suppose that my waiting time for a train is uniformly distributed on the interval $[0, \theta]$, and that the results $x_1, x_2, \ldots, x_n$ of a random sample from this distribution have been observed. If my waiting times are 4.5, 5.3, 1.2, 7.4, 3.6, 4.8, and 2.9,

   (a)   calculate the estimate of θ using the method of moments.
   (b)   calculate the estimate of θ using the method of maximum likelihood.

26. (a) Find the method of moments estimate for θ based on a random sample of size n taken from the following probability density function

$$f(x) = (\theta + 1)x^\theta \quad \text{for} \quad 0 < x < 1.$$

   (b) Calculate the estimate of θ for the sample $x_1 = 0.7$, $x_2 = 0.4$, $x_3 = 0.8$, $x_4 = 0.5$

27. At time $x = 0$, twenty identical units are put on test. Suppose that the lifetime probability density function of each unit with parameter λ is given by

$$f(x) = \lambda e^{-\lambda x} \quad \text{for } x \geq 0, \lambda > 0$$

The quality manager then leaves the test facility unmonitored. On his return 24 h later, the manager immediately terminates the test after noticing that $y = 15$ of the 20 units are still in operation (so five have failed). Determine the maximum likelihood estimate of λ.

28. At time $x = 0$, forty identical units are put on test. Suppose that the lifetime probability density function of each unit with parameter θ is given by

$$f(x) = \frac{x}{\theta^2} e^{-\frac{x^2}{2\theta^2}} \text{ for } x \geq 0, \quad \theta > 0$$

The quality manager then leaves the test facility unmonitored. On his return 8 h later, the manager immediately terminates the test after noticing that 30 of the 40 units are still in operation (so ten have failed). Determine the maximum likelihood estimate of θ.

29. The following data were drawn one observation at a time in the order records:

g g b g g b g g g b g b g g b b b g g g.

where b denotes a defective item and g denotes a good item. The experiment was performed to test the following hypothesis:

H_0: p = p_0 = 0.10 versus H_1: p = p_1 = 0.20.

where p denotes the proportion defective items in the population. It is desired to reject H_0 when it is true with probability 0.05 and to accept H_0 when H_1 is true with probability 0.20. Using sequential testing plan, determine whether you would accept or reject a lot on the basis of the observations above.

30. The following data were drawn one observation at a time in the order records:

b g b b g b g b.

where b denotes a defective item and g denotes a good item. The experiment was performed to test the following hypothesis:
H_0: $p = p_0 = 0.10$ versus H_1: $p = p_1 = 0.25$.

where p denotes the proportion defective items in the population. It is desired to reject H_0 when it is true with probability 0.05 and to accept H_0 when H_1 is true with probability 0.10. Using sequential testing plan, determine whether you would accept or reject a lot on the basis of the observations above.

31. Let X represent a random variable of service time at a certain facility, and suppose that the cumulative distribution function of X is

$$F(x) = \begin{cases} 1 - e^{-\lambda x} - \lambda x e^{-\lambda x} & \text{for } x > 0 \\ 0 & \text{for } x \leq 0 \end{cases}$$

where $\lambda > 0$. A random sample of 10 customers yielded data.

$$x_1 = 12, \quad x_2 = 18, \quad x_3 = 8, \quad x_4 = 15, \quad x_5 = 11,$$
$$x_6 = 9, \quad x_7 = 8, \quad x_8 = 10, \quad x_9 = 18, x_{10} = 21$$

(a) Use the method of moments to obtain an estimator of λ, and then compute the estimate for this data.
(b) Obtain the maximum likelihood estimator of λ, and then compute the estimate for the given data.

32. Let X represent a random variable of service time at a certain facility, and suppose that the probability density function of X is

$$f(x) = 0.5 \, \lambda^3 x^2 e^{-\lambda x} \quad \text{for } x > 0, \lambda > 0.$$

A random sample of 5 customers yielded data
$x_1 = 20, x_2 = 16, x_3 = 15, x_4 = 18, x_5 = 25$
Use the method of moments to obtain an estimator of λ, and then compute the estimate for this data.

33. A study of the relationship between facility conditions at gasoline stations and aggressiveness in the pricing of gasoline reported the accompanying data below based on a sample of n = 441 stations. At the 0.10 significance level,

does the data suggest that facility conditions and pricing policy are independent of one another?

Observed Pricing Policy

Facility conditions	Aggressive	Neutral	Nonaggressive	Total
Substandard	24	15	17	56
Standard	52	73	80	205
Modern	58	86	36	180
Total	134	174	133	441

References

Akaike H (1973) Information theory and an extension of the maximum likelihood principle. In: Petrov BN, Caski F (eds) The second international symposium proceedings on information theory. AkademiaiKiado, Budapest, Hungary, pp 267–281

Asher H, Feingold H (1984) Repairable systems reliability. Marcel Dekker, New York

Feller W (1966) An introduction to probability theory and its applications, vol II. Wiley, New York

Feller W (1968) An introduction to probability theory and its applications, vol 1, 3rd edn. Wiley, New York

Jelinski Z, Moranda PB (1972) Software reliability research. In: Freiberger W (ed) Statistical computer performance evaluation. Academic Press, New York

Musa JD, Iannino A, Okumoto K (1987) Software reliability: measurement, prediction, application. McGraw-Hill, New York

Nelson SC, Haire MJ, Schryver JC (1992) Network-simulation modeling of interactions between maintenance and process systems. In: Proceedings of the 1992 annual reliability and maintainability symposium, Las Vegas, NV, USA,1992 Jan 21–23, pp 150–156

Pham H (2000) Software reliability. Springer, Singapore

Pham H (2002) A Vtub-shaped hazard rate function with applications to system safety. Int J Reliab Appl 3(1):1–16

Pham H (2006) System software reliability. Springer

Pham H (2020) On estimating the number of deaths related to Covid-19. Mathematics 8

Pham H (2014) A new software reliability model with Vtub-shaped fault-detection rate and the uncertainty of operating environments. Optimization 63(10):1481–1490

Pham H (2019) A new criteria for model selection. Mathematics 7(12):2019

Pham H, Deng C (2003) Predictive-ratio risk criterion for selecting software reliability models. In: Proceedings of the ninth international conference on reliability and quality in design

Schwarz G (1978) Estimating the dimension of a model. Ann Stat 1978(6):461–464

Smirnov N (1948) Table for estimating the goodness of fit of empirical distributions. Ann Math Stat 19:279–281

Zhao M (2003) Statistical reliability change-point estimation models. In: Pham H (ed) Handbook of reliability engineering. Springer, pp 157–163.

Chapter 4
System Reliability Modeling

4.1 Introduction

Assess the reliability during the development of a new product especially a complex system is important to be able to predict how likely the product can be performed for its customers throughout the time the customers plan to use it. This chapter presents methods and techniques for calculating the reliability of complex systems consisting of non-repairable independent components including series–parallel, parallel–series, k-out-of-n, standby and degraded systems (Pham 1997, 2003, 2014; Pham et al. 1996, 1997). The chapter also discusses basic mathematical reliability optimization methods and the reliability of systems with multiple failure modes.

4.2 Series Systems

A series system is comprised of n components, shown in Fig. 4.1, the failure of any of which will cause a system failure. In other words, the series system fails if and only if at least one component fails. Similarly, for the series system to function then all its components must function. An example of a series system is in the area of communication that a system consists of three subsystems: a transmitter, a receiver, and a processor. The failure of any one of these three subsystems will cause a system failure.

If all n component failures are independent, then the reliability of series system is given by

$$R = \prod_{i=1}^{n} R_i. \tag{4.1}$$

If the n components are identical with reliability R_0, the system reliability is

© The Author(s), under exclusive license to Springer Nature Switzerland AG 2022
H. Pham, *Statistical Reliability Engineering*, Springer Series in Reliability Engineering,
https://doi.org/10.1007/978-3-030-76904-8_4

Fig. 4.1 A simplified series system

$$R = R_0^n.$$

If component i has failure time T_i, for $i = 1, 2, \ldots, n$ then since the system will fail as soon as one of its components fails, it follows that the system failure time is given by

$$T_s = \min_{1 \leq i \leq n} T_i$$

Let us define $f_i(t)$ and $R_i(t)$ are, respectively, the failure time density and reliability function of the ith component, the failure time distribution function of the system is

$$F(t) = P(T_s \leq t) = 1 - P(T_s > t)$$

$$= 1 - \prod_{i=1}^{n} P(T_i > t)$$

$$= 1 - \prod_{i=1}^{n} R_i(t) \tag{4.2}$$

The system failure time density is given by

$$f(t) = \frac{dF}{dt} = \sum_{i=1}^{n} f_i(t) \prod_{\substack{j=1 \\ j \neq i}}^{n} R_j(t) \tag{4.3}$$

The system reliability is also given by

$$R(t) = \prod_{i=1}^{n} R_i(t) \tag{4.4}$$

The system failure rate $\lambda_S(t)$ is

$$\lambda_S(t) = \frac{\sum_{i=1}^{n} f_i(t) \cdot \left[\prod_{j \neq i} R_j(t)\right]}{R(t)} = \frac{\sum_{i=1}^{n} f_i(t) \cdot \frac{\left[\prod_j R_j(t)\right]}{R_i(t)}}{R(t)} \tag{4.5}$$

Example 4.1 Given that each of the n independent components has an exponential failure time distribution with constant failure rate λ_i, that is $R_i(t) = e^{-\lambda_i t}$, obtain the reliability function of series system.

Solution The system reliability is

$$R(t) = e^{-\left(\sum_{i=1}^{n} \lambda_i\right)t} = e^{-\lambda t}$$

where $\lambda = \sum_{i=1}^{n} \lambda_i$ which is just the sum of the failures of the components; that is also an exponential distribution with constant failure rate λ. If each component has 99.99% reliability after a month in the field, a system consisting of 5 independent components in series will have a reliability of $(0.9999)^5$ or 0.9995. In fact, in a series system high reliability can be achieved only if the individual component has very high reliability.

In general, the fact that a system consisting of n independent components arranged in series has failure rate equal to the sum of the component failure rates is general true for any distribution other than the exponential. See Problem 1.

The mean time to failure of the system can be obtained

$$MTTF = \int_0^\infty R(t)dt = \frac{1}{\sum_{i=1}^{n} \lambda_i}.$$

Example 4.2 Consider a communication system with three independent units such as transmitting unit, receiving unit and a processing unit. Assume that the transmitter has a constant failure rate of $\lambda_T = 0.002$ per hour, the receiver has a constant failure rate of $\lambda_R = 0.005$ per hour, and the processor has a constant failure rate of $\lambda_P = 0.008$ per hour. Obtain the system reliability for a mission of 25 h.

Solution The reliability of transmitting unit is given by

$$R_T(25) = e^{-0.002(25)} = 0.9512$$

The reliability of receiving unit is

$$R_R(25) = e^{-0.005(25)} = 0.8825$$

The reliability of processing unit is

$$R_P(25) = e^{-0.008(25)} = 0.8187.$$

The reliability of this communication system is given by

$$R(25) = (0.9512) * (0.8825) * (0.8187) = 0.6872$$

4.3   Parallel Systems

A parallel system is composed of n components that perform independent functions, shown in Fig. 4.2, the success of any of which will lead to system success. In other words, a parallel system works if and only if at least one component works. Also, the parallel system can fail only when all components have failed.

Assume that a parallel system consists of n components (see Fig. 4.2) and that the failures are independent. Note that for the parallel system to fail, all n components must fail. The unreliability function, F ($= 1 - R$) can be written as

$$F = \prod_{i=1}^{n} (1 - R_i). \tag{4.6}$$

The system reliability is

$$R = 1 - \prod_{i=1}^{n} (1 - R_i). \tag{4.7}$$

If components are identical, that is $R_i = R_0$, then

$$R = 1 - (1 - R_0)^n \tag{4.8}$$

where R_0 is the reliability of a component. If R_s is of reliability requirement then the minimum number of components required can be obtained by solving the following inequality:

$$1 - (1 - R_0)^n \geq R_s$$

and, therefore, the minimum number of components is:

$$n = \frac{\ln(1 - R_s)}{\ln(1 - R_0)}.$$

Fig. 4.2 A simplified parallel system

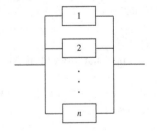

Example 4.3 If a parallel system consists of 3 independent components, each having 0.95 reliability, then the system reliability is given by

$$R = 1 - (1 - 0.95)^3 = 0.9999$$

If the ith component has failure time T_i for $i = 1, 2, \ldots, n$ and the parallel system will fail only upon failure of all of its components, thus, the system failure time is

$$T_P = \max_{1 \le i \le n} T_i$$

Let us define $f_i(t)$, $F_i(t)$, and $R_i(t)$ are, respectively, the failure time density, failure time distribution function and reliability function of the ith component. The failure time distribution function of the system is

$$F(t) = P(T_P \le t) = \prod_{i=1}^{n} F_i(t)$$

The system failure time density is

$$f(t) = \frac{dF}{dt} = \sum_{i=1}^{n} f_i(t) \prod_{\substack{j=1 \\ j \ne i}}^{n} F_j(t)$$

The system reliability is

$$R(t) = 1 - F(t) = 1 - \prod_{i=1}^{n} (1 - R_i(t)) \qquad (4.9)$$

The system failure rate can be obtained:

$$r(t) = \frac{f(t)}{R(t)} = \frac{\sum_{i=1}^{n} f_i(t) \prod_{\substack{j=1 \\ j \ne i}}^{n} F_j(t)}{1 - \prod_{\substack{j=1 \\ j \ne i}}^{n} F_j(t)}$$

If each of the n independent components has an exponential failure time distribution with constant failure rate λ_i, that is $R_i(t) = e^{-\lambda_i t}$, then the system reliability is

$$R(t) = 1 - \prod_{i=1}^{n} (1 - e^{-\lambda_i t}) \qquad (4.10)$$

The mean time to failure of the system can be obtained

$$M_{tbf} = \int\limits_0^\infty R(t)dt = \int\limits_0^\infty \left(1 - \prod_{i=1}^n (1 - e^{-\lambda_i t})\right) dt$$

After simplifications, we can obtain

$$M_{tbf} = \sum_{i=1}^n \sum_{j_i=1}^n \sum_{\substack{j_{i-1}=1 \\ \{j_1 > j_2 > ..., > j_i\}}}^n \cdots \sum_{j_2=1}^n \sum_{j_1=1}^n \frac{(-1)^{i+1}}{\left(\sum_{k=1}^i \lambda_{j_k}\right)} \tag{4.11}$$

For $n = 2$, from the above formula we obtain the system MTTF

$$M_{tbf} = \sum_{i=1}^2 \sum_{\substack{j_2=1 \\ \{j_1 > j_2\}}}^2 \sum_{j_1=1}^2 \frac{(-1)^{i+1}}{\left(\sum_{k=1}^i \lambda_{j_k}\right)}$$

$$= \sum_{\substack{j_2=1 \\ \{j_1 > j_2\}}}^2 \sum_{j_1=1}^2 \frac{(-1)^{1+1}}{\left(\sum_{k=1}^1 \lambda_{j_k}\right)} + \sum_{\substack{j_2=1 \\ \{j_1 > j_2\}}}^2 \sum_{j_1=1}^2 \frac{(-1)^{2+1}}{\left(\sum_{k=1}^2 \lambda_{j_k}\right)}$$

$$= \frac{1}{\lambda_1} + \frac{1}{\lambda_2} - \frac{1}{\lambda_1 + \lambda_2}$$

Thus,

$$MTTF = \frac{1}{\lambda_1} + \frac{1}{\lambda_2} - \frac{1}{\lambda_1 + \lambda_2}.$$

Similarly, it can be shown that for a 3-component parallel system, the system MTTF is given by

$$MTTF = \frac{1}{\lambda_1} + \frac{1}{\lambda_2} + \frac{1}{\lambda_3} - \frac{1}{\lambda_1 + \lambda_2} - \frac{1}{\lambda_1 + \lambda_3} - \frac{1}{\lambda_2 + \lambda_3} + \frac{1}{\lambda_1 + \lambda_2 + \lambda_3}$$

Example 4.4 Consider a parallel system of n identical and independent components with constant failure rate λ, then the component reliability is

$$R_0(t) = e^{-\lambda t},$$

and from Eq. (4.8), the system reliability is

$$R(t) = 1 - \left(1 - e^{-\lambda t}\right)^n. \tag{4.12}$$

The mean time to failure (MTTF) is given by

$$MTTF = \int_0^\infty R(t)dt = \frac{1}{\lambda} \sum_{i=1}^n \frac{1}{i}. \tag{4.13}$$

When $n = 2$, the 2-component system reliability and system MTTF are

$$R(t) = 1 - \left(1 - e^{-\lambda t}\right)^2 = 2e^{-\lambda t} - e^{-2\lambda t}$$

and

$$MTTF = \frac{1}{\lambda} \sum_{i=1}^2 \frac{1}{i} = \frac{3}{2}\lambda$$

respectively. In general, the MTTF of a parallel system with n components is $\sum_{i=1}^n \frac{1}{i}$ times better than that of a single-component system.

4.4 Series–Parallel Systems

Consider a series–parallel system consists of n subsystems connected in series where subsystem i with m_i components in parallel, for $i = 1, 2, ..., n$, as shown in Fig. 4.3. Define $R_{ij}(t)$ is as the reliability of component j in subsystem i for $i = 1, 2, ..., n$ and $j = 1, 2, ..., m_i$. The series–parallel system operates successfully if all of n subsystems work. In other words, there is at least one component in each subsystem must function for the successful operation of the system. The reliability of series–parallel systems can be obtained

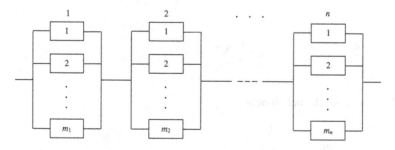

Fig. 4.3 A series–parallel system

$$R(t) = \prod_{i=1}^{n} \left[1 - \prod_{j=1}^{m_i} (1 - R_{ij}(t)) \right] \tag{4.14}$$

When all the components are identical and the numbers of components in each subsystem are the same, then system reliability from Eq. (4.14) becomes

$$R(t) = \left[1 - (1 - R_0(t))^m \right]^n \tag{4.15}$$

where $R_0(t)$ is a component reliability and m is the number of components in each subsystem.

If each component has a constant failure rate, that is $R_0(t) = e^{-\lambda t}$, then the system MTTF is given by

$$MTTF = \int_0^\infty \left(1 - (1 - R_0(t))^m \right)^n dt$$

$$= \int_0^\infty \left(1 - (1 - e^{-\lambda t})^m \right)^n dt$$

$$= \frac{1}{\lambda} \int_0^1 \frac{(1 - u^m)^n}{1 - u} du$$

where $u = 1 - e^{-\lambda t}$. After simplifications, we obtain

$$MTTF = \frac{1}{\lambda} \int_0^1 \frac{(1 - u^m)}{1 - u} (1 - u^m)^{n-1} du$$

$$= \frac{1}{\lambda} \sum_{i=0}^{m-1} \int_0^1 (1 - u^m)^{n-1} u^i du$$

$$= \frac{1}{m\lambda} \sum_{i=0}^{m-1} \int_0^1 x^{\frac{i+1}{m}-1} (1 - x)^{n-1} dx$$

Note that from the beta function

$$B(a, b) = \int_0^1 x^{a-1} (1 - x)^{b-1} dx = \frac{(a - 1)!(b - 1)!}{(a + b - 1)!}$$

we have

$$MTTF = \frac{1}{m\lambda} \sum_{i=0}^{m-1} B\left(\frac{i+1}{m}, n\right) = \frac{(n-1)!}{m\lambda} \sum_{i=0}^{m-1} \frac{\left\lfloor \frac{i+1}{m} - 1 \right\rfloor!}{\left\lfloor \frac{i+1}{m} + n - 1 \right\rfloor!} \qquad (4.16)$$

where $\lfloor x \rfloor$ denotes the largest integer not exceeding x.

4.5 Parallel–Series Systems

Consider a parallel–series system of m subsystems in parallel where subsystem i with n_i components in series, for $i = 1, 2, ..., m$, as shown in Fig. 4.4. Define $R_{ij}(t)$ is as the reliability of component j in subsystem i for $i = 1, 2, ..., m$ and $j = 1, 2, ..., n_i$. The parallel–series system operates successfully if all of the components in any subsystem m that functions. Thus, the reliability of parallel–series system is given by

$$R(t) = 1 - \prod_{i=1}^{m} \left(1 - \prod_{j=1}^{n_i} R_{ij}(t)\right). \qquad (4.17)$$

If all components in the parallel–series system are identical and the numbers of components in all subsystems are equal, the reliability of the system can be written as

$$R(t) = 1 - \left[1 - (R_0(t))^n\right]^m \qquad (4.18)$$

where $R_0(t)$ is the reliability of an individual component, and n is the number of components in each subsystem. If each component has a constant failure rate, that is $R_0(t) = e^{-\lambda t}$, then the system MTTF is given by

Fig. 4.4 A parallel–series system

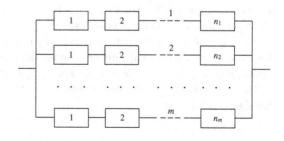

$$MTTF = \int\limits_{0}^{\infty} R(t)dt = \int\limits_{0}^{\infty} \left(1 - \left[1 - e^{-n\lambda t}\right]^{m}\right)dt$$

Let $u = 1 - e^{-n\lambda t}$ then we obtain

$$MTTF = \frac{1}{n\lambda} \int\limits_{0}^{1} \frac{1 - u^{m}}{1 - u}du = \frac{1}{n\lambda} \sum_{i=1}^{m} \frac{1}{i} \qquad (4.19)$$

4.6   k-out-of-n Systems

A common form of redundancy is a k-out-of-n system in which at least k out of n components must function properly for the successful functioning of the system (Pham 1992b, 2003; Pham and Pham 1991, 1992; Pham et al. 1996). For example, a four-engine plane in which two engines must work for the plane to fly is a 2-out-of-4 system. Similarly, consider a hot standby system consisting of two statistically identical buses. There is a primary bus and a standby bus, that is switched online upon failure of the primary bus. Therefore, the system, which operates correctly so long as one of the two buses operates correctly, is a 1-out-of-2 system. Another example is that cables for a bridge at Station M where a minimum of 250 out of 275 cables are necessary to support the structure.

An electrical power-generating system consisting of five generators will be considering to operate in a full mode if there are at least four generators to operate in full mode in order to deliver sufficient power. Such system is a 4-out-of-5 system. Furthermore, in an 8-cylinder engine automobile it may be possible to drive the car if only five cylinders are firing but if less than five fire then the automobile cannot be driven, therefore, the engine functions as a 5-out-of-8 system. This configuration of a reliability system is a generalization of a parallel redundant system with a requirement for k out of n (obviously $k \leq n$) identical and independent components to function in order for the whole system to function. When $k = 1$ the previously discussed complete redundancy occurs as parallel systems, while if $k = n$ the system is comprised of n components in series.

The k-out-of-n system functions if and only if at least k of its n components must function, where n is the total number of components in the system. By the definition, a parallel system is a 1-out-of-n while a series system is an n-out-of-n system.

The reliability of a k-out-of-n system with independently and identically distributed (i.i.d.) components is given by

$$R(k, n) = \sum_{i=k}^{n} \binom{n}{i} p^{i} q^{n-i} \qquad (4.20)$$

where p is the reliability of a component. The above equation can be rewritten as

$$R(k, n) = p^k \sum_{i=k}^{n} \binom{i-1}{k-1} q^{i-k} \tag{4.21}$$

Example 4.5 An aircraft has four engines (i.e., two on each wing) but can land using just any three or more engines. Assume that the reliability of each engine for the duration of a mission is $R_e = 0.997$ and that engine failures are independent. Calculate the mission reliability of the aircraft.

Solution Given $R_e = 0.997$. The mission reliability of the aircraft is given by

$$R_{system} = \sum_{i=3}^{4} \binom{4}{i} (0.997)^i (0.003)^{4-i} = 0.9999462$$

Time-Dependent Reliability Function

If all the components have the same failure time distribution function $F_0(t)$ and $R_0(t) = 1 - F_0(t)$, then the probability of having exactly k components operational is given by

$$P(X = k) = \binom{n}{k}(R_0(t))^k (1 - R_0(t))^{n-k}, k = 0, 1, \ldots, n$$

where $R_0(t)$ is the reliability of a component. Therefore, the reliability of the k-out-of-n system is given by

$$R(t) = \sum_{i=k}^{n} \binom{n}{i}(R_0(t))^i (1 - R_0(t))^{n-i}. \tag{4.22}$$

When $k = 1$, that is, the n components are in parallel, Eq. (4.22) becomes

$$R(t) = 1 - (1 - R_0(t))^n$$

which is the same as Eq. (4.8). When $k = n$, that is, the n components are in series, thus

$$R(t) = (R_0(t))^n$$

If the time to failure is exponential, the reliability of the system is

$$R(t) = \sum_{i=k}^{n} \binom{n}{i} e^{-\lambda i t} (1 - e^{-\lambda t})^{n-i} \tag{4.23}$$

where λ is the component constant failure rate. It can be shown that the mean time to failure (MTTF) of the k-out-of-n system is given by

$$MTTF = \int_0^\infty R(t)dt = \frac{1}{\lambda} \sum_{i=k}^n \frac{1}{i}. \tag{4.24}$$

For a special case when $k = 1$ then it becomes a parallel system. In general, it is not simple to write a reliability expression for a k-out-of-n system when the components do not have identical failure time distributions. However, it is possible to derive the desired reliability expressions when n and k are small. Let us look at the following examples.

Example 4.6 Given components with an exponential failure time distribution such that the ith component has constant failure rate λ_i then it can be shown that the reliability and MTTF of a 2 out of 3 system are, respectively,

$$R(t) = e^{-(\lambda_1+\lambda_2)t} + e^{-(\lambda_1+\lambda_3)t} + e^{-(\lambda_2+\lambda_3)t} - 2e^{-(\lambda_1+\lambda_2+\lambda_3)t}$$

and

$$M_{tbf} = \int_0^\infty (e^{-(\lambda_1+\lambda_2)t} + e^{-(\lambda_1+\lambda_3)t} + e^{-(\lambda_2+\lambda_3)t} - 2e^{-(\lambda_1+\lambda_2+\lambda_3)t})dt$$

$$= \frac{1}{\lambda_1 + \lambda_2} + \frac{1}{\lambda_1 + \lambda_3} + \frac{1}{\lambda_2 + \lambda_3} - \frac{2}{\lambda_1 + \lambda_2 + \lambda_3}$$

Similarly, one can easily obtain the reliability and MTTF of a 2 out of 4 system are, respectively,

$$R(t) = e^{-(\lambda_1+\lambda_2)t} + e^{-(\lambda_1+\lambda_3)t} + e^{-(\lambda_1+\lambda_4)t} + e^{-(\lambda_2+\lambda_3)t} + e^{-(\lambda_2+\lambda_4)t}$$
$$+ e^{-(\lambda_3+\lambda_4)t} - 2e^{-(\lambda_1+\lambda_2+\lambda_3)t} - 2e^{-(\lambda_1+\lambda_2+\lambda_4)t}$$
$$- 2e^{-(\lambda_1+\lambda_3+\lambda_4)t} - 2e^{-(\lambda_2+\lambda_3+\lambda_4)t} + 3e^{-(\lambda_1+\lambda_2+\lambda_3+\lambda_4)t}$$

and

$$MTTF = \frac{1}{\lambda_1 + \lambda_2} + \frac{1}{\lambda_1 + \lambda_3} + \frac{1}{\lambda_1 + \lambda_4} + \frac{1}{\lambda_2 + \lambda_3} + \frac{1}{\lambda_2 + \lambda_4}$$
$$+ \frac{1}{\lambda_3 + \lambda_4} - \frac{2}{\lambda_1 + \lambda_2 + \lambda_3} - \frac{2}{\lambda_1 + \lambda_2 + \lambda_4} - \frac{2}{\lambda_1 + \lambda_3 + \lambda_4}$$
$$- \frac{2}{\lambda_2 + \lambda_3 + \lambda_4} + \frac{3}{\lambda_1 + \lambda_2 + \lambda_3 + \lambda_4}$$

Similarly, one can easily obtain the reliability and MTTF of a 3-out-of-4 system are, respectively,

$$R(t) = e^{-(\lambda_1+\lambda_2+\lambda_3)t} + e^{-(\lambda_1+\lambda_2+\lambda_4)t} + e^{-(\lambda_1+\lambda_3+\lambda_4)t}$$
$$+ e^{-(\lambda_2+\lambda_3+\lambda_4)t} - 3e^{-(\lambda_1+\lambda_2+\lambda_3+\lambda_4)t}$$

and

$$M_{tbf} = \frac{1}{\lambda_1+\lambda_2+\lambda_3} + \frac{1}{\lambda_1+\lambda_2+\lambda_4} + \frac{1}{\lambda_1+\lambda_3+\lambda_4}$$
$$+ \frac{1}{\lambda_2+\lambda_3+\lambda_4} - \frac{3}{\lambda_1+\lambda_2+\lambda_3+\lambda_4}$$

4.7 *k*-to-*l*-out-of-*n* Noncoherent Systems

A *k*-to-*l*-out-of-*n* system works if and only if no fewer than *k* and no more than *l* out of its *n* components are to function for the successful operation of the system. This class of noncoherent systems was first proposed by Heidtmann (1981). The *k*-to-*l*-out-of-*n* system takes into account the failures caused by the underproduction of units or services and the failures that are caused by overproduction as well. Examples of noncoherent systems can be found in the area of avionics system. A space-shuttle computer-complex system has 5 identical computers assigned to redundant operation. During the critical mission phases such as boost, reentry, and landing, 4 of these computers operate in an N-Modular Redundancy (NMR) mode, receiving the same inputs and executing identical tasks. Computer 5 performs non-critical tasks. The shuttle can tolerate up to 2 failures. If all 5 computers are functioning during the critical mission phases, the technique to detect a failed computer might not be able to identify the failure, given that a failure has occurred. This eventually causes the space shuttle to fail (Upadhyaya and Pham 1992, 1993). It is therefore a 2-to-4-out-of-5 system.

A noncoherent system is any system that is not coherent; a system is coherent if and only if: (a) it is good (bad) when all components are good (bad), and (b) it never gets worse (better) when the number of good components increases (decreases). Such a system is noncoherent because the system reliability can decrease when the number of good components increases (Pham 1991). The generality of the *k*-to-*l*-out-of-*n* system can be demonstrated as follows: (i) when $k = l = n$, the *k*-to-*l*-out-of-*n* system becomes a series system; (ii) when $k = 1$ and $l = n$, the *k*-to-*l*-out-of-*n* system becomes a parallel system.

If we assume that all the *n* components in the *k*-to-*l*-out-of-*n* system are i.i.d., then the reliability of the *k*-to-*l*-out-of-*n* system is

$$R(t) = \sum_{i=k}^{l} \binom{n}{i}(R_0(t))^i(1 - R_0(t))^{n-i} \tag{4.25}$$

where $R_0(t)$ is the reliability of a component. When $k = 1$ and $l = n$, then the above system reliability function becomes

$$R(t) = 1 - (1 - R_0(t))^n$$

which is the same as Eq. (4.8) the parallel system. Similarly, when $k = l = n$, the above system reliability function becomes a series system and is

$$R(t) = (R_0(t))^n$$

If the time to failure is exponential, the reliability of the k-to-l-out-of-n system is

$$R(t) = \sum_{i=k}^{l} \binom{n}{i}e^{-\lambda i t}(1 - e^{-\lambda t})^{n-i}$$

where λ is the component constant failure rate. It can be shown that the MTTF of the k-to-l-out-of-n system is given by

$$MTTF = \int_0^{\infty} R(t)dt = \int_0^{\infty} \left(\sum_{i=k}^{l} \binom{n}{i}e^{-\lambda i t}(1 - e^{-\lambda t})^{n-i} \right)dt = \frac{1}{\lambda}\sum_{i=k}^{l}\frac{1}{i}. \tag{4.26}$$

4.8   Standby Systems

In general standby systems can be divided into three categories:

- Cold-standby system
- Warm-standby system
- Hot-standby system.

In a cold-standby system unit 1 is powered up (or initially "on-line") and operational, the remaining units are not powered (Pham 1992a). When unit 1 fails, the switching device switches in component 2 which remain powered up until it fails. Unlike the parallel systems where all units in the system are operating simultaneously, in the standby systems, one or more units are in standby mode waiting to take over the operation from the primary (or "on-line") operating unit as soon as the failure of the primary unit occurs. If a standby unit is fully activated when the

Fig. 4.5 A standby system

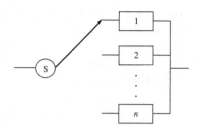

system is in use, it is said to be in *hot standby*. If a standby unit is fully activated only when the primary unit fails, it is in *cold standby*. A cold standby system needs a sensing mechanism to detect the failure of the primary unit and a switching device to activate the standby unit when the failure occurs. If the standby unit is partially activated where the system is in operating mode, the standby unit is said to be in *warm standby*. Cold standby units are not subject to failure until activated where a warm standby unit usually is subjected to a reduced level of stress, and may fail before it is fully activated.

Consider the cold standby system with a switching device. The system consisting of *n* components is shown in Fig. 4.5 where component 1 is the primary one, and S represents the switching device.

The typical assumptions in considering the standby systems are:

1. The switching device is 100% reliable.
2. If switching is necessary due to failure in an active or primary component, the time required is insignificant and will not affect the desired operation.
3. No warm up time is necessary for cold-standby components being switched in.
4. Failure can be detected by the failure sensing device with probability 1 and the subsequent is then switched in automatically.

In this section, we will discuss reliability of various aspects of standby systems including the above assumptions and others including imperfect switching device, load-sharing standby units.

4.8.1 Cold Standby Systems with Perfect Switching Device

Consider a standby system consisting of *n* components (only one active and $(n-1)$ components are in cold-standby), shown in Fig. 4.5. Here we assume that the switch S functions perfectly and that components cannot fail while they are in standby, also called as *cold-standby*. In other words, cold-standby components will not fail while waiting to be activated. The component that is in operation is called the *active* (or primary) component while the components are still in standby called *standby* (or cold standby) components.

The standby system operates as follows: Component 1, called primary component, is put into operation at time $t = 0$ and operates until failure. When it fails, component 2 is activated. If component 2 fails, component 3 is activated, and so forth. Similarly, component $(n - 1)$ operates until failure. When it fails, component n is activated. When component n fails, the system fails. Let T_i denote the time to failure of component i $(i = 1, 2, \ldots, n)$ and T be the system failure time, then we can easily see that

$$T = \sum_{i=1}^{n} T_i. \tag{4.27}$$

If T_1, T_2, $\ldots$, T_n are independent and identically exponential distributed with failure rate λ then the system failure time T follows a gamma distribution with parameters n and λ, and the corresponding pdf is

$$f(t) = \frac{\lambda^n}{\Gamma(n)} t^{n-1} e^{-\lambda t}$$

The reliability of the system is given by

$$R(t) = \int_t^{\infty} \frac{\lambda^n}{\Gamma(n)} s^{n-1} e^{-\lambda s} ds = e^{-\lambda t} \sum_{i=0}^{n-1} \frac{(\lambda t)^i}{i!}. \tag{4.28}$$

The system MTTF can be easily obtained as follows

$$MTTF = \frac{n}{\lambda}$$

or, equivalently, using the sum of the mean,

$$MTTF = E(T) = \sum_{i=1}^{n} E(T_i) = \sum_{i=1}^{n} \frac{1}{\lambda} = \frac{n}{\lambda} \tag{4.29}$$

Two-Component Standby Systems

Consider the standby system with two independent components where component 1 is the primary component and component 2 is in standby. The system will survive at time t if any of the following two events occurs.

- Event 1: The primary component (whose life T_1) does not fail at time t; that is, $T_1 \geq t$.
- Event 2: If the primary component fails at τ $(\tau < t)$, then the standby component (whose life T_2) must function at time τ and does not fail in the remaining time $(t - \tau)$. The event can be described as $\{T_1 < t \text{ and } T_2 \geq t - \tau\}$.

Since the above two events are mutually exclusive, the reliability of the system can be written as

$$R(t) = \Pr\{T_1 \geq t\} + \Pr\{T_1 < \tau \text{ and } T_2 \geq (t - \tau)\}$$

$$= R_1(t) + \int_0^t f_1(\tau) R_2(t - \tau) d\tau \tag{4.30}$$

where $R_i(t)$ and $f_i(t)$ are component i reliability and pdf, respectively. When the pdf of component 1 and 2 are given, it can be easily obtained the reliability of the system using Eq. (4.30).

For components with an exponential pdf having constant failure rate λ_i for the ith component and the switching device is 100% reliable, from Eq. (4.30) we can rewritten the system reliability as follows:

$R(t) = P\{\text{component \#1 does not fail at time t}\}$

$\qquad + P\{\text{component \#1 has failed before time t \& component \#2 was switched on and is}$

$\qquad \text{functioning until time t}\}$

Therefore,

$$R(t) = e^{-\lambda_1 t} + \int_0^t \lambda_1 e^{-\lambda_1 s} e^{-\lambda_2 (t-s)} ds$$

$$= \begin{cases} e^{-\lambda t} + \lambda t e^{-\lambda t} & \text{if } \lambda_1 = \lambda_2 = \lambda \\ \frac{\lambda_1 e^{-\lambda_2 t} - \lambda_2 e^{-\lambda_1 t}}{\lambda_1 - \lambda_2} & \text{if } \lambda_1 \neq \lambda_2 \end{cases} \tag{4.31}$$

Three-Component Standby Systems

Consider the standby system consisting of three independent components where component 1 is the primary component and two standby components (component 2 and 3) are in standby. Let $R_i(t)$ and $f_i(t)$ be reliability and pdf of component i. From Eq. (4.30), the system reliability can be obtained as follows:

$$R(t) = R_1(t) + \int_0^t f_1(t_1) R_2(t - t_1) dt_1 + \int_0^t f_1(t_1) \left(\int_0^{t-t_1} f_2(t_2) R_3(t - t_1 - t_2) dt_2 \right) dt_1 \tag{4.32}$$

If we assume that all the components are non-identical but exponentially distributed with constant failure rate parameter λ_i for component i, then from Eq. (4.32) we obtain

$$R(t) = e^{-\lambda_1 t} + \frac{\lambda_1}{\lambda_2 - \lambda_1}\left(e^{-\lambda_1 t} - e^{-\lambda_2 t}\right)$$

$$+ \lambda_1 \lambda_2 \left(\frac{e^{-\lambda_1 t}}{(\lambda_1 - \lambda_2)(\lambda_1 - \lambda_3)} + \frac{e^{-\lambda_2 t}}{(\lambda_2 - \lambda_1)(\lambda_2 - \lambda_3)} + \frac{e^{-\lambda_3 t}}{(\lambda_3 - \lambda_1)(\lambda_3 - \lambda_2)}\right)$$

$$(4.33)$$

It can be easily shown that for all non-identical n components in a standby system where component i is exponentially distributed with constant failure rate parameter λ_i, then the n-component standby system is given by

$$R(t) = \frac{\lambda_2 \lambda_3 \ldots \lambda_n\, e^{-\lambda_1 t}}{(\lambda_2 - \lambda_1)(\lambda_3 - \lambda_1) \ldots (\lambda_n - \lambda_1)} + \frac{\lambda_1 \lambda_3 \ldots \lambda_n\, e^{-\lambda_2 t}}{(\lambda_1 - \lambda_2)(\lambda_3 - \lambda_2) \ldots (\lambda_n - \lambda_2)}$$

$$+ \cdots + \frac{\lambda_1 \lambda_2 \ldots \lambda_{n-1}\, e^{-\lambda_n t}}{(\lambda_1 - \lambda_n)(\lambda_3 - \lambda_n) \ldots (\lambda_{n-1} - \lambda_n)}$$

$$= \sum_{i=1}^{n} \left(\prod_{\substack{j=1 \\ j \neq i}}^{n} \frac{\lambda_j}{(\lambda_j - \lambda_i)}\right) e^{-\lambda_i t} \tag{4.34}$$

The system MTTF then

$$MTTF = \int_{0}^{\infty} R(t)dt$$

$$= \int_{0}^{\infty} \left(\sum_{i=1}^{n} \left(\prod_{\substack{j=1 \\ j \neq i}}^{n} \frac{\lambda_j}{(\lambda_j - \lambda_i)}\right) e^{-\lambda_i t}\right) dt$$

$$= \sum_{i=1}^{n} \left\{\left(\prod_{\substack{j=1 \\ j \neq i}}^{n} \frac{\lambda_j}{(\lambda_j - \lambda_i)}\right) \frac{1}{\lambda_i}\right\} \tag{4.35}$$

If $\lambda_i = \lambda$ then from Eq. (4.34), it can be shown that the reliability of n-component standby system is given by

$$R(t) = e^{-\lambda t} \sum_{i=0}^{n-1} \frac{(\lambda t)^i}{i!} \tag{4.36}$$

Generalization of Standby System Reliability Calculations

In general, we can obtain the reliability of standby systems directly from the system pdf. Let $g_i(t)$ denotes the density function for T_i and f(t) is the corresponding system failure density function, then we can obtain the pdf of the 2-component standby system as follows:

$$f(t) = \int_0^t g_1(x)g_2(t-x)dx = (g_1 * g_2)(t) \tag{4.37}$$

where * is the convolution function.

The pdf of the 3-component standby system can also be obtained

$$f(t) = \int_0^t \int_0^{t-t_1} g_1(t_1)\, g_2(t_2)\, g_3(t-t_1-t_2)\, dt_2\, dt_1 = (g_1 * g_2 * g_3)(t) \tag{4.38}$$

Similarly, the pdf of the 4-component standby system is given by

$$f(t) = \int_0^t \int_0^{t-t_1} \int_0^{t-t_1-t_2} g_1(t_1)\, g_2(t_2)\, g_3(t_3)\, g(t-t_1-t_2-t_3)dt_3\, dt_2\, dt_1 = (g_1 * g_2 * g_3 * g_4)(t)$$

In general, the pdf of the n-component standby system is given by

$$f(t) = \int_0^t \int_0^{t-t_1} \cdots \int_0^{t-\sum_{i=1}^{n-2} t_i} g_1(t_1)g_2(t_2)\ldots g_n\left(t - \sum_{i=1}^{n-1} t_i\right) dt_{n-1} dt_{n-2} \ldots dt_1$$

$$= (g_1 * g_2 * \cdots * g_n)(t) \tag{4.39}$$

Therefore, the reliability of the n-component standby system can be written as

$$R(t) = \int_t^\infty f(x)dx$$

$$= \int_t^\infty (g_1 * g_2 * \ldots * g_n)(t)dt \tag{4.40}$$

Example 4.7 Consider a two-component standby system with an exponential pdf, that is,

$$g_i(t) = \lambda_i e^{-\lambda_i t} \quad \text{for } i = 1 \text{ and } 2$$

Using the convolution approach, then the pdf of the system can be written as

$$
f(t) = \int_0^t \lambda_1 \lambda_2 e^{-\lambda_1 s} e^{-\lambda_2(t-s)} ds
$$

$$
= \begin{cases} \lambda^2 t e^{-\lambda t} & \text{if } \lambda_1 = \lambda_2 = \lambda \\ \frac{\lambda_1 \lambda_2 (e^{-\lambda_2 t} - e^{-\lambda_1 t})}{\lambda_1 - \lambda_2} & \text{if } \lambda_1 \neq \lambda_2 \end{cases} \tag{4.41}
$$

The system reliability can be obtained as follows

$$
R(t) = \int_t^\infty f(s) ds
$$

$$
= \begin{cases} (1 + \lambda t) e^{-\lambda t} & \text{if } \lambda_1 = \lambda_2 = \lambda \\ \frac{\lambda_1 e^{-\lambda_2 t} - \lambda_2 e^{-\lambda_1 t}}{\lambda_1 - \lambda_2} & \text{if } \lambda_1 \neq \lambda_2 \end{cases} \tag{4.42}
$$

is same as Eq. (4.31).

In general, if $g_i(t) = \lambda_i e^{-\lambda_i t}$ for $i = 1, 2 \ldots, n$ then we can show that the reliability of n components standby system is given by

$$
R(t) = \int_t^\infty \left\{ \int_0^t \int_0^{t-t_1} \cdots \int_0^{t-\sum_{i=1}^{n-2} t_i} g_1(t_1) g_2(t_2) \ldots g_n\left(t - \sum_{i=1}^{n-1} t_i\right) dt_{n-1} dt_{n-2} \ldots dt_1 \right\} dt
$$

$$
= \int_t^\infty \{(g_1 * g_2 * \cdots * g_n)(t)\} dt
$$

$$
= \sum_{i=1}^n \left\{ \prod_{\substack{j=1 \\ j \neq i}}^n \frac{\lambda_j}{(\lambda_j - \lambda_i)} \right\} e^{-\lambda_i t} \tag{4.43}
$$

for if $\lambda_i \neq \lambda_j$ for $i \neq j$. If $\lambda_i = \lambda$ then from Eq. (4.43), it can be shown that

$$
R(t) = e^{-\lambda t} \sum_{i=0}^{n-1} \frac{(\lambda t)^i}{i!} \tag{4.44}
$$

4.8.2 Cold-Standby Systems with Imperfect Switching Device

Let us consider a switching device consisting of a failure detection mechanism and a switching sensor that subject to failure with a pdf $f_s(t)$ and reliability $R_s(t)$.

2-component Cold-Standby System with imperfect switching device

For a two-component cold standby system, the reliability of the system is given by

$$R(t) = R_1(t) + \int_0^t R_s(\tau)f_1(\tau)R_2(t-\tau)d\tau \qquad (4.45)$$

where $R_s(\tau)$ is the reliability of the switching device at time τ. The first term $R_1(t)$ in Eq. (4.45) represents the component 1 does not fail in (0, t]. The second term $\int_0^t R_s(\tau)f_1(\tau)R_2(t-\tau)d\tau$ indicates that component 1 fails at time τ with probability $f_1(\tau)$ and the switch S does not fail at time τ and is able to activate component 2. Component 2 then does not fail in $(\tau, t]$.

Assume that the two components are identically and exponentially distributed with constant failure rate parameter λ, and that the life of a switching device follows exponentially distributed with parameter λ_s, then from Eq. (4.45) the reliability of the 2-component standby system is given by

$$R(t) = e^{-\lambda t} + \int_0^t e^{-\lambda_s \tau}\lambda e^{-\lambda \tau}e^{-\lambda(t-\tau)}d\tau = e^{-\lambda t}\left[1 + \frac{\lambda}{\lambda_s}\left(1 - e^{-\lambda_s t}\right)\right].$$

The mean time to failure is

$$MTTF = \int_0^\infty R(t)dt = \frac{1}{\lambda} + \frac{1}{\lambda_s} - \frac{\lambda}{\lambda_s(\lambda + \lambda_s)}.$$

In case the reliability of switching device is independent of time. Let $R_s(\tau) = p_s$. Then from Eq. (4.45), we have

$$R(t) = e^{-\lambda t} + p_s \int_0^t \lambda e^{-\lambda \tau}e^{-\lambda(t-\tau)}d\tau = (1 + p_s\lambda t)e^{-\lambda t}.$$

The system MTTF is given by

$$MTTF = \int_0^\infty R(t)dt = \frac{1 + p_s}{\lambda}.$$

Example 4.8 Consider a standby system with two identical valves each with constant failure rate $\lambda = 10^{-5}$ failures/hours. The probability p_s that the switch S will activate the standby valve is 0.995. What is the reliability of the system at time $t = 8760$ h (or one year)?

Here $p_s = 0.995$, from Eq. (4.48) the reliability that the system will survive one year is,

$$
\begin{aligned}
R(8760) &= (1 + p_s \lambda t)e^{-\lambda t} \\
&= \left(1 + (0.995)(10^{-5})8760\right)e^{-(10^{-5})8760} \\
&= 0.9960
\end{aligned}
$$

The system MTTF is

$$
\begin{aligned}
MTTF &= \frac{1 + p_s}{\lambda} \\
&= \frac{1 + 0.995}{10^{-5}} = 199,500 \text{ hours}
\end{aligned}
$$

Non-identical components. Let us assume that the component 1 (active unit) and component 2 (standby unit) have constant rate λ_1 and λ_2, respectively. Let p_s is the probability that the switching is successful. This implies that the switch S will activate the standby unit with probability p_s. From Eq. (4.45), we have

$$
\begin{aligned}
R(t) &= e^{-\lambda_1 t} + p_s \int_0^t \lambda_1 e^{-\lambda_1 \tau} e^{-\lambda_2 (t - \tau)} d\tau \\
&= e^{-\lambda_1 t} + \frac{p_s \lambda_1}{\lambda_1 - \lambda_2} e^{-\lambda_2 t} - \frac{p_s \lambda_1}{\lambda_1 - \lambda_2} e^{-\lambda_1 t}
\end{aligned}
$$

The system MTTF is given by

$$
\begin{aligned}
MTTF &= \int_0^\infty R(t)dt = \int_0^\infty \left(e^{-\lambda t} + \frac{p_s \lambda_1}{\lambda_1 - \lambda_2} e^{-\lambda_2 t} - \frac{p_s \lambda_1}{\lambda_1 - \lambda_2} e^{-\lambda_1 t} \right) dt \\
&= \frac{1}{\lambda_1} + \frac{p_s}{\lambda_2}
\end{aligned}
\tag{4.46}
$$

3-component Cold-Standby System with imperfect switching device

Let $f_i(.)$ be the probability density function of the ith active component, $R_{si}(.)$ is the reliability of the switch when it is connected to the ith active component, and $R_i(.)$ is the reliability of the ith active component. Similarly, the reliability of a 3-component cold-standby system with imperfect switching device is given by

$$R(t) = R_1(t) + \int_0^t f_1(t_1)R_{s1}(t_1)\,R_2(t - t_1)dt_1$$

$$+ \int_0^t f_1(t_1)R_{s1}(t_1)\left(\int_0^{t-t_1} f_2(t_2)R_{s2}(t_2)R_3(t - t_1 - t_2)dt_2\right)dt_1 \qquad (4.47)$$

n-component Cold-Standby System with imperfect switching device

In general, one can obtain the reliability of a cold-standby system with n components as follows:

$$R(t) = R_1(t) + \int_0^t f_1(t_1)R_{s1}(t_1)\,R_2(t - t_1)dt_1$$

$$+ \int_0^t f_1(t_1)R_{s1}(t_1)\left(\int_0^{t-t_1} f_2(t_2)R_{s2}(t_2)R_3(t - t_1 - t_2)dt_2\right)dt_1$$

$$+ \cdots + \int_0^t \int_0^{t-t_1} \cdots \int_0^{t-t_1-t_2-\cdots-t_{n-2}} f_1(t_1)R_{s1}(t_1)f_2(t_2)R_{s2}(t_2)\ldots$$

$$f_{n-1}(t_{n-1})R_{s(n-1)}(t_{n-1})R_n(t - t_1 - t_2 - \cdots - t_{n-1})dt_{n-1}\ldots dt_2 dt_1 \qquad (4.48)$$

where $f_i(.)$ is the failure function of the ith active component, $R_{si}(.)$ is the reliability of the switch when it is connected to the ith active component, and $R_i(.)$ is the reliability of the ith active component. The mean time to failure of a cold-standby system with imperfect switching device can be obtained as

$$MTTF_{system} = \int_0^\infty R(t)dt$$

where $R(t)$ is given in Eq. (4.48).

Assume that all n components are identically and exponentially distributed with constant failure rate parameter λ, and that the failure time of the switching device follows an exponential distribution with parameter λ_{si} when it is connected to the ith active component. The reliability of the n-component cold standby system is given by

$$R(t) = e^{-\lambda t} + \int_0^t \lambda e^{-\lambda t_1} e^{-\lambda_{s1} t_1} e^{-\lambda(t-t_1)}dt_1 + \int_0^t \lambda e^{-\lambda t_1} e^{-\lambda_{s1} t_1}\left(\int_0^{t-t_1} \lambda e^{-\lambda t_2} e^{-\lambda_{s2} t_2} e^{-\lambda(t-t_1-t_2)}dt_2\right)dt_1$$

$$+\cdots+\int_0^t \int_0^{t-t_1} \cdots \int_0^{t-t_1-t_2-\cdots-t_{n-2}} \lambda e^{-\lambda t_1} e^{-\lambda_{s1} t_1} \lambda e^{-\lambda t_2} e^{-\lambda_{s2} t_2} \cdots$$

$$\lambda e^{-\lambda t_{n-1}} e^{-\lambda_{s(n-1)} t_{n-1}} e^{-\lambda (t-t_1-t_2-\cdots-t_{n-1})} dt_{n-1} \cdots dt_2 dt_1$$

After several calculations, we obtain

$$R(t) = e^{-\lambda t} + \left[\left(\frac{\lambda}{\lambda_{s1}}\right)e^{-\lambda t} - \left(\frac{\lambda}{\lambda_{s1}}e^{-(\lambda+\lambda_{s1})t}\right)\right]$$

$$+ \left[\left(\frac{\lambda^2}{\lambda_{s1}\lambda_{s2}}\right)e^{-\lambda t} - \left(\frac{\lambda^2}{\lambda_{s1}(\lambda_{s2}-\lambda_{s1})}e^{-(\lambda+\lambda_{s1})t} + \frac{\lambda^2}{\lambda_{s2}(\lambda_{s1}-\lambda_{s2})}e^{-(\lambda+\lambda_{s2})t}\right)\right]$$

$$+ \cdots + \left[\left(\frac{\lambda^{n-1}}{\prod_{i=1}^{n-1}\lambda_{si}}\right)e^{-\lambda t} - \left(\sum_{i=1}^{n-1}\frac{\lambda^{n-1}}{\lambda_{si}\prod_{j=1 \wedge j\neq i}^{n-1}(\lambda_{sj}-\lambda_{si})}\right)e^{-(\lambda+\lambda_{si})t}\right] \tag{4.49}$$

where λ and λ_{si} are the failure rates of all components in the active state and the failure rate of the switch when it is connected to the ith active component, respectively. From Eq. (4.49), the reliability of a non-repairable n-component cold-standby system with an imperfect switching when all the failure rates of the n components follow an exponential distribution can be rewritten as

$$R(t) = \begin{cases} e^{-\lambda t} & \text{for } n = 1 \\ \left[\left(\sum_{i=0}^{n-1}\frac{\lambda^i \lambda_{s(i+1)}}{\prod_{j=1}^{i+1}\lambda_{sj}}\right) - \left[\sum_{i=1}^{n-1}\sum_{k=1}^{i+1}\left(\frac{\lambda^i(\lambda_{s(i+1)}-\lambda_{sk})e^{-\lambda_{sk}t}}{\lambda_{sk}\left(\prod_{j=1 \wedge j\neq k}^{i+1}(\lambda_{sj}-\lambda_{sk})\right)}\right)\right]\right]e^{-\lambda t} & \text{for } n > 1 \end{cases} \tag{4.50}$$

Thus, the system MTTF is given by

$$MTTF_{system} = \int_0^\infty R(t)dt$$

$$= \int_0^\infty \sum_{i=0}^{n-1}\left(\frac{\lambda^i \lambda_{s(i+1)}e^{-\lambda t}}{\prod_{j=1}^{i+1}\lambda_{sj}}\right)dt - \int_0^\infty\left[\sum_{i=1}^{n-1}\sum_{k=1}^{i+1}\left(\frac{\lambda^i(\lambda_{s(i+1)}-\lambda_{sk})e^{-(\lambda_{sk}+\lambda)t}}{\lambda_{sk}\left(\prod_{j=1 \wedge j\neq k}^{i+1}(\lambda_{sj}-\lambda_{sk})\right)}\right)\right]dt$$

$$= \sum_{i=0}^{n-1}\left(\frac{\lambda^{i-1}\lambda_{s(i+1)}}{\prod_{j=1}^{i+1}\lambda_{sj}}\right) - \sum_{i=1}^{n-1}\sum_{k=1}^{i+1}\left(\frac{\lambda^i(\lambda_{s(i+1)}-\lambda_{sk})}{\lambda_{sk}(\lambda_{sk}+\lambda)\left(\prod_{j=1 \wedge j\neq k}^{i+1}(\lambda_{sj}-\lambda_{sk})\right)}\right) \tag{4.51}$$

From Eq. (4.51), when $\lambda_{si} = 0$, the system MTTF yields

$$MTTF_{system} = \frac{n}{\lambda}.$$

same as in Eq. (4.29).

2-component Hot-Standby System with imperfect switching device

If we assume that the standby component 2 may fail (in standby mode) before it is activated with a pdf $f_0(t)$, then from Eq. (4.45), the reliability of the two-component hot-standby system is given by

$$R(t) = R_1(t) + \int\limits_0^t R_s(\tau)R_0(\tau)f_1(\tau)R_2(t - \tau)d\tau \qquad (4.52)$$

where $R_0(\tau) = \int_\tau^\infty f_0(s)ds$ represents that component 2 does not fail in $(0, \tau]$, and $R_s(\tau)$ is the reliability of the switching device at time τ. The first term $R_1(t)$ in Eq. (4.52) represents the component 1 does not fail in $(0, t]$.

4.9 Load-Sharing Systems with Dependent Failures

A shared-load system refers to a parallel system whose units equally share the load for the system to function (Pham 1992a). Here we assume that for the shared-load parallel system to fail, all the units in the system must fail. In this section we will derive the reliability of the two-unit and three unit of load-sharing systems. Other higher number of units in the system can similarly be easily obtained the results.

Two-non-identical-unit load-sharing systems

Let $f_{if}(t)$ and $f_{ih}(t)$ be the pdf for time to failure of unit i, for $i = 1, 2$, under full load and half load condition, respectively. Also let $R_{if}(t)$ and $R_{ih}(t)$ be the reliability of unit i under full load and half load condition, respectively. Thus the system will be working if: (1) both units are working until the end of mission time t; (2) unit 1 fails before time t then unit 2 takes the full load and is operating until the end of time t; or (3) same as (2) but, instead, the unit 2 fails before time t then unit 1 takes the full load and is operating until the end of time t.

Mathematically, we can obtain the reliability function of shared-load 2-unit system as follows (Pham 2014):

$$R(t) = R_{1h}(t)R_{2h}(t) + \int\limits_0^t f_{1h}(s)R_{2h}(s)R_{2f}(t - s)ds + \int\limits_0^t f_{2h}(s)R_{1h}(s)R_{1f}(t - s)ds \qquad (4.53)$$

Three-non-identical unit load-sharing systems

Let $f_{if}(t)$, $f_{ih}(t)$, and $f_{ie}(t)$ be the pdf for time to failure of unit i, for $i = 1, 2$, and 3 under full load and half load and equal-load condition (this occurs when all three units are working), respectively. Also let $R_{if}(t)$, $R_{ih}(t)$, and $R_{ie}(t)$ be the reliability of unit i under full load and half load condition, respectively.

The following events would be considered for the three-unit load-sharing system to work:

Event 1: All the three units are working until the end of mission time t;

Event 2: All three units are working until time t_1; at time t_1 one of the three units fails. The remaining two units are working until the end of the mission.

Event 3: All three units are working until time t_1; at time t_1 one of the three units fails. Then at time t_2 the second unit fails, and the remaining unit is working until the end of mission t.

$$\text{Prob } \{event\ 1\} = \prod_{i=1}^{3} R_{ie}(t) \tag{4.54}$$

$$\text{Prob } \{event\ 2\} = \int_0^t f_{1e}(s)R_{2e}(s)R_{3e}(s)R_{2h}(t-s)R_{3h}(t-s)ds$$

$$+ \int_0^t f_{2e}(s)R_{1e}(s)R_{3e}(s)R_{1h}(t-s)R_{3h}(t-s)ds$$

$$+ \int_0^t f_{3e}(s)R_{1e}(s)R_{2e}(s)R_{1h}(t-s)R_{2h}(t-s)ds \tag{4.55}$$

and

$$\text{Prob } \{event\ 3\} = \int_0^t f_{1e}(s)R_{2e}(s)R_{3e}(s) \int_0^{t-t_1} f_{2h}(x)R_{3h}(x)R_{3f}(t-s-x)dx\,ds$$

$$+ \int_0^t f_{1e}(s)R_{2e}(s)R_{3e}(s) \int_0^{t-t_1} f_{3h}(x)R_{2h}(x)R_{2f}(t-s-x)dx\,ds$$

$$+ \int_0^t f_{2e}(s)R_{1e}(s)R_{3e}(s) \int_0^{t-t_1} f_{1h}(x)R_{3h}(x)R_{3f}(t-s-x)dx\,ds$$

$$+ \int_0^t f_{2e}(s)R_{1e}(s)R_{3e}(s) \int_0^{t-t_1} f_{3h}(x)R_{1h}(x)R_{1f}(t-s-x)dx\,ds$$

$$+ \int_0^t f_{3e}(s)R_{1e}(s)R_{2e}(s) \int_0^{t-t_1} f_{1h}(x)R_{2h}(x)R_{2f}(t-s-x)dx\,ds$$

$$+ \int_0^t f_{3e}(s)R_{1e}(s)R_{2e}(s) \int_0^{t-t_1} f_{2h}(x)R_{1h}(x)R_{1f}(t-s-x)dx\,ds. \tag{4.56}$$

The reliability of a three-unit load-sharing system is, from Eqs. (4.54–4.56):

$$R(t) = \text{Prob}\{event\ 1\} + \text{Prob}\{event\ 2\} + \text{Prob}\{event\ 3\}$$

where Prob {event 1}, Prob {event 2} and Prob {event 3} are given in Eqs. (4.54), (4.55) and (4.56), respectively.

Example 4.9 Consider a three-unit load-sharing parallel system where λ_0 be the constant failure rate of a unit when all the three units are working; λ_h be the constant failure rate of each of the two surviving units, each of which shares half of the total load; and λ_f be the constant failure rate of a unit at full load. Then we obtain

$$\text{Prob}\{event\ 1\} = \prod_{i=1}^{3} R_{ie}(t) = e^{-3\lambda_0 t} \tag{4.57}$$

$$\text{Prob}\{event\ 2\} = \int_0^t f_{1e}(s)R_{2e}(s)R_{3e}(s)R_{2h}(t-s)R_{3h}(t-s)ds$$

$$+ \int_0^t f_{2e}(s)R_{1e}(s)R_{3e}(s)R_{1h}(t-s)R_{3h}(t-s)ds$$

$$+ \int_0^t f_{3e}(s)R_{1e}(s)R_{2e}(s)R_{1h}(t-s)R_{2h}(t-s)ds$$

$$= 3 \int_0^t \lambda_0 e^{-\lambda_0 s} e^{-2\lambda_0 s} e^{-2\lambda_h(t-s)} ds$$

$$= \frac{3\lambda_0}{2\lambda_h - 3\lambda_0} \left(e^{-3\lambda_0 t} - e^{-2\lambda_h t} \right). \tag{4.58}$$

And from Eq. (4.56), we obtain

$$\text{Prob}\{event\ 3\} = 6 \int_0^t \lambda_0 e^{-\lambda_0 t_1} e^{-2\lambda_0 t_1} \int_0^{t-t_1} \lambda_h e^{-\lambda_h t_2} e^{-\lambda_h t_2} e^{-\lambda_f(t-t_1-t_2)} dt_2 dt_1$$

$$= 6\lambda_0\lambda_h \left(\frac{e^{-3\lambda_0 t}}{(3\lambda_0 - 2\lambda_h)(3\lambda_0 - \lambda_f)} + \frac{e^{-2\lambda_h t}}{(2\lambda_h - 3\lambda_0)(2\lambda_h - \lambda_f)} + \frac{e^{-\lambda_f t}}{(\lambda_f - 3\lambda_0)(\lambda_f - 2\lambda_h)} \right) \tag{4.59}$$

The reliability of a three-unit load-sharing parallel system is

$$R(t) = \text{Prob}\{event\ 1\} + \text{Prob}\{event\ 2\} + \text{Prob}\{event\ 3\}$$

$$= e^{-3\lambda_0 t} + \frac{3\lambda_0}{2\lambda_h - 3\lambda_0} \left(e^{-3\lambda_0 t} - e^{-2\lambda_h t} \right)$$

$$+ 6\lambda_0\lambda_h\left(\frac{e^{-3\lambda_0 t}}{(3\lambda_0 - 2\lambda_h)(3\lambda_0 - \lambda_f)} + \frac{e^{-2\lambda_h t}}{(2\lambda_h - 3\lambda_0)(2\lambda_h - \lambda_f)} + \frac{e^{-\lambda_f t}}{(\lambda_f - 3\lambda_0)(\lambda_f - 2\lambda_h)}\right). \quad (4.60)$$

The system MTTF can be easily obtained as follows

$$\text{MTTF} = \int_0^\infty R(t)dt = \frac{1}{3\lambda_0} + \frac{1}{2\lambda_h} + \frac{1}{\lambda_f}$$

Example 4.10 Continued from Example 4.8, given that $\lambda_0 = 0.001/hour$, $\lambda_h = 0.005/hour$, $\lambda_f = 0.05/hour$ then

$$\text{MTTF} = \frac{1}{3\lambda_0} + \frac{1}{2\lambda_h} + \frac{1}{\lambda_f} = \frac{1}{3(0.001)} + \frac{1}{2(0.005)} + \frac{1}{(0.05)} = 453.33 \text{ hours.}$$

Example 4.11 A Three-unit Load-Sharing System

A shared-load system refers to a parallel system whose units equally share the system function. Therefore, for a shared-load parallel system to fail, all the units in the system must fail. Determine the reliability and mean time to failure (MTTF) of a 3-unit shared-load parallel system where

λ_0 is the constant failure rate of a unit when all the three units are operational;
λ_h is the constant failure rate of each of the two surviving units, each of which shares half of the total load; and.
λ_f is the constant failure rate of a unit at full load.

(a)   Determine the reliability at $t = 10$ h where

$$\lambda_0 = 0.001/hour; \quad \lambda_h = 0.005/hour; \quad \text{and } \lambda_f = 0.05/hour.$$

(b)   What is the MTTF?

Solution Consider the following three events:

Event 1: All the three units are working until the end of the mission time t where each unit shares one-third of the total load.

Event 2: All the three units are working until time t_1 (each shares one-third of the total load). At time t_1, one of the units (say unit 1) fails, and the other two units (say unit 2 and 3) remain to work until the mission time t. Here, once a unit fails at time t_1, the remaining two working units would take half each of the total load and have a constant rate λ_h. As for all identical units, there are 3 possibilities under this situation.

Event 3: All the three units are working until time t_1 (each shares one-third of the total load) when one (say unit 1) of the three units fails. At time t_2, $(t_2 > t_1)$ one more unit fails (say unit 2) and the remaining unit works until the end of the mission time t. Under this event, there are 6 possibilities that the probability of two units failing before time t and only one unit remains to work until time t.

Let T_1, T_2 and T_3 be failure time of unit 1, 2 and 3, respectively. Let

$R_0(t)$ be the reliability of a unit when all the three units are operational
$R_h(t)$ be reliability of each of the two remaining working units each of which shares half of the total load
$R_f(t)$ be the reliability of a unit at full load.

Then

$$
\begin{aligned}
P_1(t) &\equiv P\{\text{Event 1}\} = P\{T_1 > t, T_2 > t, T_3 > t\} \\
&= (R_0(t))^3 \\
&= e^{-3\lambda_0 t}
\end{aligned}
\tag{4.61}
$$

Next,

$$
\begin{aligned}
P_2(t) &\equiv P\{\text{Event 2}\} \\
&= 3\, P\{\{T_1 = t_1 < t, T_2 > t_1, T_3 > t_1\} \text{ and } \{T_2 > (t-t_1) \text{ and } T_3 > (t-t_1)\}\}+ \\
&= 3 \int_0^t f_0(t_1) R_0^2(t_1) R_h^2(t-t_1)\, dt_1
\end{aligned}
\tag{4.62}
$$

where $f_0(t_1)$ is the pdf of a unit with one-third of the total load when all three units are functioning. For exponential distribution function, from Eq. (4.62) we have

$$
\begin{aligned}
P_2(t) &= 3 \int_0^t \lambda_0\, e^{-\lambda_0 t_1}\, e^{-2\lambda_0 t_1}\, e^{-2\lambda_h(t-t_1)}\, dt_1 \\
&= \frac{3\lambda_0}{2\lambda_h - 3\lambda_0} \left(e^{-3\lambda_0 t} - e^{-2\lambda_h t} \right)
\end{aligned}
\tag{4.63}
$$

Finally,

$$
\begin{aligned}
P_3(t) &\equiv P\{\text{Event 3}\} \\
&= 6\, P\{\{T_1 = t_1 < t, T_2 > t_1, T_3 > t_1\} \cap \{T_2 = t_2 > t_1 \text{ and } T_3 > t_2\} \cap \{T_3 > (t-t_1-t_2)\}\} \\
&= 6 \int_0^t f_0(t_1) R_0^2(t_1) \left\{ \int_0^{t-t_1} f_h(t_2) R_h(t_2) R_f(t-t_1-t_2)\, dt_2 \right\} dt_1
\end{aligned}
\tag{4.64}
$$

where $f_h(t_2)$ is the pdf of a unit with a half of the total load.
For exponential distribution function, from Eq. (4.64) we obtain

$$
P_3(t) = 6 \int_0^t \lambda_0 e^{-\lambda_0 t_1} e^{-2\lambda_0 t_1} \left\{ \int_0^{t-t_1} \lambda_h\, e^{-\lambda_h t_2}\, e^{-\lambda_h t_2}\, e^{-\lambda_f(t-t_1-t_2)}\, dt_2 \right\} dt_1
$$

$$= 6\lambda_0\lambda_h\left[\frac{e^{-3\lambda_0 t}}{(3\lambda_0 - 2\lambda_h)(3\lambda_0 - \lambda_f)} + \frac{e^{-2\lambda_h t}}{(2\lambda_h - 3\lambda_0)(2\lambda_h - \lambda_f)} + \frac{e^{-\lambda_f t}}{(\lambda_f - 3\lambda_0)(\lambda_f - 2\lambda_h)}\right] \quad (4.65)$$

Note that one can also calculate the $P_3(t)$ as follows:

$$P_3(t) \equiv P\{\text{Event 3}\}$$

$$= \binom{3}{1}\binom{2}{1}P\{\{T_1 = t_1 < t, T_2 > t_1, T_3 > t_1\} \cap \{T_2 = t_2 - t_1 \ \& \ T_3 > t_2 - t_1\} \cap \{T_3 > (t - t_2)\}\}$$

$$= 6\int_{t_1=0}^{t} f_0(t_1)R_0^2(t_1)\left\{\int_{t_2=t_1}^{t} f_h(t_2 - t_1)R_h(t_2 - t_1)R_f(t - t_2)dt_2\right\}dt_1 \quad (4.66)$$

which is the same as Eq. (4.64). For exponential distribution function, from Eq. (4.66) we obtain as follows:

$$P_3(t) = 6\int_0^{t} \lambda_0 e^{-\lambda_0 t_1} e^{-2\lambda_0 t_1}\left\{\int_{t_2=t_1}^{t} \lambda_h e^{-\lambda_h(t_2 - t_1)} e^{-\lambda_h(t_2 - t_1)} e^{-\lambda_f(t - t_2)}dt_2\right\}dt_1$$

$$= 6\lambda_0\lambda_h\left[\frac{e^{-3\lambda_0 t}}{(3\lambda_0 - 2\lambda_h)(3\lambda_0 - \lambda_f)} + \frac{e^{-2\lambda_h t}}{(2\lambda_h - 3\lambda_0)(2\lambda_h - \lambda_f)} + \frac{e^{-\lambda_f t}}{(\lambda_f - 3\lambda_0)(\lambda_f - 2\lambda_h)}\right]$$

$$= \frac{6\lambda_0\lambda_h}{(\lambda_f - 2\lambda_h)}\left[\frac{e^{-3\lambda_0 t}}{(2\lambda_h - 3\lambda_0)} - \frac{e^{-2\lambda_h t}}{(2\lambda_h - 3\lambda_0)} - \frac{e^{-3\lambda_0 t}}{(\lambda_f - 3\lambda_0)} + \frac{e^{-\lambda_f t}}{(\lambda_f - 3\lambda_0)}\right] \quad (4.67)$$

where Eq. (4.67) is the same as Eq. (4.65). Thus, the reliability of a three-unit shared load parallel system is

$$R(t) = P_1(t) + P_2(t) + P_3(t)$$

$$= e^{-3\lambda_0 t} + \frac{3\lambda_0}{2\lambda_h - 3\lambda_0}\left(e^{-3\lambda_0 t} - e^{-2\lambda_h t}\right)$$

$$+ 6\lambda_0\lambda_h\left[\frac{e^{-3\lambda_0 t}}{(3\lambda_0 - 2\lambda_h)(3\lambda_0 - \lambda_f)} + \frac{e^{-2\lambda_h t}}{(2\lambda_h - 3\lambda_0)(2\lambda_h - \lambda_f)} + \frac{e^{-\lambda_f t}}{(\lambda_f - 3\lambda_0)(\lambda_f - 2\lambda_h)}\right]. \quad (4.68)$$

The system mean time to failure (MTTF) is

$$MTTF = \int_0^{\infty} R(t)dt$$

$$= \int_0^{\infty}\left\{\begin{array}{l} e^{-3\lambda_0 t} + \frac{3\lambda_0}{2\lambda_h - 3\lambda_0}\left(e^{-3\lambda_0 t} - e^{-2\lambda_h t}\right) \\ + 6\lambda_0\lambda_h\left[\frac{e^{-3\lambda_0 t}}{(3\lambda_0 - 2\lambda_h)(3\lambda_0 - \lambda_f)} + \frac{e^{-2\lambda_h t}}{(2\lambda_h - 3\lambda_0)(2\lambda_h - \lambda_f)} + \frac{e^{-\lambda_f t}}{(\lambda_f - 3\lambda_0)(\lambda_f - 2\lambda_h)}\right] \end{array}\right\}dt$$

$$= \frac{1}{3\lambda_0} + \frac{1}{2\lambda_h} + \frac{1}{\lambda_f}.$$

Given $\lambda_0 = 0.001/hour$; $\lambda_h = 0.005/hour$; and $\lambda_f = 0.05/hour$. From Eq. (4.68), the reliability at time $t = 10$ h is given by

$$R(t) = e^{-3(0.001)(10)} + \frac{3(0.001)}{2(0.005) - 3(0.001)} \left(e^{-3(0.001)(10)} - e^{-2(0.005)(10)}\right)$$

$$+ 6(0.001)(0.005) \left[\begin{array}{l} \frac{e^{-3(0.001)(10)}}{(3(0.001)-2(0.005))(3(0.001)-0.05)} \\ + \frac{e^{-2(0.005)(10)}}{(2(0.005)-3(0.001))(2(0.005)-0.05)} \\ + \frac{e^{-(0.05)(10)}}{(0.05-3(0.001))(0.05-2(0.005))} \end{array} \right]$$

$$= 0.9997.$$

The system mean time to failure (MTTF) is

$$MTTF = \frac{1}{3\lambda_0} + \frac{1}{2\lambda_h} + \frac{1}{\lambda_f}$$

$$= \frac{1}{3(0.001)} + \frac{1}{2(0.005)} + \frac{1}{(0.05)}$$

$$= 453.33 \text{ hours.}$$

An application: Consider a high-voltage load-sharing system consisting of a power supply and two transmitters, say A and B, using mechanically tuned magnetrons (Pham 1992a). Two transmitters are used, each tuning one-half the desired frequency range; however, if one transmitter fails, the other can tune the entire range with a resultant change in the expected time to failure. Suppose that for this system to work, the power supply and at least one of the transmitters must work. The system diagram that characterizes this functional behavior is shown in Fig. 4.6. The transmitters operate independently when both are functional. An undetected fault in either of the transmitters causes the system to fail.

Suppose that transmitter A has constant failure rate λ_A if transmitter B is operating and failure rate λ'_A if transmitter B has failed. Similarly, suppose that transmitter B has constant failure rate λ_B if transmitter A is operating and failure rate λ'_B if transmitter A has failed. Denote λ_P and λ_C be the constant failure rate of power supply and fault coverage, respectively.

The system works if and only if the power supply works and at least one transmitters operate. Let

Event E_1: Both transmitters and power supply operate by time T

Fig. 4.6 A simplified high-voltage load-sharing system

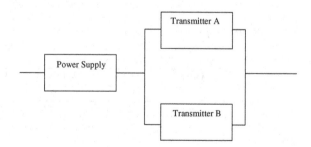

Event E_2: The power supply works by time T, transmitter A fails at time t for $0 \leq t \leq T$, and transmitter B does not fail by by time T.

Event E_3: The power supply works at time T, transmitter B fails at time t for $0 \leq t \leq T$, and transmitter A does not fail by time T.

Then we have

$$P(E_1) = e^{-(\lambda_A + \lambda_B + \lambda_P)t}$$

$$P(E_2) = \int_0^T \lambda_A e^{-\lambda_A t} e^{-\lambda_B t} e^{-\lambda_C t} e^{-\lambda_B'(T-t)} e^{-\lambda_P T} dt$$

$$= \begin{cases} \frac{\lambda_A e^{-(\lambda_P + \lambda_B')T}}{(\lambda_A + \lambda_B + \lambda_C - \lambda_B')} \left(1 - e^{-(\lambda_A + \lambda_B + \lambda_C - \lambda_B')T}\right) & \text{if } \lambda_A + \lambda_B + \lambda_C - \lambda_B' \neq 0 \\ \lambda_A T e^{-(\lambda_P + \lambda_B')T} & \text{if } \lambda_A + \lambda_B + \lambda_C - \lambda_B' = 0 \end{cases}$$

Similarly, we can obtain

$$P(E_3) = \int_0^T \lambda_B e^{-\lambda_B t} e^{-\lambda_A t} e^{-\lambda_C t} e^{-\lambda_A'(T-t)} e^{-\lambda_P T} dt$$

$$= \begin{cases} \frac{\lambda_B e^{-(\lambda_P + \lambda_A')T}}{(\lambda_A + \lambda_B + \lambda_C - \lambda_A')} \left(1 - e^{-(\lambda_A + \lambda_B + \lambda_C - \lambda_A')T}\right) & \text{if } \lambda_A + \lambda_B + \lambda_C - \lambda_A' \neq 0 \\ \lambda_B T e^{-(\lambda_P + \lambda_A')T} & \text{if } \lambda_A + \lambda_B + \lambda_C - \lambda_A' = 0 \end{cases}$$

It can be shown (see Problem 3) that the reliability of this high-voltage load-sharing system with dependent failures is given by (Pham 1992a):

$$R(t) = e^{-(\lambda_A + \lambda_B + \lambda_P)t} + \frac{\lambda_A e^{-(\lambda_P + \lambda_B')t}}{(\lambda_A + \lambda_B + \lambda_C - \lambda_B')} \left(1 - e^{-(\lambda_A + \lambda_B + \lambda_C - \lambda_B')t}\right)$$

$$+ \frac{\lambda_B e^{-(\lambda_P + \lambda_A')t}}{(\lambda_A + \lambda_B + \lambda_C - \lambda_A')} \left(1 - e^{-(\lambda_A + \lambda_B + \lambda_C - \lambda_A')t}\right) \qquad (4.69)$$

The system MTTF is

$$MTTF = \int_0^\infty R(t) dt$$

$$= \frac{1}{\lambda_A + \lambda_B + \lambda_P} + \frac{\lambda_A}{(\lambda_A + \lambda_B + \lambda_C - \lambda_B')} \left(\frac{1}{\lambda_P + \lambda_B'} - \frac{1}{\lambda_A + \lambda_B + \lambda_C + \lambda_P}\right)$$

$$+ \frac{\lambda_B}{(\lambda_A + \lambda_B + \lambda_C - \lambda_A')} \left(\frac{1}{\lambda_P + \lambda_A'} - \frac{1}{\lambda_A + \lambda_B + \lambda_C + \lambda_P}\right) \qquad (4.70)$$

t	R_1	R_2	R_3
0	1.0000	1.0000	1.0000
5	0.9986	0.9985	0.9985
10	0.9950	0.9949	0.9948
15	0.9897	0.9895	0.9892
20	0.9829	0.9825	0.9820
25	0.9748	0.9742	0.9734
30	0.9656	0.9648	0.9638
35	0.9555	0.9544	0.9531
40	0.9446	0.9433	0.9416
45	0.9330	0.9314	0.9294
50	0.9209	0.9190	0.9166
55	0.9082	0.9060	0.9033
60	0.8952	0.8926	0.8895
65	0.8817	0.8789	0.8754
70	0.8680	0.8649	0.8610
75	0.8541	0.8506	0.8463
80	0.8399	0.8361	0.8314
85	0.8256	0.8215	0.8164
90	0.8112	0.8067	0.8013
95	0.7967	0.7919	0.7861
100	0.7821	0.7770	0.7708

Table 4.1 Reliability values R_1, R_2, R_3 for $\lambda_C = 0.0001$, $\lambda_C = 0.0005$, and $\lambda_C = 0.001$, respectively

Consider a high-voltage load-sharing system where, for example,

$$\lambda_A = 0.0005/h \quad \lambda_B = 0.0001/h \quad \lambda_A' = 0.005/h \quad \lambda_B' = 0.001/h$$
$$\lambda_C = 0.000005/h \quad \lambda_P = 0.000001/h$$

From Eqs. (4.69) and (4.70), Table 4.1 and Fig. 4.7 show the values of system reliability for various mission time t (in hours). The system MTTF values are 249.4 h, 244.3 h and 238.7 h for $\lambda_C = 0.0001$, $\lambda_C = 0.0005$, and $\lambda_C = 0.001$, respectively.

4.10 Reliability Evaluation of Complex Systems Using Conditional Probability

In practice many systems, i.e., electric power station, telephone network, communication and control systems, are complicated that perhaps cannot by easily decomposed into the well-system structures such as series, parallel, series–parallel, parallel–series,

Fig. 4.7 System reliability values for $\lambda_C = 0.0001$, $\lambda_C = 0.0005$, and $\lambda_C = 0.001$, respectively

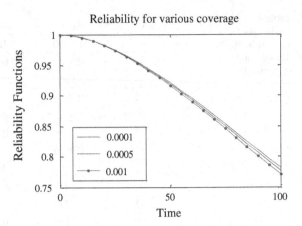

standby, or k-out-of-n. Reliability calculation of those complex systems would require more complicate approach. We now discuss the conditional probability approach, also called the total probability theorem or Bayes' method. The approach starts with choosing a keystone component, say component X, which hopefully appears to bind the system together. In general, an expression for the reliability of the system, R, can be derived using the form of the total probability theorem as translated into the language of reliability. Let X is the event that keystone component X is working, $\bar{X}$ is the event that keystone component X has failed. Then

$$\text{Pr}\{\text{system works}\} = \text{Pr}\{\text{system works when unit X is good}\} \cdot \text{Pr}\{\text{unit } X \text{ is good}\}$$
$$+ \text{Pr}\{\text{system works when unit } X \text{ fails}\} \cdot \text{Pr}\{\text{unit } X \text{ is failed}\}$$

or, equivalently,

$$R = \text{Pr}(\text{system works}|\text{X}) \ \text{Pr}(X) + \text{Pr}(\text{system works}|\overline{\text{X}}) \ \text{Pr}(\overline{X}) \tag{4.71}$$

For example in Fig. 4.8, we can consider component 5 as a keystone component. Note that one can just use any component in the system as keystone component, not has to be component 5. The efficiency of the approach depends on the selection of the keystone component. The reliability results obviously should be the same whatever the selection of the keystone component. However, a careful choice of keystone components would be of help to save lot of computational times.

Fig. 4.8 A bridge system

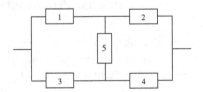

The reliability of the bridge system (Fig. 4.8) can be written as

$$R = \Pr(\text{system works}|\text{component 5 works}) \; \Pr(\text{component 5 works})$$
$$+ \Pr(\text{system works}|\text{comp. 5 has failed}) \; \Pr(\text{comp. 5 has failed}) \quad (4.72)$$

$\Pr(\text{system works}|\text{component 5 works})$ is the probability that the system is working given that component 5 works, and $\Pr(\text{system works}|\text{component 5 has failed})$ is the probability that the system is working given that component 5 has failed.

Example 4.12 Consider the bridge system shown in Fig. 4.8. Suppose that the reliability of component i is R_i, $i = 1, 2, \ldots, 5$. To obtain the reliability of the system, let us consider component 5 as the keystone component. From Eq. (4.72), the reduced system is a series–parallel structure as shown in Fig. 4.9, given that the component 5 is working, and that

$$\Pr(\text{system works}/\text{component 5 works}) = [1 - (1 - R_1)(1 - R_3)] \cdot [1 - (1 - R_2)(1 - R_4)].$$

Similarly, the reduced system is a parallel–series structure as shown in Fig. 4.10, given that the component 5 has failed. That is,

$$\Pr(\text{system works}|\text{component 5 has failed}) = 1 - (1 - R_1 R_2)(1 - R_3 R_4).$$

Hence, the reliability of the bridge system is

$$R = \left[1 - (1 - R_1)(1 - R_3)\right]\left[1 - (1 - R_2)(1 - R_4)\right]R_5 + \left[1 - (1 - R_1 R_2)(1 - R_3 R_4)\right](1 - R_5)$$
$$= R_1 R_2 + R_3 R_4 + R_1 R_4 R_5 + R_2 R_3 R_5 - R_1 R_2 R_4 R_5 - R_2 R_3 R_4 R_5 - R_1 R_2 R_3 R_5$$
$$- R_1 R_3 R_4 R_5 - R_1 R_2 R_3 R_4 + 2 R_1 R_2 R_3 R_4 R_5.$$

Fig. 4.9 The bridge system when component 5 never fails

Fig. 4.10 The bridge system when component 5 has failed

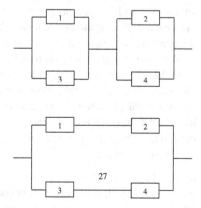
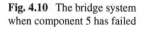

4.10.1   Applications of Fault-Tolerant Systems Using Decomposition Method

In many critical applications of digital systems, fault tolerance has been an essential architectural attribute for achieving high reliability. It is universally accepted that computers cannot achieve the intended reliability in operating systems, application programs, control programs, or commercial systems, such as in the space shuttle, nuclear power plant control, etc., without employing redundancy. Several techniques can achieve fault tolerance using redundant hardware or software. Typical forms of redundant hardware structures for fault-tolerant systems are of two types: fault masking and standby. Masking redundancy is achieved by implementing the functions so that they are inherently error correcting, e.g. triple-modular redundancy (TMR), N-modular redundancy (NMR), and self-purging redundancy. In standby redundancy, spare units are switched into the system when working units break down.

This section presents a fault-tolerant architecture to increase the reliability of a special class of digital systems in communication (Pham and Upadhyaya 1989). In this system, a monitor and a switch are associated with each redundant unit. The switches and monitors can fail. The monitors have two failure modes: failure to accept a correct result, and failure to reject an incorrect result. The scheme can be used in communication systems to improve their reliability.

Consider a digital circuit module designed to process the incoming messages in a communication system Pham and Upadhyaya (1989). This module consists of two units: a converter to process the messages, and a monitor to analyze the messages for their accuracy. For example, the converter could be decoding or unpacking circuitry, whereas the monitor could be checker circuitry. To guarantee a high reliability of operation at the receiver end, n converters are arranged in "parallel". All, except converter n, have a monitor to determine if the output of the converter is correct. If the output of a converter is not correct, the output is cancelled and a switch is changed so that the original input message is sent to the next converter. The architecture of such a system has been proposed by Pham and Upadhyaya (1989). Systems of this kind have useful applications in communication and network control systems and in the analysis of fault-tolerant software systems.

We assume that a switch is never connected to the next converter without a signal from the monitor, and the probability that it is connected when a signal arrives is p_s. We next present a general expression for the reliability of the system consisting of n non-identical converters arranged in "parallel". An optimization problem is formulated and solved for the minimum average system cost. Let us define the following notation, events, and assumptions.

The notation is as follows:

p_i^c Pr{converter i works}

p_i^s Pr{switch i is connected to converter $(i + 1)$ when a signal arrives}

p_i^{m1} Pr {monitor i works when converter i works} $=$ Pr{not sending a signal to the switch when converter i works}

p_i^{m2} Pr{i monitor works when converter i has failed} $=$ Pr{sending a signal to the switch when converter i has failed}

R_{n-k}^k reliability of the remaining system of size $n-k$ given that the first k switches work

R_n reliability of the system consisting of n converters.

The events are:

C_i^w, C_i^f converter i works, fails

M_i^w, M_i^f monitor i works, fails

S_i^w, S_i^f switch i works, fails

W system works.

The assumptions are:

1. The system, the switches, and the converters are two-state: good or failed.
2. The module (converter, monitor, or switch) states are mutually statistical independent.
3. The monitors have three states: good, failed in mode 1, failed in mode 2.
4. The modules are not identical.

Reliability Evaluation

The reliability of the system is defined as the probability of obtaining the correctly processed message at the output. To derive a general expression for the reliability of the system, we use an adapted form of the total probability theorem as translated into the language of reliability. Let A denote the event that a system performs as desired. Let X_i and X_j be the event that a component X(e.g. converter, monitor, or switch) is good or failed respectively. Then

Pr{system works} $=$ Pr{system works when unit X is good} $\cdot$ Pr{ unit X is good}

$+$ Pr{system works when unit X fails} $\cdot$ Pr{unit X is failed}

The above equation provides a convenient way of calculating the reliability of complex systems. Notice that $R_1 = p_i^c$ and for $n \geq 2$, the reliability of the system can be calculated as follows:

$$R_n = \text{Pr}\{W|C_1^w \text{ and } M_1^w\} \text{Pr}\{C_1^w \text{ and } M_1^w\} + \text{Pr}\{W|C_1^w \text{ and } M_1^f\} \text{Pr}\{C_1^w \text{ and } M_1^f\}$$
$$+ \text{Pr}\{W|C_1^f \text{ and } M_1^w\} \text{Pr}\{C_1^f \text{ and } M_1^w\} + \text{Pr}\{W|C_1^f \text{ and } M_1^f\} \text{Pr}\{C_1^f \text{ and } M_1^f\}$$

In order for the system to operate when the first converter works and the first monitor fails, the first switch must work and the remaining system of size $n-1$ must work:

$$\text{Pr}\{W|C_1^w \text{ and } M_1^f\} = p_1^s R_{n-1}^1$$

Similarly:

$$\Pr\{W|C_1^f \text{ and } M_1^w\} = p_1^s R_{n-1}^1$$

then

$$R_n = p_1^c p_1^{m_1} + [p_1^c(1 - p_1^{m_1}) + (1 - p_1^c)p_1^{m_2}]p_1^s R_{n-1}^1$$

The reliability of the system consisting of n non-identical converters can be easily obtained:

$$R_n = \sum_{i=1}^{n-1} p_i^c p_i^{m1} \pi_{i-1} + \pi_{n-1} p_n^c \quad \text{for } n > 1 \tag{4.73}$$

and

$$R_1 = p_1^c$$

where

$$\pi_k^j = \prod_{i=j}^{k} A_i \text{ for } k \geq 1$$

$$\pi_k \equiv \pi_k^1 \quad \text{for all } k, \text{ and } \pi_0 = 1$$

and

$$A_i \equiv [p_i^c(1 - p_i^{m1}) + (1 - p_i^c)p_i^{m2}] \text{ for all } i = 1, 2, \dots, n$$

Assume that all the converters, monitors, and switches have the same reliability, that is:

$$p_i^c = p^c, p_i^{m1} = p^{m1}, p_i^{m2} = p^{m2}, \ p_i^s = p^s \quad \text{for all } i$$

then we obtain a closed form expression for the reliability of system as follows:

$$R_n = \frac{p^c p^{m1}}{1 - A}(1 - A^{n-1}) + p^c A^{n-1} \tag{4.74}$$

where

$$A = [p^c(1 - p^{m1}) + (1 - p^c)p^{m2}]p^s$$

Redundancy Optimization

Assume that the system failure costs d units of revenue, and that each converter, monitor, and switch module costs a, b, and c units respectively. Let T_n be system cost for a system of size n. The average system cost for size n, $E[T_n]$, is the cost incurred when the system has failed, plus the cost of all n converters, $n - 1$ monitors, and $n - 1$ switches. Therefore:

$$E[T_n] = an + (b + c)(n - 1) + d(1 - R_n) \qquad (4.75)$$

where R_n is given in Eq. (4.74). The minimum value of $E[T_n]$ is attained at

$$n* = \begin{cases} 1 & \text{if } A \le 1 - p^{m1} \\ \lfloor n_0 \rfloor & otherwise \end{cases}$$

where

$$n_0 = \frac{\ln(a + b + c) - \ln[dp^c(A + p^{m1} - 1)]}{\ln A} + 1$$

Example 4.13 (Pham and Upadhyaya 1989) Given a system with $p^c = 0.8$, $p^{m1} = 0.90$, $p^{m2} = 0.95$, $p^s = 0.90$, and $a = 2.5$, $b = 2.0$, $c = 1.5$, $d = 1200$. The optimal system size is $n* = 4$, and the corresponding average cost (81.8) from Eq. (4.75) is minimized.

4.11 Degradable Systems

In many applications, parallel redundant components are employed in order to guarantee a certain service level. In reality with complex dynamic changing environments, the state of reliability of the components can change due to the changing conditions of complex environments (Pham 1992b, 2010; Pham and Galyean 1992; Pham et al. 1996, 1997). To many applications, a minimum number of units is needed to function in order to guarantee the functioning of the systems, particular for the fault-tolerant computing systems.

Consider a network of electric generators for example, there are at least 80% of all generators must work in order to ensure a certain service level of the entire station under normal conditions, called "good mode". Under a stress condition, called "overloaded mode" or "degradation mode", there are at least 90% of all generators must work in order to ensure a full service level of the station. In the latter case, due to a higher work load of each generator, the reliability of working generator will be considerably decreased.

In this section, we consider that the system consists of n statistically independent and non-repairable units. The system and the units are either in a successful state (working) or in a failed state (not working). What separates this system from other two state systems is that the successful state contains two modes: a good mode and a degradation mode.

Consider a power plant with four generators to provide electricity for town X for example (Pham 2009). To fully power the town, it is recommended that at least three generators must be working; though the plant could provide all necessary power with only at least two working generators. Under non-peak (normal conditions) hours of the day, that means any time of the day except from 5 p.m. to 8 p.m., the power plant is in a good working mode if only two or more generators out of the four are working. However during the peak (stress conditions) hours of the day between 5 and 8 p.m., there are at least three out of four generators must work for the plant to be able to provide the town the full capacity with electricity; resulting in a degradation in generator reliability (a different generator reliability than earlier) and thus being in an overloaded mode. The power plant fails if any three generators break down, as it will no longer be able to provide the town with electricity. This is an example of degradable dual-system with competing operating modes: a 2-out-of-4 system under a good mode and 3-out-of-4 system under overloaded mode.

An application of the power substations is another example. A faulty control that improperly restricted power output was at the root of a massive power outage on the May 25, 2006 incident that shut down the Washington-to-New York City rail corridor for more than two hours during the morning rush hours. The events leading up to the outage began several days before when the station reduced the power capacity at one of the substations in order to perform the maintenance. The system was designed to be able to keep operating during non-rush hours with one or two substations in the maintenance status. After the problem was fixed, station employees tried to restore output to full capacity to be ready for the morning rush hours. Unfortunately a computerized control never implemented the command. So when they got into the rush hours, the other substations, however, attempted to compensate for the one that was limited, but were severely overloaded during peak-hour demand, which led them to shut down.

Here, the system is composed of n independent units; the units and the system are in one of two states: successful or failed; a successful system and all successful units are either in good mode or in an overloaded mode. Define

n	number of units in the system
$p_1^{(i)}$	probability for the ith component to be successful when the system is in good mode
$p_2^{(i)}$	probability for the ith component to be successful when the system is in overloaded mode $\left(p_1^{(i)} \geq p_2^{(i)}\right)$
w_i	weight of unit i

$$w(J) = \sum_{i \in J} w_i$$

$\alpha(1 - \alpha)$ probability for the system to be in good (overloaded) mode
R(n)   system reliability
t_1, t_2 threshold for the cumulative weight subject to good mode and overloaded mode, respectively.

The system works in a good mode if and only if the cumulative weight of all working units exceeds or equals t_1 w(I). Similarly, the system works in overloaded mode if and only if the cumulative weight of all working units exceeds or equals t_2 w(I) where I is the set of indices representing all units.

Conditioning on the mode of the system, the system reliability can be expressed as follows:

$R(n) = P\{\text{system works/system is in good mode}\} \cdot P\{\text{system is in good mode}\}$

$\quad + P\{\text{system works/system is in overloaded mode}\} \cdot P\{\text{system is in overloaded mode}\}$

The dual-system reliability can be obtained as follows (Pham 2009):

$$R(n) = \alpha \sum_{\substack{S_w \subseteq I \\ w(S_w) \geq t_1 w(I)}} \prod_{i \in S_w} p_1^{(i)} \prod_{i \in I \backslash S_w} \left(1 - p_1^{(i)}\right)$$

$$+ (1 - \alpha) \sum_{\substack{S_w \subseteq I \\ w(S_w) \geq t_2 w(I)}} \prod_{i \in S_w} p_2^{(i)} \prod_{i \in I \backslash S_w} \left(1 - p_2^{(i)}\right) \qquad (4.76)$$

Systems with Identical Components

Assume that all the components in the system are statistically independent and identical and are defined as $p_1^{(i)} = p_1; p_2^{(i)} = p_2$; and $w_i = 1$ for all i = 1, 2, ..., n. We also define as follows: $t_1 w(I) = k$ and $t_2 w(I) = m$. Again the system works in a good (overloaded) mode if and only if at least k (m) units are to function for the successful operation of the system. Let $R_g(n)$ and $R_o(n)$ are the reliability functions when the system is in good mode and overloaded mode, respectively. The reliability of the system which is in good mode and in overloaded mode can be expressed as, respectively,

$$R_g(n) = \alpha \left[\sum_{i=k}^{n} \binom{n}{i} p_1^i (1 - p_1)^{n-i} \right] \qquad (4.77)$$

$$R_o(n) = (1 - \alpha) \left[\sum_{i=m}^{n} \binom{n}{i} p_2^i (1 - p_2)^{n-i} \right] \qquad (4.78)$$

Combining Eqs. (4.77) and (4.78), the reliability of degradable dual-systems is given by

$$R(n) = \alpha \left[\sum_{i=k}^{n} \binom{n}{i} p_1^i (1 - p_1)^{n-i} \right] + (1 - \alpha) \left[\sum_{i=m}^{n} \binom{n}{i} p_2^i (1 - p_2)^{n-i} \right] \quad (4.79)$$

where $k \leq m \leq n$ and $p_1 \geq p_2$.

Example 4.14 Given k = 5, m = 7, $\alpha = 0.7, p_1 = 0.5$ and $p_2 = 0.4$. From Eq. (4.79), we obtain

n	R(n)
8	0.2569
9	0.3575
10	0.4525
15	0.7757

Example 4.15 Given k = 5, m = 7, $\alpha = 0.7, p_1 = 0.8$ and $p_2 = 0.7$. From Eq. (4.79), we have

n	R(n)
10	0.8908
15	0.9955
20	1.0000
25	1.0000

Example 4.16 Given k = 5, n = 10, $\alpha = 0.6, p_1 = 0.8$ and $p_2 = 0.7$. From Eq. (4.79),

m	R(m)
5	0.9776
6	0.9364
7	0.8564
8	0.7496
9	0.6560
10	0.6076

4.12   Dynamic Redundant Systems with Imperfect Coverage

Consider a system with dynamic redundancy which consists of several modules but with only one operating at a time. Various fault detection schemes are used to determine when a module has become faulty, and fault location is used to determine exactly which module, if any, is fault. If a fault is detected and located, the faulty module is removed from operation and replaced by a spare. This section discusses a dynamic redundancy requires consecutive actions of fault detection and fault recovery (Pham and Pham 1992). The modules could be a processor in a multiprocessor system, a generator in an electrical power system or a logic gate.

The system consists of $(S + 1)$ modules where S is the total number of spare modules. Examples of duplex (when $S = 1$) configurations are the UDET 7116 telephone switch system, the COMTRAC railway traffic controller, and the Bell ESS (Pham and Pham 1992). It is assumed that the modules are i.i.d., both modules and the system are two-state, i.e. either successful or unsuccessful (failed) and the reliability of each module, active or spare, is p.

The reliability of the dynamic redundant system with S spares is given by

$$R(S) = 1 - (1 - p)^{S+1} \qquad (4.80)$$

The above reliability function is obtained assuming that the fault detection and the switch over mechanism are perfect. This implies that the function R(S) is, of course, an increasing function of the number of spare modules. Clearly adding spare modules makes it more likely that the reliability of the system will increase, while at the same time it also increases the cost of the system. In practice, the dynamic redundant system may not be able to use the redundancy because it cannot identify that a module is faulty, remove that fault module, and replace it with a fault-free one. This section addresses the impact of the fault coverage in the dynamic redundant system with S spares.

Denote c is the probability that the fault detection and switching mechanisms operate correctly, given the failure of one of the modules. Then the reliability of a dynamic redundant system with S spares is given by

$$R_c(S) = \sum_{i=1}^{S+1} \binom{S + 1}{i} p^i \left[(1 - p)c\right]^{S+1-i} \qquad (4.81)$$

where c is the probability that the fault detection and switching mechanisms operate correctly, given the failure of one of the modules, and p is the reliability of each module.

Theorem 4.1 (Pham and Pham 1992) *For fixed p and c, the maximum value of $R_c(S)$ is attained at $S^* = S_1 + 1$ where*

$$S_1 = \frac{\ln\left(\frac{(1-p)(1-c)}{1-(1-p)c}\right)}{\ln\left(\frac{(1-p)c}{p+(1-p)c}\right)} \qquad (4.82)$$

If S_1 is an integer, S_1 and $S_1 + 1$ both maximize $R_c(S)$.

Fail-Safe Systems

A fail-safe system (FSS) consists of a fail-safe redundancy (FSR) core with an associated bank of S spare units such that when one of the $(N - 1)$ operating units fails, the redundant unit immediately starts operating. The failed unit is then replaced by a spare, which then becomes the new redundant unit restoring the FSR core to the all-perfect state. The system survives provided at least $(N - 1)$ out of $(N + S)$ units are operable. Therefore, the FSS can tolerate up to $(S + 1)$ unit failures (i.e., all spares plus the active redundant unit). Applications of such systems can be found in the areas of safety monitoring and reactor trip function systems. For example, in the areas of safety monitoring systems, nuclear plants often use three identical and independently functioning counters to monitor the radioactivity of the air in ventilation systems, with the aim to initiating reactor shutdown when a dangerous level of radioactivity is present. When two or more counters register a dangerous level of radioactivity, the reactor automatically shuts down (Pham and Galyean 1992).

The system is characterized by the following properties:

1. The system consists of $(N + S)$ independent and identical units
2. Each unit of the system has two states, operable and failed.
3. The system works if and only if at least $(N - 1)$ out of $(N + S)$ units are operable.

Suppose that each unit costs c dollars and system failure costs d dollars of revenue. Let us define the average total system cost $E(T_s)$ is as follows:

$$E(T_s) = c(N + S) + d[1 - R(S)] \qquad (4.83)$$

where

$$R(S) = \sum_{i=N-1}^{N+S} \binom{N + S}{i} p^i (1 - p)^{N+S-i} = \sum_{i=0}^{S+1} \binom{N + S}{i} (1 - p)^i p^{N+S-i} \quad (4.84)$$

and p is a unit reliability.

The average total system cost of system size $(N + S)$ is the cost incurred when the system has failed plus the cost of all $(N + S)$ modules. We determine the optimal value of S, say S*, which minimizes the average total system cost.

Theorem 4.2 (Pham and Galyean 1992) *Fix p, N, d and c. The optimal value S^* such that the FSS minimizes the average total system cost is as follows:*

(1)   *If $f(S_0) < A$ then $S^* = 0$*
(2)   *If $f(S_0) \geq A$ and $f(0) \geq A$ then $S^* = S_a$*
(3)   *If $f(S_0) \geq A$ and $f(0) < A$ then*

$$S^* = \begin{cases} 0 & \text{if } E(T_0) \leq E(T_{S_a}) \\ S_a & \text{if } E(T_0) > E(T_{S_a}) \end{cases} \qquad (4.85)$$

where

$$S_0 = \max\{0, [S_1] + 1\}; \quad S_1 = \frac{(N+1)q - 3}{p}; \quad q = 1 - p$$

$$A = \frac{c}{dp^{N-1}q^2}; \quad S_a = \inf\{S \in [S_0, \infty) : f(S) < A\}$$

$$f(S) = \binom{N+S}{S+2} sq^S. \tag{4.86}$$

Example 4.17 Consider a FSS with $N = 5$, $p = 0.85$, $c = 10$, and $d = 300$. Then from Eq. (4.20), A = 2.84, $S_1 = -2.47$, therefore, $S_0 = 0$ and $f(0) = 10$. Because $f(S0) = f(0) > A$, and from Theorem 4.2, it can be easily obtained that the optimal value of S is $S^* = 2$ and the corresponding average total system cost is 73.6.

Application 4.1 (Pham and Galyean 1992) A cooling water system for a commercial nuclear power plant (either service water or component cooling) will typically be a two-train system, each with a dedicated pump. The system will usually have a third 'swing' pump that provides automatics backup to either of the two trains. Failure to maintain both colling water trains operating cold result in a plant shutdown and require the utility to buy replacement power at a cost of $1,000,000 per day. The cost of a redundant pump is estimated at $20,000. The constant failure rate (from random, independent faults) of the pumps is estimated at $\lambda = 10^{-4}/hour$, a 30-day pump postulated system failure (i.e. failure of second pump), if it were to occur, is assumed midway in the replacement/repair time of the first failed pump (i.e. $\frac{T_r}{2}$), and would result in the plant being shutdown for 15 days. We have

$$N = 3 \text{ pumps}, \ c = \$20,000/\text{pump} \quad d = \$1,000,000 \times 15 \text{ days} = \$15 \times 10^6$$

$$\lambda = 10^{-4}/\text{hour}; \quad T_r = 30 \text{ days}$$

$$q = 1 - e^{-\lambda T_r} = 0.07$$

From Eq. (4.86), we have $A = 0.315$; $S_1 = -2.9$; $S_0 = 0$; $f(S_0) = f(0) = 3$. Then $f(S_0) = f(0) > A$ and $S^* = S_a$. For $S = 0$ then $f(0) = 3$, $S = 1$ then $f(1) = 0.28$. Since $f(0) > A > f(1)$, then $S^* = 1$.

Application 4.2 Same situation as in Application 4.1 above, except the time to replace/repair (T_r) a failed pumps is 7 days, and the average consequence of a system failure is the plant being shutdown for 3.5 days. Therefore, $q = 0.017$, $d = \$3.5 \times 10^6$. We obtain

$$A = 20.5, \ S_1 = -3, \ S_0 = 0, f(S_0) = f(0) = 3.$$

Then $f(S_0) < A$ and therefore, $S^* = 0$.

4.13   Noncoherent *k*-to-*l*-out-of-*n* Systems

A *k*-to-*l*-out-of-*n* system is a noncoherent system in which "no fewer than *k* and no more than *l*" out of its *n* components are to function for the successful operation of the system. This class of noncoherent systems was first proposed by Heidtmann (1981). A noncoherent system is any system that is not coherent; a system is coherent if and only if: (a) it is good (bad) when all components are good (bad), and (b) it never gets worse (better) when the number of good components increases (decreases). Such a system is noncoherent because the system reliability can decrease when the number of good components increases. The generality of the *k*-to-*l*-out-of-*n* system can be demonstrated as follows: (i) when $k = l = n$, the *k*-to-*l*-out-of-*n* system becomes a series system; (ii) when $k = 1$ and $l = n$, the *k*-to-*l*-out-of-*n* system becomes a parallel system.

Examples of noncoherent systems can be found in multiprocessor and communication (Pham 1991; Upadhyaya and Pham 1993). A multiprocessor system, for example, shares resources such as memory, I/O units, and buses, among n processors. If less than k processors are being used, the system does not work to its maximum capacity and the system efficiency is poor. On the other hand, if more than l processors are being used then efficiency again is poor due to the traffic congestion caused by the too many processors in use. Both of these cases need to be avoided; this can be accomplished by modeling the system as failed for these cases. This multiprocessor system is thus a k-to-l-out-of-n system. Another example is in the area of avionics system. A space-shuttle computer-complex system has 5 identical computers assigned to redundant operation. During the critical mission phases such as boost, reentry, and landing, 4 of these computers operate in an N-Modular Redundancy (NMR) mode, receiving the same inputs and executing identical tasks. Computer 5 performs non-critical tasks. The shuttle can tolerate up to 2 failures. If all 5 computers are functioning during the critical mission phases, the technique to detect a failed computer might not be able to identify the failure, given that a failure has occurred. This eventually causes the space shuttle to fail. It is therefore a 2-to-4-out-of-5 system.

Reliability analysis

Consider a system composed of n components. We assume that both components and the system are 2-state, i.e., either successful or failed and the component states are independent and the component reliabilities are not necessarily equal (non-identical components). Let p_i (q_i) be the reliability (unreliability) of component i. Let $S(x, n)$ be the probability that exactly x units out of the n units are successful. Let $R(k: l, n)$ denote the reliability of the *k*-to-*l*-out-of-*n* system. By definition, the reliability of the *k*-to-*l*-out-of-*n* system is given by

$$R(k; l, n) = \sum_{x=k}^{l} S(x, n)$$

Let us assume that the units are independent and identically distributed (i.i.d. units) with $p_i = p$ *for* $1 \leq i \leq n$.. The reliability of the k-to-l-out-of-n system becomes

$$R(n) = \sum_{i=k}^{l} \binom{n}{i} p^i (1-p)^{n-i}$$

For fixed k, l and p, there exists an optimal value n^* for n that maximizes the system reliability R(n). The value n^* is given by (Upadhyaya and Pham 1993):

$$n^* = \inf\{n : f(n) < a^{l+1-k}\}$$

where

$$f(n) = \frac{\binom{n}{k-1}}{\binom{n}{l}} \text{ and } a = \frac{p}{1-p}$$

Similarly, for fixed values of k, l and n, there exists an optimal value p^* for p that maximizes the system reliability R(p). The value p^* is given by (Upadhyaya and Pham 1993):

$$p^* = \frac{1}{1 + \frac{1}{b^{k-l-1}}}$$

where

$$b = \frac{l!(n-l-1)!}{(k-1)!(n-k)!} \text{ and } m! = 0 \text{ for } m < 0.$$

Profit Optimization

Let us consider a computer installation where a certain number of terminals are to be connected to the system. We will look at what is the optimal number of terminals that maximizes the mean profit of the terminal subsystem. If the number of terminals is greater than, say l, then the response time might not be tolerable, causing loss of profit. This is over-performance of work. If, on the other hand, the number of terminals is less than, say k, then the computer is not being used to its maximum capacity. This is under-performance of work and lower profit. Both cases are to be avoided.

Notation

$f_n(i)$ profit when i units are functioning
$M(n)$ mean system-profit
S_n binomially distributed random variable with parameters p, n
p reliability of a unit.

Assumptions

(a)   The units and the system are 2-state: either good or failed
(b)   The states of the units are i.i.d.
(c) $f_n(i) < f_n(j)$ and $f_n(r) < f_n(j)$ for $0 < i < k$, $k \leq j \leq l$, $l < r \leq n$.

The mean system-profit is given by (Pham 1991):

$$M(n) = \sum_{i=0}^{n} \binom{n}{i} p^i (1-p)^{n-i} f_n(i)$$

$$= E\{f_n(S_n)\}. \tag{4.87}$$

Theorem 4.3 (Pham 1991) *Fix k, l and p. If the profit function f_n is concave, then there is an optimal value of n such that M(n) is maximized.*

Proof Let f_n be concave. Use the property of conditional expectation, we obtain

$$E\{f_n(S_{n+1}) - f_n(S_n)\} = E\{E\{f_n(S_{n+1}) - f_n(S_n)/S_n\}\}$$
$$= E\{pf_n(S_n + 1) + (1-p)f_n(S_n) - f_n(S_n)\}$$
$$= E\{p[f_n(S_n + 1) - f_n(S_n)]\} = E\{g_n(S_n)\} \tag{4.88}$$

where $g_n \equiv g_n(i) = p[f_n(i+1) - f_n(i)]$. From Eq. (4.88), we have

$$E\{[f_n(S_{n+2}) - f_n(S_{n+1})] - [f_n(S_{n+1}) - f_n(S_n)]\} = E\{g_n(S_{n+1}) - g_n(S_n)\} = E\{h_n(S_n)\}$$

where $h_n(i) = p[g_n(i+1) - g_n(i)]$. Hence,

$$E\{[f_n(S_{n+2}) - f_n(S_{n+1})] - [f_n(S_{n+1}) - f_n(S_n)]\} = p^2[f_n(i+2) - 2f_n(i+2) + f_n(i)] \leq 0$$

for all i since f_n is concave. This implies that $M(n)$ is concave.

Example 4.18 Given $k = 5$, $l = 8$ and the profit function

$$f_n(i) = \begin{cases} 100i & for \ \ 0 \leq i < k \\ 500 & for \ \ k \leq i \leq l \\ 500 - 10(i - l) & for \ \ l < i \leq n \end{cases}$$

From the Theorem 4.3, the optimal number of units to maximize the mean system-profit is 9 for $p = 0.9$. The mean system-profit corresponding to this optimum value is 496. For a lower value of p, e.g., $p = 0.7$, the optimal number of units, 11, is higher. If p is close enough to 1, it seems that the optimal n is very close to l (Pham 1991).

4.14 Mathematical Optimization

Let $y = f(x_1, x_2, \ldots, x_n)$ be defined and differentiable in an n-dimensional domain D. Define $\widetilde{X}^a = \left(X_1^a, X_2^a, \ldots, X_n^a\right)$ be a point at which $f(x_1, x_2, \ldots, x_n)$ attains its optimal value is commonly referred to as a critical point of $f(\tilde{x})$, that is, the point $\widetilde{X}$ at which $f(\widetilde{X}) = 0$. For every point $\widetilde{X} \in D, f\left(\widetilde{X}^a\right)$ is called the global maximum of f in $\tilde{x} \in D$ if $f(\tilde{x}) \leq f(\tilde{x}^a)$. Similarly, $f\left(\widetilde{X}^a\right)$ is called the global minimum of f in $\tilde{x} \in D$ if $f(\tilde{x}) \geq f(\tilde{x}^a)$.

The function $f(\tilde{x})$ is said to have a local minimum (maximum) at $\tilde{x}^a$ if $f(\tilde{x}) \geq$ ($\leq$)$f(\tilde{x}^a)$ for all $\tilde{x}$ sufficiently close to $\tilde{x}^a$. Note that while every local minimum and maximum occurs at a critical point, a critical point need not to give either minimum or maximum. One can easily show that

$$\max_D -f(\tilde{x}) = -\min_D f(\tilde{x})$$

$$\min_D \{f(\tilde{x}) + g(\tilde{x})\} \geq \min_D f(\tilde{x}) + \min_D g(\tilde{x})$$

$$\max_D \{f(\tilde{x}) + g(\tilde{x})\} \leq \max_D f(\tilde{x}) + \max_D g(\tilde{x})\theta$$

Assume that $f(x)$ is differentiable and has a local minimum or a local maximum at a point x_0 then $f'(x_0) = 0$. However, when $f'(x_0) = 0$ it is not sufficient to determine whether $f(x)$ has a maximum or minimum at x_0. From elementary differential calculus we have the well-known second derivative method to determine the sufficiency condition. That is, if $f(x)$ is a differential function of x and $f'(x_0) = 0$, then if $f''(x_0)$ exists and is positive (negative), $f(x)$ has a local minimum (maximum) at x_0.

Example 4.19 A manager has determined that, for a given product, the average cost function $\overline{C}(x)$ (in dollars per unit) is given by

$$\overline{C}(x) = 2x^2 - 36x + 210 - \frac{200}{x} \quad \text{for } 2 \leq x \leq 10,000$$

where x represents the number of units produced.

(a) For which values of x in the interval [2, 10000] is the total cost function increasing?

(b) At what level within the interval [2, 10000] should production be fixed in order to minimize total cost? What is the minimum total cost?

(c)   If production were required to lie within the interval [1000, 10000], what value
      of x would minimize total cost? Show your work.

Solution

(a)   Let x be the number of units produced and $C(x)$ be the total cost function. The
      total cost function $C(x)$ can be written as

$$C(x) = x\,\overline{C}(x) = 2x^3 - 36x^2 + 210x - 200 \quad \text{for} \ \ 2 \le x \le 10{,}000$$

We have

$$\frac{\partial C(x)}{\partial x} = 6x^2 - 72x + 210 = 6(x - 5)(x - 7)$$

It can be shown that the total cost function increases when $2 \le x < 5$ and
$7 < x \le 10000$.

(b)   We have

$$\frac{\partial^2 C(x)}{\partial x^2} = 12x - 72 = 12(x - 6)$$

This implies that

$$\frac{\partial^2 C(x)}{\partial x^2} = 12x - 72 = 12(x - 6)\begin{cases} < 0 & \text{when } x = 5\text{: local maximum} \\ > 0 & \text{when } x = 7\text{: local minimum.} \end{cases}$$

Thus, we just need to compare the values of $C(2)$ and $C(7)$ where $C(2) = 92$ and
$C(7) = 192$. So $x = 2$ and the minimum total cost is $C(2) = 92$.

(c)   Since the cost function $C(x)$ is increasing in (7, 10000], the minimal total cost
      would occur at $x = 1000$ and $C(1000) = 1{,}964{,}209{,}800$.

A function $f(\tilde{x})$ defined on an n-dimensional domain D is monotone increasing
(decreasing) in x_j iff

$$\frac{\partial f(\tilde{x})}{\partial x_j} \ge 0(\le 0) \text{ for all } x_j, \quad 1 \le j \le n.$$

If the monotonicity of the derivative holds on the entire domain, then the extremum
(maximum or minimum) is global.

A function $f(\tilde{x})$ defined on an n-dimensional domain D is said to be convex on D
if the following two conditions are satisfied:

(i)   D is convex set
(ii)   For any two points $\tilde{x}$ and $\tilde{y}$ in D, we have

$$f(\lambda x_1 + (1 - \lambda)y_1); \lambda x_2 + (1 - \lambda)y_2); \ldots, \lambda x_n + (1 - \lambda)y_n))$$
$$\leq \lambda f(x_1, x_2, \ldots, x_n) + (1 - \lambda)f(y_1, y_2, \ldots, y_n)$$

For all $\lambda, 0 \leq \lambda \leq 1$. Similarly, a function $f(\tilde{x})$ is called concave if $-f(\tilde{x})$ is convex.

Discrete Derivative Variables

Let $f(n)$ be defined only for positive integers of n and define

$$\Delta f(n) = f(n+1) - f(n) \text{ and } \Delta^2 f(n) = \Delta(f(n+1) - f(n)) = f(n+2) - 2f(n+1) + f(n)$$

The function $f(n)$ has a local minimum at $n = n_0$ if

$$f(n_0 + 1) \geq f(n_0) \text{ and } f(n_0) \leq f(n_0 - 1) \tag{4.89}$$

or, equivalently, if $\Delta f(n_0) \geq 0$ and $\Delta f(n_0 - 1) \leq 0$. Similarly, $f(n)$ has a local maximum at $n = n_0$ if

$$\Delta f(n_0) \leq 0 \leq \Delta f(n_0 - 1) \leq 0. \tag{4.90}$$

Note that a sufficient condition for $f(n)$ to have a global minimum at $n = n_0$ if Eq. (4.89) hold and $\Delta^2 f(n) \geq 0$.

Similarly, a sufficient condition for $f(n)$ to have a global maximum at $n = n_0$ is that Eq. (4.90) hold and $\Delta^2 f(n) \leq 0$. In other words, the function $f(n)$ will have a global minimum at $n = n_0$ if $f(n_0) \leq f(n)$ for all n in the domain of the function, and a global maximum at $n = n_0$ if $f(n_0) \geq f(n)$ for all n in the domain of the function.

Lagrange Multiplier Method

Suppose $f(\tilde{x})$ and $g_j(\tilde{x})$ for $j = 1, 2, \ldots, m$ are functions having continuous first partial derivatives on a domain D. The maximization or minimization of an objective function of n variables, say $f(\tilde{x})$, subject to m constraints of the forms $g_j(\tilde{x}) = a_j$ for $j = 1, 2, \ldots, m$ is discussed here called the Lagrange multiplier method,

$$H(\tilde{x}) = f(\tilde{x}) - \sum_{j=1}^{m} \lambda_j g_j(\tilde{x}) \tag{4.91}$$

where λ_j are arbitrary real numbers and called Lagrange multipliers. If the point $\tilde{x}$ is a solution of the extreme problem, then it will satisfy the following system of $(n + m)$ equations:

$$\begin{cases} \frac{\partial H(\tilde{x})}{\partial x_i} = 0 \text{ for } i = 1, 2, \ldots, n \\ g_j(\tilde{x}) = a_j \text{ for } j = 1, 2, \ldots, m \end{cases} \tag{4.92}$$

In other words, one attempts to solve the above system for the $(n + m)$ unknown. They are: $x_1, x_2, \ldots, x_n$ and $\lambda_1, \lambda_2, \ldots, \lambda_m$. The points $(x_1, x_2, \ldots, x_n)$ obtained must need to test to determine whether they yield a maximum, a minimum or not. Again, the parameters $\lambda_1, \lambda_2, \ldots, \lambda_m$ which are introduced only help solve the system for $x_1, x_2, \ldots, x_n$, are known as Lagrange's multipliers.

Example 4.20 Consider a series–parallel system consisting of n subsystems in series where subsystem i consists of m_i components. The problem here is to determine the system design that will yield a desired level of the entire system reliability but at a minimum cost.

Let p_i be the reliability of a component of type i and suppose that the number of components of type i in the parallel subsystem is m_i. The reliability of the ith subsystem is given by

$$r_i = 1 - (1 - p_i)^{m_i}$$

The system reliability is given by

$$R_s = \prod_{i=1}^{n} r_i = \prod_{i=1}^{n} \left[1 - (1 - p_i)^{m_i} \right] \tag{4.93}$$

Let C_i be the cost of a component of type i. Consider the problem of minimizing the total system cost

$$C(m_1, m_2, \ldots, m_n) = \sum_{i=1}^{n} m_i c_i \tag{4.94}$$

subject to a fixed system reliability level R_s

$$R_s = \prod_{i=1}^{n} \left[1 - (1 - p_i)^{m_i} \right]$$

We can use the Lagrange multiplier method to solve this optimization problem with care since the variables $m_1, m_2, \ldots, m_n$ are all positive integers. Denote

$$a_i = \frac{\log r_i}{\log R_s}$$

then we have $\sum_{i=1}^{n} a_i = 1$. From Eq. (4.93), we can rewrite

$$r_i = R_s^{a_i} \text{ for } 0 \leq a_i \leq 1 \text{ and } m_i = \frac{\log\left(1 - R_s^{a_i}\right)}{\log(1 - p_i)}$$

Then Eq. (4.94) can be rewritten the minimization function as

$$C(m_1, m_2, ..., m_n) = \sum_{i=1}^{n} \frac{c_i \log(1 - R_s^{a_i})}{\log(1 - p_i)}$$

where $0 \leq a_i \leq 1$ and $\sum_{i=1}^{n} a_i = 1$.

Let's assume that R_s to be close to 1, say $R_s = 1 - \varepsilon$ where ε is very small. So,

$$\log R_s = \log(1 - \varepsilon) \simeq -\varepsilon$$

and

$$1 - R_s^{a_i} = -a_i \log R_s = a_i \varepsilon$$

The Lagrange function can be written as

$$H(a_1, a_2, ..., a_n) \simeq \sum_{i=1}^{n} \frac{C_i \log(a_i \varepsilon)}{\log(1 - p_i)} - \lambda_1 \sum_{i=1}^{n} a_i$$

Then

$$\frac{\partial H}{\partial a_i} = \frac{C_i}{a_i \log(1 - p_i)} - \lambda_1 \quad \text{for } i = 1, 2, \ldots, n$$

$$\frac{\partial H}{\partial a_i} = 0 \Leftrightarrow a_i = \frac{C_i}{\lambda_1 \log(1 - p_i)}$$

Since the constraint $\sum_{i=1}^{n} a_i = 1$ we have

$$\lambda_1 = \frac{1}{\sum_{i=1}^{n} \frac{C_i}{\log(1 - p_i)}}.$$

From Eq. (4.94), we obtain

$$a_i = \frac{C_i \sum_{j=1}^{n} \frac{C_j}{\log(1 - p_j)}}{\log(1 - p_i)}. \tag{4.95}$$

Therefore, the solution is,

$$m_i = \begin{cases} k_i & \text{if } k_i \text{ is an integer} \\ \lfloor k_i \rfloor + 1 & \text{if } k_i \text{ is not integer} \end{cases}$$

where

$$k_i = \frac{\log(1 - R_s^{a_i})}{\log(1 - p_i)}$$

where a_i is given in Eq. (4.95).

Example 4.21 As given the reliability of series–parallel system is given by

$$R(m_1, m_2, ..., m_n) = \prod_{i=1}^{n} \left(1 - (1 - p_i)^{m_i}\right)$$

Here is a reliability optimization that we will consider

$$\text{Maximize} \quad R(m_1, m_2, \ldots, m_n)$$

subject to the cost constraint $\sum_{i=1}^{n} c_i m_i \leq C$. Note that to maximize $R(m_1, m_2, \ldots, m_n)$ is the same as to maximize its logarithm given by

$$\log R(m_1, m_2, \ldots, m_n) = \sum_{i=1}^{n} \log(1 - (1 - p_i)^{m_i}).$$

Therefore, we now can rewrite the maximization problem is as follows:

$$H(m_1, m_2, \ldots, m_n) = \sum_{i=1}^{n} \log(1 - (1 - p_i)^{m_i}) - \lambda \sum_{i=1}^{n} c_i m_i$$

$$= \sum_{i=1}^{n} \left\{ \log(1 - (1 - p_i)^{m_i}) - \lambda c_i m_i \right\}.$$

To maximize $H(m_1, m_2, \ldots, m_n)$ is the same as to maximize f(m$_i$) where

$$f(m_i) = \log(1 - (1 - p_i)^{m_i}) - \lambda c_i m_i$$

Using an approximation as a continuous variable, we obtain

$$\frac{\partial f(m_i)}{\partial m_i} = 0 \quad \text{iff} \quad m_i = \frac{\log \lambda c_i - \log(\lambda c_i - \log(1 - p_i))}{\log(1 - p_i)}$$

So the solution can be considered as

Table 4.2 Optimal solutions and their total system costs

λ	m_1	m_2	m_3	m_4	Reliability R_s	System cost
0.0003	5	6	5	4	0.997470	54.8
0.0004	5	6	5	4	0.997470	54.8
0.0005	5	6	5	4	0.997470	54.8
0.0006	5	6	5	4	0.997470	54.8
0.0007	5	6	5	3	0.994608	50.3
0.0008	5	5	5	3	0.992914	48.0
0.0009	5	5	4	3	0.990003	44.6
0.0002	6	7	6	4	0.998967	61.7
0.000210	6	7	6	4	0.998967	61.7
0.000215	5	7	6	4	0.998712	60.5
0.000220	5	7	5	4	0.997980	57.1
0.000225	5	6	5	4	0.997470	54.8

$$m_i = \lfloor k_i \rfloor \quad \text{or} \quad \lfloor k_i \rfloor + 1$$

where

$$k_i = \frac{\log \lambda c_i - \log(\lambda c_i - \log(1 - p_i))}{\log(1 - p_i)}$$

Example 4.22 Given $n = 4$ and

n	Cost C_i	p_i
1	1.2	0.80
2	2.3	0.70
3	3.4	0.75
4	4.5	0.85

If we assume that the cost constraint is at most at 55, then the solution using Lagrange approximation approach is $m_1 = 5$, $m_2 = 6$, $m_3 = 5$, $m_4 = 4$ at a total cost of 54.8 with system reliability is 0.99747. See Table 4.2 for details.

4.15 Reliability of Systems with Multiple Failure Modes

A component is subject to failure in either open or closed modes. Networks of relays, fuse systems for warheads, diode circuits, fluid flow valves, etc. are a few examples of such components. Redundancy can be used to enhance the reliability of a system without any change in the reliability of the individual components that form the system. However, in a two-failure mode problem, redundancy may either increase or decrease the system's reliability (Pham 1989). For example, a network consisting of n relays in series has the property that an open-circuit failure of any one of the relays would cause an open-mode failure of the system and a closed-mode failure of the system. (The designations "closed mode" and "short mode" both appear in this chapter, and we will use the two terms interchangeably.) On the other hand, if the n relays were arranged in parallel, a closed-mode failure of any one relay would cause a system closed-mode failure, and an open-mode failure of all n relays would cause an open-mode failure of the system. Therefore, adding components in the system may decrease the system reliability. Diodes and transistors also exhibit open-mode and short-mode failure behavior (Pham 1989, 1999, 2003).

For instance, in an electrical system having components connected in series, if a short circuit occurs in one of the components, then the short-circuited component will not operate but will permit flow of current through the remaining components so that they continue to operate. However, an open-circuit failure of any of the components will cause an open-circuit failure of the system. As an example, suppose we have a number of 5 W bulbs that remain operative in satisfactory conditions at voltages ranging between 3 and 6 V. Obviously, on using the well-known formula in physics, if these bulbs are arranged in a series network to form a two-failure mode system, then the maximum and the minimum number of bulbs at these voltages are $n = 80$ and $k = 40$, respectively, in a situation when the system is operative at 240 V. In this case, any of the bulbs may fail either in closed or in open mode till the system is operative with 40 bulbs. Here, it is clear that, after each failure in closed mode, the rate of failure of a bulb in open mode increases due to the fact that the voltage passing through each bulb increases as the number of bulbs in the series decreases.

System reliability where components have various failure modes is covered in (Barlow and Proschan 1965, 1975; Barlow et al. 1963; Pham 1989; Pham and Pham 1991; Pham and Malon 1994). Barlow and Proschan (1965) studied various system configurations including series–parallel and parallel–series systems, where the size of each subsystem was fixed, but the number of subsystems was varied as well as k out of n systems to maximize the reliability of such systems. Ben-Dov (1986) determined a value of k that maximizes the reliability of k-out-of-n systems. Jenney and Sherwin (1986) considered systems in which the components are i.i.d. and subject to mutually exclusive open and short failures. Sah and Stiglitz (1988) obtained a necessary and sufficient condition for determining a threshold value that maximizes the mean profit of k-out-of-n systems. Pham and Pham (1991) further studied the effect of system parameters on the optimal k or n and showed that there does not exist a (k, n) maximizing the mean system profit.

This section discusses in detail the aspects of the reliability optimization of systems subject to two types of failure. It is assumed that the system component states are statistically independent and identically distributed, and that no constraints are imposed on the number of components to be used. Reliability optimization of series, parallel, parallel–series, series–parallel, and k-out-of-n systems subject to two types of failure will be discussed next.

In general, the formula for computing the reliability of a system subject to two kinds of failure is (Pham 1989):

$$\text{System reliability} = \text{Pr\{system works in both modes\}}$$
$$= \text{Pr\{system works in open mode\}} - \text{Pr\{system fails in closed mode\}}$$
$$+ \text{Pr\{system fails in both modes\}} \quad (4.96)$$

When the open- and closed-mode failure structures are dual of one another, i.e. Pr{system fails in both modes} $= 0$, then the system reliability given by Eq. (4.96) becomes

$$\text{System reliability} = 1 - \text{Pr\{system fails in open mode\}} - \text{Pr\{system fails in closed mode\}} \quad (4.97)$$

Notation

q_0	the open-mode failure probability of each component ($p_0 = 1 - q_0$)
q_s	the short-mode failure probability of each component ($p_s = 1 - q_s$)
Φ	implies $1 - \Phi$ for any Φ
$\lfloor x \rfloor$	the largest integer not exceeding x
*	implies an optimal value.

4.15.1 The Series System

Consider a series system consisting of n components. In this series system, any one component failing in an open mode causes system failure, whereas all components of the system must malfunction in short mode for the system to fail. The probabilities of system fails in open mode and fails in short mode are

$$F_0(n) = 1 - (1 - q_0)^n$$

and

$$F_s(n) = q_s^n$$

respectively. From Eq. (4.97), the system reliability is:

$$R_s(n) = (1 - q_0)^n - q_s^n \quad (4.98)$$

where n is the number of identical and independent components. In a series arrangement, reliability with respect to closed system failure increases with the number of components, whereas reliability with respect to open system failure decreases.

Let q_0 and q_s be fixed. There exists an optimum number of components, say n^*, that maximizes the system reliability. If we define

$$n_0 = \frac{\log\left(\frac{q_0}{1-q_s}\right)}{\log\left(\frac{q_s}{1-q_0}\right)}$$

then the system reliability, $R_s(n^*)$, *is* maximum for (Barlow and Proschan 1965)

$$n* = \begin{cases} \lfloor n_0 \rfloor + 1 & \text{if } n_0 \text{ is not an integer} \\ n_0 \text{ or } n_0 + 1 & \text{if } n_0 \text{ is not an integer} \end{cases} \tag{4.99}$$

Example 4.23 A switch has two failure modes: fail-open and fail-short. The probability of switch open-circuit failure and short-circuit failure are 0.1 and 0.2 respectively. A system consists of n switches wired in series. That is, given $q_0 = 0.1$ and $q_s = 0.2$. From Eq. (4.99)

$$n_0 = \frac{\log\left(\frac{0.1}{1-0.2}\right)}{\log\left(\frac{0.2}{1-0.1}\right)} = 1.4$$

Thus, $n^* = \lfloor 1.4 \rfloor + 1 = 2$. Therefore, when $n^* = 2$ the system reliability $R_s(n)$ $= 0.77$ is maximized.

4.15.2   The Parallel System

Consider a parallel system consisting of n components. For a parallel configuration, all the components must fail in open mode or at least one component must malfunction in short mode to cause the system to fail completely. The system reliability is

$$R_p(n) = (1 - q_s)^n - q_0^n \tag{4.100}$$

where n *is* the number of components connected in parallel. In this case, $(1 - q_s)^n$ represents the probability that no components fail in short mode, and $q_0{}^n$ represents the probability that all components fail in open mode.

Let q_0 and q_s be fixed. If we define

$$n_0 = \frac{\log\left(\frac{q_s}{1-q_0}\right)}{\log\left(\frac{q_0}{1-q_s}\right)} \tag{4.101}$$

then the system reliability $R_p(n^*)$ is maximum for (Barlow and Proschan 1965):

$$n* = \begin{cases} \lfloor n_0 \rfloor + 1 & \text{if } n_0 \text{ is not an integer} \\ n_0 \text{ or } n_0 + 1 & \text{if } n_0 \text{ is not an integer} \end{cases} \qquad (4.102)$$

It is observed that, for any range of q_0 and q_s, the optimal number of parallel components that maximizes the system reliability is one, if $q_s > q_0$. For most other practical values of q_0 and q_s the optimal number turns out to be two. In general, the optimal value of parallel components can be easily obtained using Eq. (4.102).

Suppose that each component costs d dollars and system failure costs c dollars of revenue. We now wish to determine the optimal system size n that minimizes the average system cost given that the costs of system failure in open and short modes are known. Let T_n be a total of the system. The average system cost is given by

$$E[T_n] = dn + c[1 - R_p(n)]$$

where $R_p(n)$ is defined as in Eq. (4.100). For given q_0, q_s, c, and d, we can obtain a value of n, say n^*, minimizing the average system cost.

Theorem 4.4 (Pham 1989) *Fix q_0, q_s, c, and d. There exists a unique value n^* that minimizes the average system cost, and*

$$n* = \inf\left\{ n \leq n_1 : (1 - q_0)q_0^n - q_s(1 - q_s)^n < \frac{d}{c} \right\} \qquad (4.103)$$

where $n_1 = \lfloor n_0 \rfloor + 1$ and n_0 is given in Eq. (4.101).

Suppose that each component costs d dollars and system failure in open mode and short mode costs c_1 and c_2 dollars of revenue respectively. Then the average system cost is given by

$$E[T_n] = dn + c_1 q_0^n + c_2[1 - (1 - q_s)^n] \qquad (4.104)$$

In other words, the average system cost of system size n is the cost incurred when the system has failed in either open mode or short mode plus the cost of all components in the system. We can determine the optimal value of n, say n^*, which minimizes the average system cost as shown in the following theorem.

Theorem 4.5 (Pham 1989) *Fix q_0, q_s, c_1, c_2, and d. There exists a unique value n^* that minimizes the average system cost, and*

$$n* = \begin{cases} 1 & \text{if } n_a \leq 0 \\ n_0 & \text{otherwise} \end{cases}$$

where

Table 4.3 The function $h(n)$ versus n

n	$h(n)$	$R_p(n)$	$E[T_n]$
1	9.6	0.6	490.0
2	2.34	0.72	212.0
3	0.216	0.702	151.8
4	4.373	0.648	155.3
5	−0.504	0.588	176.5
6	−0.506	0.531	201.7

$$n_0 = \inf\left\{ n \le n_a : h(n) \le \frac{d}{c_2 q_s} \right\}$$

and

$$h(n) = q_0^n \left[\frac{1 - q_0}{q_s} \frac{c_1}{c_2} - \left(\frac{1 - q_s}{q_0} \right)^n \right]$$

$$n_a = \frac{\log\left(\frac{1 - q_0}{q_s} \frac{c_1}{c_2} \right)}{\log\left(\frac{1 - q_s}{q_0} \right)}$$

Example 4.24 Suppose $d = 10$, $c_1 = 1500$, $c_2 = 300$, $q_s = 0.1$, $q_0 = 0.3$. Then

$$\frac{d}{c_2 q_s} = 0.333$$

From Table 4.3, $h(3) = 0.216 < 0.333$; therefore, the optimal value of n is $n^* = 3$. That is, when $n^* = 3$ the average system cost (151.8) from Eq. (4.104) is minimized.

4.15.3   The Parallel–Series System

Consider a system of components arranged so that there are m subsystems operating in parallel, each subsystem consisting of n identical components in series. Such an arrangement is called a parallel–series arrangement. The components could be a logic gate, a fluid-flow valve, or an electronic diode, and they are subject to two types of failure: failure in open mode and failure in short mode. Applications of the parallel–series systems can be found in the areas of communication, networks, and nuclear power systems. For example, consider a digital circuit module designed to process the incoming message in a communication system. Suppose that there are, at most, m ways *of* getting a message through the system, depending on which of the branches with n modules are operable. Such a system is subject to two failure modes: (1) a failure in open circuit of a single component in each subsystem would render the system unresponsive; or (2) a failure in short circuit of all the components in any subsystem would render the entire system unresponsive.

Notation

m number *of* subsystems in a system (or subsystem size)

n number *of* components in each subsystem

$F_0(m)$ probability of system failure in open mode

$F_s(m)$ probability *of* system failure in short mode.

The systems are characterized by the following properties:

1. The system consists of m subsystems, each subsystem containing n i.i.d. components.
2. A component is either good, failed open, or failed short. Failed components can never become good, and there are no transitions between the open and short failure modes.
3. The system can be (a) good, (b) failed open (at least one component in each subsystem fails open), or (c) failed short (all the components in any subsystem fail short).
4. The unconditional probabilities of component failure in open and short modes are known and are constrained: $q_0, q_s > 0; q_0 + q_s < 1$.

The probabilities of a system failing in open mode and failing in short mode are given by

$$F_0(m) = [1 - (1 - q_0)^n]^m \tag{4.105}$$

and

$$F_s(m) = [1 - (1 - q_s)^m]^m \tag{4.106}$$

respectively. The system reliability is

$$R_{ps}(n, m) = (1 - q_s^n)^m - [1 - (1 - q_0)^n]^m \tag{4.107}$$

where m is the number of identical subsystems in parallel and n is the number of identical components in each series subsystem. The term $\left(1 - q_s^n\right)^m$ represents the probability that none of the subsystems has failed in closed mode. Similarly, $\left[1 - (1 - q_0)^n\right]^m$ represents the probability that all the subsystems have failed in open mode.

An interesting example in Barlow and Proschan (1965) shows that there exists no pair n, m maximizing system reliability, since R_{ps} be made arbitrarily close to one by appropriate choice of m and n. To see this, let

$$a = \frac{\log q_s - \log(1 - q_0)}{\log q_s + \log(1 - q_0)} \quad M_n = q_s^{-n/(1+a)} \quad m_n = \lfloor M_n \rfloor$$

For given n, take $m = m_n$; then one can rewrite Eq. (4.107) as:

$$R_{ps}(n, m_n) = (1 - q_s^n)^{m_n} - [1 - (1 - q_0)^n]^{m_n}.$$

A straightforward computation yields

$$\lim_{n \to \infty} R_{ps}(n, m_n) = \lim_{n \to \infty} \{(1 - q_s^n)^{m_n} - [1 - (1 - q_0)^n]^{m_n}\} = 1$$

For fixed n, q_0, and q_s, one can determine the value of m that maximizes R_{ps}, and this is given in Barlow and Proschan (1965).

Theorem 4.6 (Barlow and Proschan 1965) *Let n, q_0, and q_s be fixed. The maximum value of $R_{ps}(m)$ is attained at $m^* = \lfloor m_0 \rfloor + 1$, where*

$$m_0 = \frac{n(\log p_0 - \log q_s)}{\log(1 - q_s^n) + \log(1 - p_0^n)} \tag{4.108}$$

If m_o is an integer, then m_o and $m_o + 1$ both maximize $R_{ps}(m)$.

The Profit Maximization Problem

We now wish to determine the optimal subsystem size m that maximizes the average system profit. We study how the optimal subsystem size m depends on the system parameters. We also show that there does not exist a pair (m, n) maximizing the average system profit.

Notation

$A(m)$	average system profit
β	conditional probability that the system is in open mode
$1 - \beta$	conditional probability that the system is in short mode
c_1, c_3	gain from system success in open, short mode
c_2, c_4	gain from system failure in open, short mode; $c_1 > c_2, c_3 > c_4$.

The average system profit is given by

$$A(m) = \beta\{c_1[1 - F_0(m)] + c_2 F_0(m)\} + (1 - \beta)\{c_3[1 - F_s(m)] + c_4 F_s(m)\} \tag{4.109}$$

Define

$$a = \frac{\beta(c_1 - c_2)}{(1 - \beta)(c_3 - c_4)} \quad \text{and} \quad b = \beta c_1 + (1 - \beta)c_4 \tag{4.110}$$

We can rewrite Eq. (4.109) as

$$A(m) = (1 - \beta)(c_3 - c_4)\{[1 - F_s(m)] - aF_0(m)\} + b \tag{4.111}$$

When the costs of the two kinds of system failure are identical, and the system is in the two modes with equal probability, then the optimization criterion becomes the same as maximizing the system reliability. Here, the following analysis deals with cases that need not satisfy these special restrictions. For a given value of n, one wishes to find the optimal number of subsystems m (m^*) that maximizes the average system profit. Of course, we would expect the optimal value of m to depend on the values of both q_0 and q_s. Define

$$m_0 = \frac{\ln a + n \ln\left(\frac{1-q_0}{q_s}\right)}{\ln\left[\frac{1-q_s^n}{1-(1-q_0)^n}\right]} \tag{4.112}$$

Theorem 4.7 (Pham 1989) *Fix β, n, q_0, q_s, and c_i for $i = 1, 2, 3, 4$. The maximum value of $A(m)$ is attained at*

$$m_0 = \begin{cases} 1 & \text{if } m_0 < 0 \\ \lfloor m_0 \rfloor + 1 & \text{if } m_0 \geq 0 \end{cases} \tag{4.113}$$

If m_o is a non-negative integer, both m_o and $m_o + 1$ maximize $A(m)$.

The proof is straightforward. When m_0 is a non-negative integer, the lower value will provide the more economical optimal configuration for the system. It is of interest to study how the optimal subsystem size m^* depends on the various parameters q_0 and q_s.

Theorem 4.8 (Pham 1989) *For fixed n, c_1, c_2, c_3, and c_4.*

(a) *If $a \geq 1$, then the optimal subsystem size m^* is an increasing function of q_0.*
(b) *If $a \leq 1$, then the optimal subsystem size m^* is a decreasing function of q_0.*
(c) *The optimal subsystem size m^* is an increasing function of β.*

The proof is left for an exercise. It is worth noting that we cannot find a pair (m, n) maximizing average system profit $A(m)$. Let

$$x = \frac{\ln q_s - \ln p_0}{\ln q_s + \ln p_0} \quad M_n = q_s^{-n/(1+x)} \quad m_n = \lfloor M_n \rfloor \tag{4.114}$$

For given n, take $m = m_n$. From Eq. (4.111), the average system profit can be rewritten as

$$A(m_n) = (1 - \beta)(c_3 - c_4)\{[1 - F_s(m_n)] - aF_0(m_n)\} + b \tag{4.115}$$

Theorem 4.9 *For fixed q_0, and q_s*

$$\lim_{n \to \infty} A(m_n) = \beta c_1 + (1 - \beta)c_3 \tag{4.116}$$

This result shows that we cannot seek a pair (m, n) maximizing the average system profit A(m_n), since A(m_n) can be made arbitrarily close to $\beta c_1 + (1 - \beta)c_3$.

Minimization Problem

We show how design policies can be chosen when the objective is to minimize the average total system cost given that the costs of system failure in open mode and short mode may not necessarily be the same.

Notation

d	cost of each component
c_1	cost when system failure in open
c_2	cost when system failure in short
$T(m)$	total system cost
$E[T(m)]$	average total system cost.

Suppose that each component costs d dollars, and system failure in open mode and short mode costs c_1 and c_2 dollars of revenue, respectively. The average total system cost is

$$E[T(m)] = dnm + c_1 F_0(m) + c_2 F_s(m) \tag{4.117}$$

In other words, the average system cost is the cost incurred when the system has failed in either the open mode or the short mode plus the cost of all components in the system. Define

$$h(m) = [1 - (p_0)^n]^m \left[c_1 p_0^n - c_2 q_s^n \left(\frac{1 - q_s^n}{1 - p_0^n} \right)^m \right] \tag{4.118}$$

$$m_1 = \inf\{m < m_2 : h(m) < dn\}$$

and

$$m_2 = \left\lceil \frac{\ln\left[\frac{c_1}{c_2} \left(\frac{p_0}{q_s} \right)^n \right]}{\ln\left(\frac{1 - q_s^n}{1 - p_0^n} \right)} \right\rceil + 1 \tag{4.119}$$

From Eq. (4.118), $h(m) > 0$ if and only if

$$c_1 p_0^n > c_2 q_s^n \left(\frac{1 - q_s^n}{1 - p_0^n} \right)^m$$

or equivalently, that $m < m_2$. Thus, the function $h(m)$ is decreasing in m for all $m < m_2$. For fixed n, we determine the optimal value of m, m^*, that minimizes the expected system cost, as shown in the following theorem.

Table 4.4 The values of m versus expected total system cost

m	$h(m)$	$E[T_m]$
1	110.146	386.17
2	74.051	326.02
3	49.784	301.97
4	33.469	302.19
5	22.499	318.71
6	15.124	346.22
7	10.166	381.10
8	6.832	420.93
9	4.591	464.10
10	3.085	509.50

Theorem 4.10 (Pham 1989) *Fixed q_0, and q_s, d, c_1, and c_2. There exists a unique value m^* such that the system minimizes the expected cost, and*

(a)   *If $m_2 > 0$ then*

$$m^* = \begin{cases} m_1 & \text{if } E[T(m_1)] \leq E[T(m_1)] \\ m_2 & \text{if } E[T(m_1)] > E[T(m_1)] \end{cases} \tag{4.120}$$

(b)   *If $m_2 \leq 0$ then $m^* = 1$.*

Since the function $h(m)$ is decreasing in m for $m < m_2$, again the resulting optimization problem in Eq. (4.118) is easily solved in practice.

Example 4.25 Suppose $n = 5$, $d = 10$, $c_1 = 500$, $c_2 = 700$, $q_s = 0.1$, and $q_0 = 0.2$. From Eq. (4.119), we obtain $m_2 = 26$. Since $m_2 > 0$, we determine the optimal value of m by using Theorem 4.10(a). The subsystem size m, $h(m)$, and the expected system cost $E[T(m)]$ are listed in Table 4.4; from this table, we have

$$m_1 = \inf\{m < 26 : h(m) < 50\} = 3 \quad \text{and} \quad E[T(m_1)] = 301.97$$

For $m_2 = 26$, $E[T(m_2)] = 1300.20$. From Theorem 4.10(a), the optimal value of m required to minimize the expected total system cost is 3, and the expected total system cost corresponding to this value is 301.97.

4.15.4   The Series–Parallel Systems

The series–parallel structure is the dual of the parallel–series structure in subsection C. We study a system of components arranged so that there are m subsystems operating in series, each subsystem consisting of n identical components in parallel. Such an arrangement is called a series–parallel arrangement. Applications of such systems can be found in the areas of communication, networks, and nuclear power systems. For example, consider a digital communication system consisting of in substations in series. A message is initially sent to substation 1, is then relayed to substation 2, etc., until the message passes through substation m and is received. The message consists of a sequence of 0's and 1's and each digit is sent separately through the series of m substations. Unfortunately, the substations are not perfect and can transmit as output a different digit than that received as input. Such a system is subject to two failure modes: errors in digital transmission occur in such a manner that either (1) a one appears instead of a zero, or (2) a zero appears instead of a one.

Failure in open mode of all the components in any subsystem makes the system unresponsive. Failure in closed (short) mode of a single component in each subsystem also makes the system unresponsive. The probabilities of system failure in open and short mode are given by

$$F_0(m) = 1 - (1 - q_0^n)^m \qquad (4.121)$$

and

$$F_s(m) = [1 - (1 - q_s^n)]^m \qquad (4.122)$$

respectively. The system reliability is

$$R(m) = (1 - q_0^n)^m - [1 - (1 - q_s^n)]^m \qquad (4.123)$$

where m is the number of identical subsystems in series and n is the number of identical components in each parallel subsystem.

Barlow and Proschan (1965) show that there exists no pair (m, n) maximizing system reliability. For fixed m, q_0, and q_s however, one can determine the value of n that maximizes the system reliability.

Theorem 4.11 (Barlow and Proschan 1965) *Let n, q_0, and q_s be fixed. The maximum value of $R(m)$ is attained at $m^* = \lfloor m_0 \rfloor + 1$, where*

$$m_0 = \frac{n(\log p_s - \log q_0)}{\log(1 - q_0^n) - \log(1 - p_s^n)} \qquad (4.124)$$

If m_0 is an integer, then m_o and $m_o + 1$ both maximize $R(m)$.

Maximizing the Average System Profit

The effect of the system parameters on the optimal m is now studied. We also determine the optimal subsystem size that maximizes the average system profit subject to a restricted type I (system failure in open mode) design error.

Notation

β conditional probability (given system failure) that the system is in open mode

$1 - \beta$ conditional probability (given system failure) that the system is in short mode

c_1 gain from system success in open mode

c_2 gain from system failure in open mode ($c_1 > c_2$)

c_3 gain from system success in short mode

c_4 gain from system failure in short mode ($c_3 > c_4$).

The average system-profit, $P(m)$, is given by

$$P(m) = \beta\{c_1[1 - F_0(m)] + c_2 F_0(m)\} + (1 - \beta)\{c_3[1 - F_s(m)] + c_4 F_s(m)\} \tag{4.125}$$

where $F_0(m)$ and $F_s(m)$ are defined as in Eqs. (4.121) and (4.122), respectively. Let

$$a = \frac{\beta(c_1 - c_2)}{(1 - \beta)(c_3 - c_4)}$$

and

$$b = \beta c_1 + (1 - \beta)c_4$$

We can rewrite Eq. (4.125) as

$$P(m) = (1 - \beta)(c_3 - c_4)[1 - F_s(m) - a F_0(m)] + b \tag{4.126}$$

For a given value of n, one wishes to find the optimal number of subsystems m, say m^*, that maximizes the average system-profit. We would anticipate that m^* depends on the values of both q_0 and q_s. Let

$$m_0 = \frac{n \ln\left(\frac{1 - q_s}{q_0}\right) - \ln a}{\ln\left[\frac{1 - q_0^n}{1 - (1 - q_s)^n}\right]} \tag{4.127}$$

Theorem 4.12 (Pham 1989) *Fix β, n, q_0, q_s, and c_i for $i = 1, 2, 3, 4$. The maximum value of $P(m)$ is attained at*

$$m* = \begin{cases} 1 & \text{if } m_0 < 0 \\ \lfloor m_0 \rfloor + 1 & \text{if } m_0 \geq 0 \end{cases}$$

If $m_o \geq 0$ and m_o is an integer, both m_o and $m_o + 1$ maximize $P(m)$.

When both m_o and $m_o + 1$ maximize the average system profit, the lower of the two values costs less. It is of interest to study how m^* depends on the various parameters q_0 and q_s.

Theorem 4.13 *For fixed n, c_i for $i = 1, 2, 3, 4$.*

(a)   *If $a \geq 1$, then the optimal subsystem size m^* is a decreasing function of q_0.*
(b)   *If $a \leq 1$, the optimal subsystem size m^* is an increasing function of q_s.*

This theorem states that when q_0 increases, it is desirable to reduce m as close to one as is feasible. On the other hand, when q_s increases, the average system-profit increases with the number of subsystems.

4.15.5   The k-out-of-n Systems

Consider a model in which a k-out-of-n system is composed of n identical and independent components that can be either good or failed. The components are subject to two types of failure: failure in open mode and failure in closed mode. The system can fail when k or more components fail in closed mode or when ($n - k + 1$) or more components fail in open mode. Applications of k-out-of-n systems can be found in the areas of target detection, communication, and safety monitoring systems, and, particularly, in the area of human organizations. The following is an example in the area of human organizations. Consider a committee with n members who must decide to accept or reject innovation-oriented projects. The projects are of two types: "good" and "bad". It is assumed that the communication among the members is limited, and each member will make a yes–no decision on each project. A committee member can make two types of error: the error of accepting a bad project and the error of rejecting a good project. The committee will accept a project when k or more members accept it, and will reject a project when ($n - k + 1$) or more members reject it. Thus, the two types of potential error of the committee are: (1) the acceptance of a bad project (which occurs when k or more members make the error of accepting a bad project); (2) the rejection of a good project (which occurs when ($n - k + 1$) or more members make the error of rejecting a good project). This section determines the

- optimal k that minimizes the expected total system cost
- optimal n that minimizes the expected total system cost
- optimal k and n that minimizes the expected total system cost.

We also study the effect of the system's parameters on the optimal k or n. The system fails in closed mode if and only if at least k of its n components fail in closed mode, and we obtain

$$F_s(k, n) = \sum_{i=k}^{n} \binom{n}{i} q_s^i p_s^{n-i} = 1 - \sum_{i=0}^{k-1} \binom{n}{i} q_s^i p_s^{n-i} \qquad (4.128)$$

The system fails in open mode if and only if at least $(n - k + 1)$ of its n components fail in open mode, that is:

$$F_0(k, n) = \sum_{i=n-k+1}^{n} \binom{n}{i} q_0^i p_0^{n-i} = \sum_{i=0}^{k-1} \binom{n}{i} p_0^i q_0^{n-i} \tag{4.129}$$

Hence, the system reliability is given by

$$R(k, n) = 1 - F_0(k, n) - F_s(k, n) = \sum_{i=0}^{k-1} \binom{n}{i} q_s^i p_s^{n-i} - \sum_{i=0}^{k-1} \binom{n}{i} p_0^i q_0^{n-i} \tag{4.130}$$

Let

$$b(k; p, n) = \binom{n}{k} p^k (1 - p)^{n-k}$$

and

$$b\inf(k; p, n) = \sum_{i=0}^{k} b(i; p, n)$$

We can rewrite Eqs. (4.128–4.130) as

$$F_s(k, n) = 1 - b\inf(k - 1; q_s, n)$$

$$F_s(k, n) = b\inf(k - 1; p_0, n)$$

$$R(k, n) = 1 - b\inf(k - 1; q_s, n) - b\inf(k - 1; p_0, n)$$

respectively. For a given k, we can find the optimum value of n, say n^*, that maximizes the system reliability.

Theorem 4.14 (Pham 1989) *For fixed k, q_0, and q_s, the maximum value of $R(k, n)$ is attained at $n^* = \lfloor n_0 \rfloor$ where*

$$n_0 = k \left[1 + \frac{\log\left(\frac{1-q_0}{q_s}\right)}{\log\left(\frac{1-q_s}{q_0}\right)} \right]$$

If n_0 is an integer, both n_0 and $n_0 + 1$ maximize $R(k, n)$.

This result shows that when n_0 is an integer, both $n^* - 1$ and n^* maximize the system reliability $R(k, n)$. In such cases, the lower value will provide the more economical optimal configuration for the system. If $q_0 = q_s$ the system reliability $R(k, n)$ is maximized when $n = 2k$ or $2k - 1$. In this case, the optimum value of n does not depend on the value of q_0 and q_s and the best choice for a decision voter is a majority voter; this system is also called a majority system (Pham 1997).

From the above Theorem 4.14 we understand that the optimal system size n^* depends on the various parameters q_0 and q_s. It can be shown the optimal value n^* is an increasing function of q_0 and a decreasing function of q_s. Intuitively, these results state that when q_s increases it is desirable to reduce the number of components in the system as close to the value of threshold level k as possible. On the other hand, when q_0 increases, the system reliability will be improved if the number of components increases.

Theorem 4.15 (Ben-Dov 1980) *For fixed n, q_0, and q_s, it is straightforward to see that the maximum value of $R(k, n)$ is attained at $k^* = \lfloor k_0 \rfloor + 1$, where*

$$k_0 = n \frac{\log\left(\frac{q_0}{p_s}\right)}{\log\left(\frac{q_s q_0}{p_s p_0}\right)}$$

If k_0 is an integer, both k_0 and $k_0 + 1$ maximize $R(k, n)$.
We now discuss how these two values, k^* and n^*, are related to one another. Define α by

$$\alpha = \frac{\log\left(\frac{q_0}{p_s}\right)}{\log\left(\frac{q_s q_0}{p_s p_0}\right)}$$

then, for a given n, the optimal threshold k is given by $k^* = \lceil n\alpha \rceil$ and for a given k the optimal n is $n^* = \lfloor k/\alpha \rfloor$. For any given q_0 and q_s, we can easily show that

$$q_s < \alpha < p_0$$

Therefore, we can obtain the following bounds for the optimal value of the threshold k:

$$nq_s < k* < np_0$$

This result shows that for given values of q_0 and q_s, an upper bound for the optimal threshold k^* is the expected number of components working in open mode, and a lower bound for the optimal threshold k^* is the expected number of components failing in closed mode.

Minimizing the Average System Cost

Notation

d each component cost
c_1 cost when system failure is in open mode
c_2 cost when system failure is in short mode.

$$b \inf(k; q_s, n) = 1 - b \inf(k - 1; q_s, n)$$

The average total system cost $E[T(k, n)]$ is

$$E[T(k, n)] = dn + [c_1 F_0(k, n) + c_2 F_s(k, n)] \qquad (4.131)$$

In other words, the average total system cost is the cost of all components in the system (dn), plus the average cost of system failure in the open mode $(c_1 F_0(k, n))$ and the average cost of system failure in the short mode $(c_2 F_s(k, n))$. We now study the problem of how design policies can be chosen when the objective is to minimize the average total system cost when the cost of components, the costs of system failure in the open, and short modes are given. We wish to find the

- optimal $k(k*)$ that minimizes the average system cost for a given n
- optimal $n(n*)$ that minimizes the average system cost for a given k
- optimal k and n $(k*, n*)$ that minimize the average system cost.

Define

$$k_0 = \frac{\log\left(\frac{c_2}{c_1}\right) + n \log\left(\frac{p_s}{q_0}\right)}{\log\left(\frac{p_0 p_s}{q_0 q_s}\right)} \qquad (4.132)$$

Theorem 4.16 *Fix n, q_0, q_s, c_1, c_2 and d. The minimum value of $E[T(k, n)]$ is attained at*

$$k* = \begin{cases} \max\{1, \lfloor k_0 \rfloor + 1\} & \text{if } k_0 < n \\ n & \text{if } k_0 \geq n \end{cases}$$

If k_0 is a positive integer, both k_0 and $k_0 + 1$ minimize $E[T(k, n)]$.

It is of interest to study how the optimal value of k, $k*$, depends on the probabilities of component failure in the open mode (q_0) and in the short mode (q_s).

It is worth noting that for fix n, f $c_1 \geq c_2$, then $k*$ is decreasing in q_0; if $c_1 \leq c_2$, then $k*$ is increasing in q_s.

Intuitively, this result states that if the cost of system failure in the open mode is greater than or equal to the cost of system failure in the short mode, then, as q_0 increases, it is desirable to reduce the threshold level k as close to one as is feasible.

Similarly, if the cost of system failure in the open mode is less than or equal to the cost of system failure in the short mode, then, as q_s increases, it is desirable to increase k as close to n as is feasible. Define

$$a = \frac{c_1}{c_2}$$

$$n_0 = \left\lfloor \frac{\log a + k \log\left(\frac{p_0 p_s}{q_0 q_s}\right)}{\log\left(\frac{p_s}{q_0}\right)} - 1 \right\rfloor$$

$$n_1 = \left\lceil \frac{k-1}{1-q_0} - 1 \right\rceil$$

$$f(n) = \left(\frac{p_s}{q_0}\right)^n \frac{(n+1)q_s - (k-1)}{(n+1)p_0 - (k-1)}$$

$$B = a\left(\frac{p_0 p_s}{q_0 q_s}\right)^k \frac{q_0}{p_s}$$

and

$$n_2 = f^{-1}(B) \quad for \quad k \le n_2 \le n_1.$$

Let

$$n_3 = \inf\left\{ n \in [n_2, n_0] : h(n) < \frac{d}{c_2} \right\}$$

where

$$h(n) = \binom{n}{k-1} p_0^k q_0^{n-k+1} \left[a - \left(\frac{q_0 q_s}{p_s p_0}\right)^k \left(\frac{p_s}{q_0}\right)^{n+1} \right] \tag{4.133}$$

It is easy to show that the function $h(n)$ is positive for all $k \le n \le n_0$, and is increasing in n for $n \in [k, n_2]$ and is decreasing in n for $n \in [n_2, n_0]$. This result shows that the function $h(n)$ is unimodal and achieves a maximum value at $n = n_2$. Since $n_2 \le n_1$, and when the probability of component failure in the open mode q_0 is quite small, then $n_1 \approx k$; so $n_2 \approx k$. On the other hand, for a given arbitrary q_0, one can find a value n_2 between the values of k and n_1 by using a binary search technique.

Theorem 4.17 *Fix q_0, q_s, k, d, c_1, and c_2. The optimal value of n, say n^*, such that the system minimizes the expected total cost is $n^* = k$ if $n_0 \le k$. Suppose $n_0 > k$ Then*

1. *if $h(n_2) < d/c_2$, then $n^* = k$*
2. *if $h(n_2) \ge d/c_2$ and $h(k) \ge d/c_2$, then $n^* = n_3$*
3. *if $h(n_2) \ge d/c_2$ and $h(k) < d/c_2$, then*

$$n* = \begin{cases} k & \text{if} \quad E[T(k, k)] \le E[T(k, n_3)] \\ n_3 & \text{if} \quad E[T(k, k)] > E[T(k, n_3)] \end{cases}$$

Proof Let $\Delta E[T(n)] = E[T(k, n + 1)] - \Delta E[T(k, n)]$. From Eq. (4.131), we obtain

$$\Delta E[T(n)] = d - c_1 \binom{n}{k-1} p_0^k q_0^{n-k+1} + c_2 \binom{n}{k-1} q_s^k p_s^{n-k+1} \qquad (4.134)$$

Substituting $c_1 = ac_2$ into Eq. (4.134), and after simplification, we obtain

$$\Delta E[T(n)] = d - c_2 \binom{n}{k-1} p_0^k q_0^{n-k+1} \left[a - \left(\frac{q_0 q_s}{p_0 p_s} \right)^k \left(\frac{p_s}{q_0} \right)^{n+1} \right]$$

$$= d - c_2 h(n)$$

The system of size $n + 1$ is better than the system of size n if, and only if, $h(n) \ge d/c_2$. If $n_0 \le k$, then $h(n) \le 0$ for all $n \ge k$, so that $E[T(k, n)]$ is increasing in n for all $n \ge k$. Thus $n* = k$ minimizes the expected total system cost. Suppose $n_0 > k$. Since the function $h(n)$ is decreasing in n for $n_2 \le n \le n_0$, there exists an n such that $h(n) < d/c_2$ on the interval $n_2 \le n \le n_0$. Let n_3 denote the smallest such n. Because $h(n)$ is decreasing on the interval $[n_2, n_0]$ where the function $h(n)$ is positive, we have $h(n) \ge d/c_2$ for $n_2 \le n \le n_3$ and $h(n) < d/c_2$ for $n > n_3$. Let $n*$ be an optimal value of n such that $E[T(k, n)]$ is minimized.

(a) If $h(n_2) < d/c_2$, then $n_3 = n_2$ and $h(k) < h(n_2) < d/c_2$, since $h(n)$ *is* increasing in $[k, n_2]$ and is decreasing in $[n_2, n_0]$.

Note that incrementing the system size reduces the expected system cost only when $h(n) \ge d/c_2$. This implies that $n* = k$ such that $E[T(k, n)]$ is minimized.

(b) Assume $h(n_2) \ge d/c_2$ and $h(k) \ge d/c_2$. Then $h(n) \ge d/c_2$ for $k \le n < n_2$, since $h(n)$ is increasing in n for $k < n < n_2$. This implies that $E[T(k, n + 1)] \le E[T(k, n)]$ for $k \le n < n_2$. Since $h(n_2) \ge d/c_2$, then $h(n) \ge d/c_2$ for $n_2 < n < n_3$ and $h(n) < d/c_2$ for $n > n_3$. This shows that $n* = n_3$ such that $E[T(k, n)]$ is minimized.

(c) Similarly, assume that $h(n_2) \ge d/c_2$ and $h(n) < d/c_2$. Then, either $n = k$ or $n* = n_3$ is the optimal solution for n.

Thus, $n* = k$ if $E[T(k, k)] \le E[T(k, n_3)]$; on the other hand, $n* = n_3$ if $E[T(k, k)] > E[T(k, n_3)]$.

In practical applications, the probability of component failure in the open mode q_0 is often quite small, and so the value of n_j is close to k. Therefore, the number of computations for finding a value of n_2 is quite small. Hence, the result of the Theorem 4.17 is easily applied in practice.

In the remaining section, we assume that the two system parameters k and n are unknown. It is of interest to determine the optimum values of (k, n), say $(k*, n*)$, that minimize the expected total system cost when the cost of components and the costs of system failures are known. Define

$$\alpha = \frac{\log\left(\frac{p_s}{q_0}\right)}{\log\left(\frac{p_0 p_s}{q_0 q_s}\right)} \quad \beta = \frac{\log\left(\frac{c_2}{c_1}\right)}{\log\left(\frac{p_0 p_s}{q_0 q_s}\right)} \tag{4.135}$$

We need the following lemma.

Lemma 4.1 *For* $0 \le m \le n$ *and* $0 \le p \le 1$:

$$\sum_{i=0}^{m} \binom{n}{i} p^i (1-p)^{n-i} < \sqrt{\frac{n}{2\pi m(n-m)}}$$

Theorem 4.18 (Pham and Malon 1994) *Fix* q_0, q_s, d, c_1, *and* c_2. *There exists an optimal pair of values* (k_n, n), *say* $(k_{n*}, n*)$, *such that average total system cost is minimized at* $(k_{n*}, n*)$, *and*

$$k_{n*} = \lfloor n * \alpha \rfloor \quad and \quad n* \le \frac{\frac{(1-q_0-q_s)}{2\pi}\left(\frac{c_1}{d}\right)^2 + 1 + \beta}{\alpha(1-\alpha)} \tag{4.136}$$

Proof Define $\Delta E[T(n)] = E[T(k_{n+1}, n+1)] - E[T(k_n, n)]$. From Eq. (4.134), we obtain

$$\Delta E[T(n)] = d + c_1[b\inf(k_{n+1}-1; p_0, n+1) - b\inf(k_n - 1; p_0, n)]$$
$$- c_2[b\inf(k_{n+1}-1; q_s, n+1) - b\inf(k_n - 1; q_s, n)]$$

Let $r = c_2/c_1$, then

$$\Delta E[T(n)] = d - c_1 g(n)$$

$$g(n) = r[b\inf(k_{n+1}-1; q_s, n+1) - b\inf(k_n - 1; q_s, n)] - [b\inf(k_{n+1}-1; p_0, n+1) - b\inf(k_n - 1; p_0, n)] \tag{4.137}$$

Case 1. Assume $k_{n+1} = k_n + 1$. We have

$$g(n) = \binom{n}{k_n} p_0^{k_n} q_0^{n-k+1} \left[r\left(\frac{q_0 q_s}{p_0 p_s}\right)^{k_n} \left(\frac{p_s}{q_0}\right)^{n+1} - 1 \right]$$

Recall that

$$\left(\frac{p_0 p_s}{q_0 q_s}\right)^{\beta} = r$$

then

$$\left(\frac{q_0 q_s}{p_0 p_s}\right)^{n\alpha+\beta}\left(\frac{p_s}{q_0}\right)^{n+1} = \frac{1}{r}\frac{p_s}{q_0}$$

since $n\alpha + \beta \le k_n \le (n+1)\alpha + \beta$, we obtain

$$r\left(\frac{q_0 q_s}{p_0 p_s}\right)^{k_n}\left(\frac{p_s}{q_0}\right)^{n+1} \le r\left(\frac{q_0 q_s}{p_0 p_s}\right)^{n\alpha+\beta}\left(\frac{p_s}{q_0}\right)^{n+1} = \frac{p_s}{q_0}$$

Thus

$$g(n) \le \binom{n}{k_n} p_0^{k_n} q_0^{n-k_n+1}\left(\frac{p_s}{q_0} - 1\right) = \binom{n}{k_n} p_0^{k_n} q_0^{n-k_n}(p_s - q_0)$$

From Lemma 4.1, and $n\alpha + \beta \le k_n \le (n+1)\alpha + \beta$, we obtain

$$g(n) \le (p_s - q_0)\frac{1}{\sqrt{2\pi n \frac{k_n}{n}\left(1 - \frac{k_n}{n}\right)}}$$

$$\le (p_s - q_0)\frac{1}{\sqrt{2\pi n\left(\alpha + \frac{\beta}{n}\right)\left[1 - \alpha\left(\frac{n+1}{n}\right) - \frac{\beta}{n}\right]}}$$

$$\le (1 - q_s - q_s)\frac{1}{\sqrt{2\pi n[n\alpha(1-\alpha) - \alpha(\alpha+\beta)]}}. \tag{4.138}$$

Case 2. Similarly, if $k_{n+1} = k_n$, then from Eq. 2.43, we have

$$g(n) = \binom{n}{k_n - 1} q_s^{k_n} p_s^{n-k_n+1}\left[\left(\frac{p_0 p_s}{q_0 q_s}\right)^{k_n}\left(\frac{q_0}{p_s}\right)^{n+1} - r\right]$$

since $k_n = [n\alpha + \beta] \le n\alpha + \beta + 1$, and from Eq. (4.137) and Lemma 4.1, we have

$$g(n) = q_s\binom{n}{k_n - 1} q_s^{k_n-1} p_s^{n-k_n+1}\left[\left(\frac{p_0 p_s}{q_0 q_s}\right)^{n\alpha+\beta+1}\left(\frac{q_0}{p_s}\right)^{n+1} - r\right]$$

$$= q_s\sqrt{\frac{n}{2\pi(k_n - 1)[n - (k_n - 1)]}}\left[\left(\frac{p_0 p_s}{q_0 q_s}\right)^{n\alpha+\beta}\left(\frac{q_0}{p_s}\right)^{n+1}\left(\frac{p_0 p_s}{q_0 q_s}\right) - r\right]$$

$$\le q_s\sqrt{\frac{n}{2\pi(k_n - 1)[n - (k_n - 1)]}}\left[r\left(\frac{q_0}{p_s}\right)\left(\frac{p_0 p_s}{q_0 q_s}\right) - r\right]$$

$$\leq \sqrt{\frac{n}{2\pi (k_n - 1)[n - (k_n - 1)]}} (1 - q_0 - q_s)$$

Note that $k_{n+1} = k_n$, then $n\alpha - (1 - \alpha - \beta) \leq k_n - 1 \leq n\alpha + \beta$. *After simplifications,*
we have

$$g(n) \leq (1 - q_0 - q_s) \frac{1}{\sqrt{2\pi n \left(\frac{k_n - 1}{n}\right)\left(1 - \frac{k_n - 1}{n}\right)}}$$

$$\leq (1 - q_0 - q_s) \frac{1}{\sqrt{2\pi n \left(\alpha - \frac{1 - \alpha - \beta}{n}\right)\left[1 - \alpha - \frac{\beta}{n}\right]}}$$

$$\leq \frac{1 - q_0 - q_s}{\sqrt{2\pi [n\alpha(1 - \alpha) - (1 - \alpha)^2 - (1 - \alpha)\beta]}} \tag{4.139}$$

From the inequalities in Eqs. (4.138) and (4.139), set

$$(1 - q_s - q_0) \frac{1}{\sqrt{2\pi [n\alpha(1 - \alpha) - \alpha(\alpha + \beta)]}} \leq \frac{d}{c_1}$$

and

$$\frac{(1 - q_s - q_0)}{\sqrt{2\pi [n\alpha(1 - \alpha) - (1 - \alpha)^2 - \alpha(\alpha + \beta)]}} \leq \frac{d}{c_1}$$

we obtain

$$(1 - q_0 - q_s)^2 \left(\frac{c_1}{d}\right)^2 \frac{1}{2\pi} \leq \min\{n\alpha(1 - \alpha) - \alpha(\alpha + \beta), n\alpha(1 - \alpha) - (1 - \alpha)^2 - (1 - \alpha)\beta\}$$

$$\Delta E[T(n)] \geq 0 \quad when \quad n \geq \frac{\frac{(1 - q_0 - q_s)^2}{2\pi}\left(\frac{c_1}{d}\right) + 1 + \beta}{\alpha(1 - \alpha)}$$

Hence

$$n* \leq \frac{\frac{(1 - q_0 - q_s)^2}{2\pi}\left(\frac{c_1}{d}\right) + 1 + \beta}{\alpha(1 - \alpha)}$$

The result in Eq. (4.139) provides an upper bound for the optimal system size.

4.16 Problems

1. Shows that a system consisting of n independent components arranged in series has failure rate equal to the sum of the component failure rates is general true for any distribution other than the exponential.

2. Given components with an exponential failure time distribution such that the ith component has constant failure rate λ_i, determine the reliability and MTTF of a 2-out-of-4 system. (Hints: see Example 4.5 in Sect. 4.6)

3. Consider a high-voltage load-sharing system consisting of a power supply and two transmitters, say A and B, using mechanically tuned magnetrons (Pham 1992a). Two transmitters are used, each tuning one-half the desired frequency range; however, if one transmitter fails, the other can tune the entire range with a resultant change in the expected time to failure. Suppose that for this system to work, the power supply and at least one of the transmitters must work. The transmitters operate independently when both are functional. An undetected fault in either of the transmitters causes the system to fail.

    Suppose that transmitter A has constant failure rate λ_A if transmitter B is operating and failure rate λ'_A if transmitter B has failed. Similarly, suppose that transmitter B has constant failure rate λ_B if transmitter A is operating and failure rate λ'_B if transmitter A has failed. Denote λ_P and λ_C be the constant failure rate of power supply and fault coverage, respectively. Obtain the reliability and MTTF of the high-voltage load-sharing system with dependent failures.

4. System XYZ consists of five water pumps for cooling a reactor. We assume that the pumps function independently and the reliability of each pump, over a period of 100,000 h, is 0.995. The system XYZ functions if at least four pumps must function. What is the reliability of the system?

5. An instrument which consists of k units operates for a time t. The reliability of each unit during the time t is p. When the time t passes, the instrument stops functioning, a technician inspects it and replaces the units that failed. He needs a time s to replace one unit.

    (a) Find the probability that in the time $2s$ after the stop the instrument will be ready for work.

    (b) Calculate (a) when $p = 0.9$ and $k = 10$.

6. A device consists of m units of type I and n units of type II. The reliability for an assigned time t of each type I unit is p_1 and that of a type II unit is p_2. The units fail independently of one another. For the device to operate, no less than two out of m type I units and no less than two out of n type II units must operate simultaneously for time t.

    (a) Find the reliability of the device.

    (b) Calculate (a) when $m = 5$, $p_1 = 0.9$, $n = 7$, $p_2 = 0.8$.

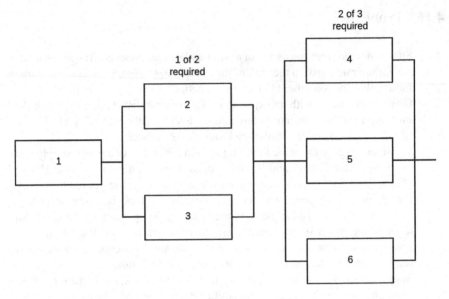

Fig. 4.11 Reliability block diagram for Problem 7

7.   A system is composed of six independent components as shown in Fig. 4.11.

   (a)   If p_i denotes the reliability of component i, find the reliability of the
         system.
   (b)   Calculate (a) when $p_i = 0.95$ for all $i = 1, 2, 3, 4, 5$, and 6.
   (c)   Suppose that each component in Fig. 4.11 has an exponential life
         distribution with failure rate λ. Find the MTTF of the system.

8.   A system is composed of three identical components, two of which are required
     to operate. The third is initially in standby. Assume that all components operate
     independently. If we assume perfect sensing and switching, that the active
     failure rate of each component is λ, and that the standby failure rate is zero:

   (a)   Find the reliability of the system.
   (b)   Determine the MTTF of the system.
   (c)   Calculate (a) and (b) when $\lambda = 0.005$ per hour and $t = 10$ h.

9.   A cooling system for a reactor has four identical cooling loops. Each cooling
     loop has two identical pumps connected in parallel. The cooling system
     requires that at least 3 of the 4 cooling loops operate successfully. The reli-
     ability of a pump over the life span of the plant is $R_p = 0.75$. Calculate the
     reliability of the cooling system.

10. An aircraft has four engines (i.e., two engines on each wing) but can land
     using just at least any two engines. Assume that the reliability of each engine
     for the duration of a flight mission is $R_e = 0.98$ and that engine failures are
     independent.

(a) Calculate the mission reliability of the aircraft.
(b) If at least one functioning engine must be on each wing, calculate the mission reliability.

4.17 Projects

Project #1

Project Description

Today, computers are used in many areas for various applications including airlines, hospitals, banks, security, automobiles, schools, etc. The impact of any such failure is significant. In airlines for example, on Wednesday July 8, 2015, United Airlines grounded 3,500 flights worldwide for less than an hour in the morning because of the computer problems that were related to network connectivity issue, according to NBC News. Delays affected 235 domestic and 138 international destinations, NBC reports (http://www.nbcnews.com/business/travel). Form teams of 3 members and select one as the team leader.

The tasks of your GROUP include:

Task 1: Discuss in detail, at least 15 computer related-failures or -problems since 2010, similar to the example given.
Task 2: Perform a comprehensive reliability/availability case study including

Distribution functions, failure rate function, MTTF, mean residual life based on a failure data set from Task 1.

Group 1:   Airlines
Group 2:   Hospitals
Group 3:   Banks
Group 4:   Security Systems
Group 5:   Automobiles

A project proposal (typed, no more than one-page) report is due in two weeks, which includes names of your group members, and a brief direction of your project.

A typed final REPORT (no more than 10 double-space pages) is due in two months.

PART I: Your Project Report must include the following:
           Introduction; Objectives
           System Failure Case Studies
           Reliability Case Study; Results, Findings and Remarks
PART II: Project Presentations (to be determined)

Project #2

Form teams of three members and select one as the team leader. Following is a brief project description.

Project Description

The system consists of n statistically identical, independent, and non-repairable units. The system and the units are either in a successful state (working) or in a failed state (not working). What separates this system from other two state systems is that the successful state contains two modes: a good mode and a degradation mode (see Sect. 4.11).

The system operates in a good mode with probability α; conversely, the system operates in a degradation mode with probability $(1 - \alpha)$. For the system to be working in a good mode, at least k out of n units must work. In a good mode, the reliability of the working units is p_1. For the system to be working in a degradation mode, at least m out of n units must work. In degradation mode, the reliability of the working units is p_2. Units in degradation mode are generally put under more stressed working conditions than normal.

Your group has been asked to answer the following questions

1. Describe the problem.
2. Provide at least 3 applications in real life about this two-state degradable system.
3. Derive the system reliability function $R(n)$.
4. Given arbitrary p_1, p_2, k, m, and α values, determine the optimal value of n, say n^*, that maximizes the system reliability function $R(n)$.
5. The cost model for this system includes the cost of a unit having reliabilities p_1 and p_2, represented as $c(p_1, p_2)$; and two penalty costs: d_g when the system fails in good mode and d_s when the system fails in degradation mode. $F_s(n)$ and $F_g(n)$ are the cumulative distribution functions for the system in degradation and good mode, respectively. The total system cost function can be written as follows

$$E(n) = nc(p_1, p_2) + d_s F_s(n) + d_g F_g(n)$$

   Given arbitrary p_1, p_2, d_s, d_g, k, m, and α values, determine the optimal value of n that minimizes the expected total system cost $E(n)$.
6. Given arbitrary $p_2, d_s, d_g, k, m, n, \alpha$ values and any specific function $c(p_1, p_2)$, obtain the optimal unit reliability value p_1, say p_1^*, that minimizes expected total cost of the system. You need to define a specific function $c(p_1, p_2)$.
7. Provide numerical examples and sensitivity analysis for all the problems above $(3 - 6)$. Draw the conclusions. Show your work.

A typed final REPORT (no more than 12 double-space pages) is due in eight weeks.

PART I: Your Project Report must include the following:
Introduction; Objectives
System Failure Case Studies
Reliability Analysis; Results, Findings and Remarks
PART II: Project Presentations (to be determined).

Project #3

Form teams of three members and select one as the team leader. Following is a brief project description.

Project Description

A manager is interested in purchasing a new cooling system for a reactor if it can meet the reliability requirement. The system information is as follows:
A cooling system for a reactor has four identical cooling loops. Each cooling loop has two identical pumps connected in parallel. The cooling system requires that at least 3 of the 4 cooling loops operate successfully.
A two-component parallel system is defined as if both components are working then the system is working. If either component 1 or 2 fails, the system is still working. The system will fail if and only if both components fail.
The following values represent the lifetimes (in years) of the pump based on a sample of historical data from the manufacture:

38.3 32.9 26.6 22.5 31.6 28.9 31.6 35.5 26.3 27.7 24.4 29.3
28.5 33.8 18.1 32.4 22.9 29.9 27.8 32.5 41.1 21.5 28.3 30.7

In this project, your group is asked to carefully study and obtain the analytical solutions for the following questions based on the mean squared error, Akai information criterion (AIC), and Bayesian information criterion (BIC) (Show all your works in detail):

1. Compute the reliability of the cooling system for a mission of 5 years.
2. Obtain the expected lifetime of the cooling system.

A typed final REPORT (no more than 5 double-space pages) is due in four weeks.

PART I: Your Project Report must include the following:
Introduction; Objectives
Methodologies, Reliability Analysis; Results, Findings and Remarks
PART II: Project Presentations (to be determined).

Project #4

Form teams of four members and select one as the team leader. Following is a brief project description.

Project Description—How Coffee Makers Work

A.   What exactly happens after adding the coffee and water into your coffee maker
     (see Fig. 4.12).

     In general, all the coffee makers consist of:

- A reservoir to hold the water
- A basic heating element
- Another smaller tube leading from the reservoir to the heating element
- A white tube leading up to the reservoir base, connecting the water to the drip
  area.

The bottom of the coffee maker is home to the coffee machine's electrical
equipment and the heating element. The heating element has two functions:

(i)   When you first put the water in the coffee maker, the heating element heats it.
(ii)   Once the coffee is made, the heating element keeps the coffee warm.

Fig. 4.12 A sample coffee machine

B. What can go wrong with your coffee machines?

Some of the most common problems that can cause your coffee machines to stop working include, but are not limited to the following:

- The power cord or switch: The power cord or on/off switch can go bad
- Clogged tubes: The tubes can get clogged with calcium
- Heating element failure.

Your group has been asked to study and answer the following questions:

1. Determine how reliable is the coffee machines with respect to reliability, performability and risk factors, and
2. To buy or not to buy this brand of coffee machine? Conduct the test by making the coffee of your own machine several times per day for a consecutive of 14 days based on your own criteria (i.e., Is the coffee machine noisy? How well does the coffee machine make coffee? etc.). Analyze the data obtained from your 14-day experiment and determine the reliability, mean time to failure and mean residual life as well as the effects of various possible inspection intervals on the reliability of this coffee machine brand

then present your results, findings as well as recommendations in two months. A typed final Report (no more than 10 pages) is due in two months.

PART I: Your Final Project Report must include the following:
Introduction; Objectives/Goals;
System description (how the machine works etc.)
Reliability block diagram or Markov diagram
Methodologies/Approaches/Modeling; Data collection
Reliability analysis, Optimization, and Examples
Results, Findings, and Conclusions/Remarks

PART II: Project Presentations (in two months, TBD).

References

Barlow RE, Proschan F (1965) Mathematical theory of reliability. Wiley, New York

Barlow RE, Proschan F (1975) Statistical theory of reliability and life testing. Holt, Renehart & Winston, New York

Barlow RE, Hunter LC, Proschan F (1963) Optimum redundancy when components are subject to two kinds of failure. J Soc Ind Appl Math 11(1):64–73

Ben-Dov Y (1980) Optimal reliability design of k-out-of-n systems subject to two kinds of failure. J Opt Res 31:743–748

Heidtmann KD (1981) A class of noncoherent systems and their reliability analysis. In: Proceedings of 11th annual symposium fault tolerant computing, June 1981, pp 96–98

Jenney BW, Sherwin DJ (1986) Open and short circuit reliability of systems of identical items. IEEE Trans Reliab R-35:532–538.

Pham H (1989) Optimal designs of systems with competing failure modes. PhD dissertation, State University of New York, Buffalo, February 1989 (unpublished)

Pham H (1991) Optimal design for a class of noncoherent systems. IEEE Trans Reliab 40(3):361–363

Pham H (1992a) Reliability analysis of a high voltage system with dependent failures and imperfect coverage. Reliab Eng Syst Saf 37(1):25–28

Pham H (1992b) On the optimal design of k-out-of-n subsystems. IEEE Trans Reliab 41(4):572–574

Pham H (1997) Reliability analysis of digital communication systems with imperfect voters. Math Comput Model J 1997(26):103–112

Pham H (1999) Reliability analysis of dynamic configurations of systems with three failure modes. Reliab Eng Syst Saf 1999(63):13–23

Pham H (2003) Handbook of reliability engineering. Springer, London

Pham H (2009) Reliability analysis of degradable dual-systems subject to competing operating-modes. Int J Reliab Qual Perform 1(1):39–51

Pham H (2010) On the estimation of reliability of k-out-of-n systems. Int J Syst Assur Eng Manage 1(1):32–35

Pham H (2014) Loglog fault-detection rate and testing coverage software reliability models subject to random environments. Vietnam J Comput Sci 1(1)

Pham H, Galyean WJ (1992) Reliability analysis of nuclear fail-safe redundancy. Reliab Eng Syst Saf 37(2):109–112

Pham H, Pham M (1991) Optimal designs of (k, n - k + 1) out-of-n: F systems (subject to 2 failure modes). IEEE Trans Reliab 1991(40):559–562

Pham H, Pham M (1992) Reliability analysis of dynamic redundant systems with imperfect coverage. Reliab Eng Syst Saf J 35(2):173–176

Pham H, Malon DM (1994) Optimal designs of systems with competing failure modes. IEEE Trans Reliab 1994(43):251–254

Pham H, Upadhyaya SJ (1989) Reliability analysis of a class of fault-tolerant systems. IEEE Trans Reliab 1989(38):3337

Pham H, Suprasad A, Misra RB (1996) Reliability and MTTF prediction of k-out-of-n complex systems with components subjected to multiple stages of degradation. Int J Syst Sci 27(10):995–1000

Pham H, Suprasad A, Misra RB (1997) Availability and mean life time prediction of multi-stage degraded system with partial repairs. Reliab Eng Syst Saf 56:169–173

Sah RK, Stiglitz JE (1988) Qualitative properties of profit making k-out-of-n systems subject to two kinds of failures. IEEE Trans Reliab 37:515–520

Upadhyaya SJ, Pham H (1992) Optimal design of k-to-l-out-of-n systems. Int J Model Simulat 12(3):69–72

Upadhyaya SJ, Pham H (1993) Analysis of noncoherent systems and an architecture for the computation of the system reliability. IEEE Trans Comput 42(4):484–493

Chapter 5
System Reliability Estimation

This chapter discusses a brief order statistics and various approaches such as MLE, UMVUE and bias-corrected estimate to obtain the reliability estimation of some form of systems including series, parallel and k-out-of-n systems for the uncensored and censored failure time data (Pham 2010, Pham and Pham 2010). The operating environments are, however, often unknown and yet different due to the uncertainties of environments in the field (Pham and Xie 2002). The *systemability* is the probability that the system must work in the uncertainty of the operational environments. The chapter discusses a concept of systemability function and its calculation for some system configurations and its applications in software reliability engineering. The chapter also discusses some life testing cost model in determining the optimum sample size on test that minimizes the expected total cost. Finally, the chapter discusses the stress-strength reliability estimation of k-out-of-n interval-system. Several applications are also discussed.

5.1 Order Statistics and Its Application in Reliability

Let $X_1, X_2, \ldots, X_n$ be a random sample of size n, assuming that they are independent and identically distributed, each with the cdf $F(x)$ and pdf $f(x)$. From Chap. 1, Sect. 12, the rth order statistic is the rth smallest value in the sample, say $X_{(r)}$, and can be written as

$$X_{(1)} \leq X_{(2)} \leq \ldots \leq X_{(r-1)} \leq X_{(r)} \leq \ldots \leq X_{(n)}$$

Let $F_{(r)}(x)$ denote the cdf of the rth order statistic $X_{(r)}$. Note that the order statistics $X_{(1)}, X_{(2)}, \ldots, X_{(n)}$ are neither independent nor identically distributed. Let $X_{(r)}$ be the rth order statistic. From Chap. 1, Eq. (71), the cdf of $X_{(r)}$ is given by

$$F_{(r)}(x) = P(X_{(r)} \leq x)$$

© The Author(s), under exclusive license to Springer Nature Switzerland AG 2022 307
H. Pham, *Statistical Reliability Engineering*, Springer Series in Reliability Engineering,
https://doi.org/10.1007/978-3-030-76904-8_5

$$= \sum_{i=r}^{n} \binom{n}{i} F^i(x)[1 - F(x)]^{n-i}$$

$$= F^r(x) \sum_{j=0}^{n-r} \binom{r+j-1}{r-1} [1 - F(x)]^j$$

$$= r \binom{n}{r} \int_{0}^{F(x)} t^{r-1}(1-t)^{n-r} dt \tag{5.1}$$

Using the integration by parts, from Eq. (5.1) it can be rewritten as:

$$r \binom{n}{r} \int_{0}^{F(x)} t^{r-1}(1-t)^{n-r} dt = \binom{n}{r} F^r(x)[1 - F(x)]^{n-r}$$

$$+ (r+1) \binom{n}{r+1} \int_{0}^{F(x)} t^r(1-t)^{n-r-1} dt$$

Thus, the cdf of $X_{(r)}$ also can be written as

$$F_{(r)}(x) = r \binom{n}{r} \int_{0}^{F(x)} t^{r-1}(1-t)^{n-r} dt$$

$$= \binom{n}{r} F^r(x)[1 - F(x)]^{n-r} + (r+1) \binom{n}{r+1} \int_{0}^{F(x)} t^r(1-t)^{n-r-1} dt$$

It is easy to show that the pdf of $X_{(r)}$ is given by

$$f_{(r)}(x) = \frac{\partial}{\partial x} F_{(r)}(x) = r \binom{n}{r} F^{r-1}(x)[1 - F(x)]^{n-r} f(x) \tag{5.2}$$

Let $X_{(1)}$ be the smallest order statistic, that is, $X_{(1)} = \min\{X_1, X_2, \ldots, X_n\}$. The cdf of $X_{(1)}$ can be obtained as follows:

$$F_{(1)}(x) = P(\min\{X_1, X_2, \ldots, X_n\} \le x) = P(X_{(1)} \le x)$$

$$= 1 - P(\min\{X_1, X_2, \ldots, X_n\} > x)$$

$$= 1 - \prod_{i=1}^{n} P(X_i > x) \quad \text{by independence}$$

$$= 1 - (1 - F(x))^n. \tag{5.3}$$

The cdf of $X_{(1)}$ also can be obtained from Eq. (5.1), we have

$$F_{(1)}(x) = P(X_{(1)} \leq x)$$
$$= \sum_{i=1}^{n} \binom{n}{i} F^i(x)[1 - F(x)]^{n-i}$$
$$= 1 - [1 - F(x)]^n$$

which is the same as Eq. (5.3). Likewise,

$$F_{(n)}(x) = P(X_{(n)} \leq x) = P(\text{all } X_i \leq x; i = 1, 2, \ldots, n)$$
$$= F^n(x) \quad \text{by independence.} \tag{5.4}$$

Let $R_i(x)$ be a reliability of component ith. As a definition from Chap. 4, the series system works if and only if all of its components work. The reliability of a series system of n components is given by

$$R_{(n:n)}(x) = P(X_{(1)} > x) = \prod_{i=1}^{n} R_i(x) \tag{5.5}$$

Also from Chap. 4, the parallel system works if and only if at least one component works. The reliability of the parallel system of n components is given by

$$R_{(1:n)}(x) = P(X_{(n)} > x) = 1 - \prod_{i=1}^{n} (1 - R_i(x)) \tag{5.6}$$

Similarly, a k-out-of-n system, in which the system functions iff at least k components function, has reliability

$$R_{(k:n)}(x) = P(X_{(n-k+1)} > x) = \sum_{i=k}^{n} \binom{n}{i} R^i(x)(1 - R(x))^{n-i} \tag{5.7}$$

Example 5.1 A life test on n items with independent lifetime X_i having exponential pdf

$$f(x) = \frac{1}{\mu} e^{-\frac{x}{\mu}} \quad x > 0 \tag{5.8}$$

is terminated as soon as N items ($N \leq n$) have failed or after time T_0, whichever comes first. It can be shown that:

(a)   The expected number of failures is

$$np \sum_{i=0}^{N-2} \binom{n-1}{i} p^i (1-p)^{n-i-1} + N \sum_{i=N}^{n} \binom{n}{i} p^i (1-p)^{n-i} \qquad (5.9)$$

where $p = 1 - e^{-\frac{T_0}{\mu}}$.

(b)   The expected duration of the test is

$$\sum_{i=1}^{N-1} \binom{n}{i} p^i (1-p)^{n-i} E(X_{(i)}) + \sum_{i=N}^{n} \binom{n}{i} p^i (1-p)^{n-i} E(X_{(N)}) \qquad (5.10)$$

where

$$E(X_{(r)}) = \mu \left(\sum_{i=0}^{r-1} \frac{1}{(n-i)} \right). \qquad (5.11)$$

and $p = 1 - e^{-\frac{T_0}{\mu}}$

Example 5.2 Under the Type II censoring, a $100(1 - \alpha)\%$ confidence interval for a reliability, with an exponential reliability function $R(x) = e^{-\frac{x}{\mu}}$, is given by $(e^{-\frac{x}{\mu_L}}, e^{-\frac{x}{\mu_U}})$ where

$$\mu_L = \frac{2N \sum_{i=1}^{N} \frac{X_{(i)}}{N}}{\chi^2_{2N, \frac{\alpha}{2}}} \quad \text{and} \quad \mu_U = \frac{2N \sum_{i=1}^{N} \frac{X_{(i)}}{N}}{\chi^2_{2N, 1-\frac{\alpha}{2}}} \qquad (5.12)$$

Similarly, let $X_{(n)}$ be the largest order statistic, that is, $X_{(n)} = \max\{X_1, X_2, \ldots, X_n\}$. We can show that

$$E(X_{(n)}) = E(X_{(n-1)}) + \int_0^{\infty} [F^{n-1}(x)][1 - F(x)]dx \quad \text{for n} = 2, 3, 4, \ldots \qquad (5.13)$$

where $X_1, X_2, \ldots, X_n$ is a random sample of size n from a continuous distribution function $F(x)$. Note that

$$E(X_{(n)}) = \int_{-\infty}^{\infty} nx[F(x)]^{n-1} dF(x) \qquad (5.14)$$

The mean of the rth order statistic $(r \leq n)$ is given by:

$$\mu_{(r)} = \int_{-\infty}^{\infty} x\, f_{(r)}(x)\, d(x) = \frac{n!}{(r-1)!(n-r)!} \int_{-\infty}^{\infty} x[F(x)]^{r-1}[1-F(x)]^{n-r} f(x) d(x) \quad (5.15)$$

It can also be shown that

$$E\big[X_{(r+1)} - X_{(r)}\big] = \binom{n}{r} \int_{-\infty}^{\infty} [F(x)]^r [1-F(x)]^{n-r} d(x) \quad \text{for } r = 1, 2, \ldots, n-1$$

Let $\mu_{r,n}$ be the mean of the rth order statistic of the sample of size n. It can be shown that

$$\mu_{r,n-1} = \mu_{r,n} + \binom{n-1}{r-1} \int_{-\infty}^{\infty} [F(x)]^r [1-F(x)]^{n-r} d(x)$$

$$\mu_{r+1,n} = \mu_{r,n-1} + \binom{n-1}{r} \int_{-\infty}^{\infty} [F(x)]^r [1-F(x)]^{n-r} d(x)$$

Order Statistics for Independent Non-identically distributed

Suppose that $X_1, X_2, \ldots, X_n$ are independent random variables, X_i having cdf $F_i(x)$. Then the cdf of $X_{(r)}$ is given by

$$F_{(r)} = \sum_{i=r}^{n} \sum_{j_1} \sum_{j_2} \sum_{j_i} \left(\prod_{k=1}^{i} [F_{j_k}(x)] \right) \left(\prod_{k=i+1}^{n} [1 - F_{j_k}(x)] \right)$$

$$\left\{ \begin{aligned} &j_1 < j_2 < \ldots < j_i \\ &\&\, j_{i+1} < j_{i+2} < \ldots < j_n \end{aligned} \right\} \quad (5.16)$$

5.2 The k-Out-of-n Systems

Consider a system consisting of n identical components having distribution function $F(x)$ and pdf $f(x)$. The k out of n system works if at least k of the n components are working; that is, it works iff at most $(n-k)$ components have failed. In particular, the parallel and series systems are 1 out of n and n out of n systems, respectively. The pdf of the k out of n system is the pdf of $(n-k+1)$th order statistics in a sample of size n and is given by

$$f_{(n-k+1)}(x) = \frac{n!}{(k-1)!(n-k)!} [F(x)]^{n-k} [1-F(x)]^{k-1} f(x) \quad (5.17)$$

The reliability of k out of n i.i.d. components system, $R_s(t)$, is given by

$$R_s(t) = \frac{n!}{(k-1)!(n-k)!} \int_t^\infty [F(x)]^{n-k}[1 - F(x)]^{k-1} f(x)\, d(x)$$

$$= \sum_{i=k}^n \binom{n}{i}[1 - F(t)]^i [F(t)]^{n-i}$$

$$= \frac{n!}{(k-1)!(n-k)!} \sum_{j=0}^{n-k} \frac{\binom{n-k}{j}(-1)^{n-k-j}}{(n-j)}[R(t)]^{n-j} \qquad (5.18)$$

where $R(t) = 1 - F(t)$. From Eq. (5.18), when $k = n$ then it becomes a series system and

$$R_{series}(t) = [R(t)]^n \qquad (5.19)$$

Similarly, when $k = 1$, it is a parallel system and

$$R_{parallel}(t) = 1 - [1 - R(t)]^n$$

$$= n \sum_{j=0}^{n-1} \frac{\binom{n-1}{j}(-1)^{n-j-1}}{(n-j)}(R(t))^{n-j} \qquad (5.20)$$

5.2.1   k *Out of* n *System Failure Rate*

The failure rate of k out of n system, $h_s(t)$, is given by

$$h_s(t) = -\frac{\partial}{\partial t}\{\ln[R_s(t)]\}$$

$$= -\frac{\partial}{\partial t}\left\{ \ln\left[\frac{n!}{(n-k)!\,(k-1)!} \sum_{j=0}^{n-k} \frac{\binom{n-k}{j}(-1)^{n-k-j}}{(n-j)}[R(t)]^{n-j} \right] \right\}$$

$$
= \frac{\lambda(t) \sum_{j=0}^{n-k} \binom{n}{j}\binom{n-j-1}{k-1}(n-j)(-1)^{n-k-j}[R(t)]^{n-j}}{\sum_{j=0}^{n-k} \binom{n}{j}\binom{n-j-1}{k-1}(-1)^{n-k-j}[R(t)]^{n-j}}
\tag{5.21}
$$

where $\lambda(t) = \frac{f(t)}{R(t)}$ is the failure rate corresponding to the cdf F.
Special cases:
If $k = n$ then $h_s(t) = n\,\lambda(t)$.
If $k = 1$ then

$$
h_s(t) = \frac{\lambda(t) \sum_{j=0}^{n-1} \binom{n}{j}(n-j)(-1)^{n-j-1}[R(t)]^{n-j}}{\sum_{j=0}^{n-1} \binom{n}{j}(-1)^{n-j-1}[R(t)]^{n-j}}
\tag{5.22}
$$

Theorem 5.1 (Abouammoh & El-Neweihi, 1986) Let X be a r.v. with the cdf F. For a random sample of size n from the pdf f. Let $F_{(k)}$ denote the distribution function of the kth order statistic. Then

(a)   F is IFR (IFRA) $\Rightarrow F_{(k)}$ is IFR (IFRA)
(b)   F is DFR (DFRA) $\neq\Rightarrow F_{(k)}$ is DFR (DFRA)
(c)   F is NBUE $\Rightarrow F_{(n)}$ is NBUE
(d)   F is DMRL $\Rightarrow F_{(n)}$ is DMRL

Theorem 5.2 (Abouammoh & El-Neweihi, 1986) Let X be a r.v. with the cdf F. For a random sample of size n from the pdf f.

(a)   If F is IFR (IFRA), then the distribution of the k out of n system is IFR (IFRA)
(b)   If F is DFR (DFRA), it is not necessary that the distribution of the k out of n system is DFR (DFRA)
(c)   If F is NBUE, then the distribution of the parallel system is also NBUE.
(d)   If F is DMRL, then the distribution of the parallel system is also DMRL.

5.3   Reliability Estimation of k-Out-of-n Systems

Consider a k out of n system which consists of n two-state (i.e., working or failed) components and that the component failures are independent where $R(t)$ denotes the reliability of the component. The k-out-of-n system functions if and only if at least k of its n components must function. In other words, as soon as $(n-k+1)$ components fail the system fails. This section discusses various approaches such as MLE, UMVUE and bias-corrected estimate (also called as delta method) for estimating the reliability of a k out of n system when the failure times are distributed as an exponential function for both the uncensored and censored failure cases.

Notation

R(t):   component reliability
f(t):   component failure probability density function
$R_s(t)$:   reliability of k out of n system
$f_s(t)$:   system failure time density function
$\widehat{R}_S(t)$ MLE of $R_s(t)$
$\widetilde{R}_S(t)$ UMVUE of $R_s(t)$
$\underset{\sim}{\hat{R}}_s(t)$ biased-corrected estimator (BCE) of $R_s(t)$
^:   implies a maximum likelihood estimate (MLE)
~ :   implies a uniformly minimum variance unbiased estimate (UMVUE)

Consider a k out of n system which consists of n subsystems and that the subsystem failures are independent where $R(t)$ denotes a subsystem reliability. The k-out-of-n system functions if and only if at least k of its n subsystems function. The reliability of the k-out-of-n system, $R_s(t)$, is:

$$R_s(t) = \sum_{i=k}^{n} \binom{n}{i}(R(t))^i(1 - R(t))^{n-i}. \tag{5.23}$$

The k-out-of-n system reliability function above can be rewritten as:

$$R_S(t) = \sum_{j=k}^{n} (-1)^{j-k} \binom{j-1}{k-1} \binom{n}{j} R^j(t) \tag{5.24}$$

For example, when $k = 2$ and $n = 4$, from Eq. (5.23)

$$R_s(t) = \sum_{i=2}^{4} \binom{4}{i}(R(t))^i(1 - R(t))^{4-i}$$
$$= \binom{4}{2}(R(t))^2(1 - R(t))^2 + \binom{4}{3}(R(t))^3(1 - R(t)) + (R(t))^4$$
$$= 6R^2(t) - 8R^3(t) + 3R^4(t)$$

is the same as when using from Eq. (5.24), that is

$$R_S(t) = \sum_{j=2}^{4} (-1)^{j-2} \binom{j-1}{1} \binom{4}{j} R^j(t)$$
$$= 6R^2(t) - 8R^3(t) + 3R^4(t)$$

If the lifetime, X, of each component has the exponential density function

$$f(x) = \frac{1}{\theta} e^{-\frac{x}{\theta}} \quad \text{for} \quad x > 0, \theta > 0 \tag{5.25}$$

where θ is the mean lifetime of the component, the component reliability, $R(t)$, is

$$R(t) = e^{-\frac{t}{\theta}} \quad \text{for} \quad t > 0, \theta > 0 \tag{5.26}$$

From Eq. (5.24), the reliability of the k out of n system is given by

$$R_S(t) = \sum_{j=k}^{n} (-1)^{j-k} \binom{j-1}{k-1} \binom{n}{j} e^{-\frac{jt}{\theta}}. \tag{5.27}$$

5.3.1 Failure-Uncensored Case of One-Parameter Exponential Distribution

Assuming that there are m units to put on test and then terminate the test when all the units have failed (uncensored case). Let $X_1, X_2, \ldots, X_m$ be the random failure times and suppose the failure times are exponentially distributed with density function as given in Eq. (5.25), the MLE of θ can be easily obtained

$$\hat{\theta} = \frac{\sum_{i=1}^{m} x_i}{m} = \bar{x} = \frac{S_m}{m} \tag{5.28}$$

where $S_m = \sum_{i=1}^{m} x_i$. The fact that (Laurent 1963) $\hat{\theta} = \bar{x} = \frac{S_m}{m}$ is also the UMVUE of θ. Thus, the MLE and UMVUE of one-unit component reliability are given by, respectively (Basu 1964):

$$\widehat{R}(t) = e^{-\frac{t}{\bar{x}}} \tag{5.29}$$

and

$$\widetilde{R}(t) = \begin{cases} \left(1 - \frac{t}{S_m}\right)^{m-1} & \text{if } t < S_m \\ 0 & \text{if } t \geq S_m \end{cases} \tag{5.30}$$

It is worth to note that the MLE has an important invariance property of the maximum likelihood estimator. However, the minimum variance unbiased estimator does not, in general, posses the invariance property of the MLE (Lehmann and Casella 1998). From Eq. (5.24) and using the invariance property, the MLE of the reliability function of a k out of n system, $R_s(t)$, are given by

$$\widehat{R}_S(t) = \sum_{j=k}^{n} (-1)^{j-k} \binom{j-1}{k-1} \binom{n}{j} e^{-\frac{jt}{\bar{x}}} \tag{5.31}$$

The UMVUE of function $R^j(t)$ is given by (Rutemiller 1966):

$$\widetilde{R}^j(t) = \begin{cases} \left(1 - \frac{jt}{S_m}\right)^{m-1} & \text{if } S_m \geq jt \\ 0 & \text{Otherwise} \end{cases} \tag{5.32}$$

Based on the polynomial combination of UMVUE functions, we can easily show that the UMVUE of reliability function of the k out of n systems, $R_s(t)$, can be obtained as follows (Pham 2010):

$$\widetilde{R}_S(t) = \begin{cases} \sum_{j=k}^{n} (-1)^{j-k} \binom{j-1}{k-1} \binom{n}{j} \left(1 - \frac{jt}{S_m}\right)^{m-1} & \text{if } S_m \geq jt \\ 0 & \text{Otherwise} \end{cases} \tag{5.33}$$

Bias-corrected Estimator (BCE) Approach

We wish to estimate the reliability using a bias-corrected estimator approach (Pham 2010) with $\hat{R}(t) = e^{-\frac{t}{\bar{x}}}$ where $\bar{x} = \frac{\sum_{i=1}^{m} x_i}{m}$. We can see that $E(\hat{R}(t)) \neq e^{-\frac{t}{\theta}}$ So $\hat{R}(t)$ is a biased estimate. Denote $\theta = E(X)$.

Delta method. Suppose the random variable Y has mean μ and variance σ^2 (i.e., $Y \sim (\mu, \sigma^2)$) and suppose we want the distribution of some function g(Y). Then

$$g(Y) = g(\mu) + (Y - \mu)\, g'(\mu) + \frac{(Y-\mu)^2}{2!}\, g'(\mu) + \dots \tag{5.34}$$

Using the Delta method and from Eq. (5.34), we obtain (Pham and Pham 2010):

$$\hat{R}(t) = e^{-\frac{t}{\bar{x}}} = e^{-\frac{t}{\theta}} + (\overline{X} - \theta)\frac{t}{\theta^2} e^{-\frac{t}{\theta}} + \frac{1}{2}(\overline{X} - \theta)^2 \left(\left(\frac{t}{\theta^2}\right)^2 - \frac{2t}{\theta^3} \right) e^{-\frac{t}{\theta}} + \dots$$

Hence,

$$E\left(\hat{R}(t)\right) \simeq \left[1 + \frac{1}{2m}\left(\frac{t^2}{\theta^2} - \frac{2t}{\theta}\right) \right] e^{-\frac{t}{\theta}}$$

So, the biased-corrected estimator of $\hat{R}(t)$ is (Pham 2010):

$$\hat{\underset{\sim}{R}}(t) = \frac{e^{-\frac{t}{\bar{x}}}}{1 + \frac{1}{2m}\left(\frac{t^2}{(\bar{x})^2} - \frac{2t}{\bar{x}}\right)}$$

Therefore, the reliability estimation function of the k out of n systems using the biased-corrected estimator approach is given by (Pham 2010):

$$\hat{\underset{\sim}{R}}_s(t) = \sum_{j=k}^{n}(-1)^{j-k}\binom{j-1}{k-1}\binom{n}{j}\left(\frac{e^{-\left(\frac{jt}{\bar{x}}\right)}}{1 + \frac{jt}{2m\bar{X}}\left(\frac{jt}{\bar{X}} - 2\right)}\right) \qquad (5.35)$$

where

$$\bar{x} = \frac{\sum_{i=1}^{m}x_i}{m} = \frac{S_m}{m}$$

5.3.2   Failure-Censored Case of One-Parameter Exponential Distribution

Consider the case where we put m units on test and terminate the test when a pre-assigned number of units, say r, of course $r \leq m$, have failed. Then the MLE of $R_s(t)$ is given by:

$$\hat{R}_S(t) = \sum_{j=k}^{n}(-1)^{j-k}\binom{j-1}{k-1}\binom{n}{j}e^{-\frac{jt}{\bar{x}}} \qquad (5.36)$$

where

$$\bar{x} = \frac{S_r}{r}$$

$$S_r = \sum_{i=1}^{r}x_{(i)} + (m-r)x_{(r)}$$

The UMVUE of $R_s(t)$ can be obtained:

$$\tilde{R}_S(t) = \begin{cases} \sum_{j=k}^{n}(-1)^{j-k}\binom{j-1}{k-1}\binom{n}{j}\left(1 - \frac{jt}{S_r}\right)^{r-1} & if \ S_r \geq jt \\ 0 & otherwise \end{cases} \qquad (5.37)$$

where S_r is given in Eq. (5.36).

The biased-corrected estimator of $R_s(t)$ is given by

$$\hat{\underset{\sim}{R}}_S(t) = \sum_{j=k}^{n} (-1)^{j-k} \binom{j-1}{k-1} \binom{n}{j} \left(\frac{e^{-\left(\frac{jt}{X}\right)}}{1 + \frac{jt}{2r\overline{X}}\left(\frac{jt}{X} - 2\right)} \right). \tag{5.38}$$

5.3.3 Applications

The power usages (in watts) of a computer system, with quad-core, 8 GB or ram, and a GeForce 9800GX-2 graphics card, performing various from simple- to-complex computing applications were recorded (every thirty-minute interval time for eleven and half hours) as follows by (Pham and Pham 2010):

253 274 285 293 257 277 251 251 252 256 293 295 285 280 259
253 253 279 273 292 306 296 259

Assuming the power usage to be distributed exponentially with mean live θ. We now calculate the MLE and UMVUE of reliability of the 2 out of 4 systems. From Eq. (5.28), we obtain the MLE and UMVUE of θ as follows:

$$\hat{\theta} = \tilde{\theta} = \frac{\sum_{i=1}^{23} x_i}{23} = \frac{6272}{23} = 272.7$$

The reliability estimation function of a 2 out of 4 system using the MLE is given by

$$\widehat{R}_S(t) = \sum_{j=2}^{4} (-1)^{j-2} \frac{4!}{j!(4-j)!}(j-1)e^{-\frac{jt}{272.7}}$$

The reliability estimation function of a 2 out of 4 system using the UMVUE is

$$\widetilde{R}_S(t) = \begin{cases} \sum_{j=2}^{4} (-1)^{j-2}\left(\frac{4!}{j!(4-j)!}\right)(j-1)\left(1 - \frac{jt}{6272}\right)^{22} & if \ \ 6272 \geq jt \\ 0 & \text{otherwise} \end{cases}$$

The reliability estimation function of a 2 out of 4 system using the Biased-Corrected estimator is

Table 5.1 Reliability estimation of a 2-out-of-4 system

t	$\widehat{R}_s(t)$	$\underset{s}{\hat{R}}(t)$	$\widetilde{R}_s(t)$	$\left(\widetilde{R}_s(t) - \widehat{R}_s(t)\right)100\%$
50	0.9836	0.9867	0.9867	0.31
100	0.9109	0.9237	0.9236	1.27
150	0.7932	0.8144	0.8144	2.12
200	0.6573	0.6807	0.6811	2.38

$$\underset{s}{\hat{R}}(t) = \sum_{j=2}^{4} (-1)^{j-2} \left(\frac{4!}{j!(4-j)!}\right) \left(\frac{e^{-\left(\frac{jt}{272.7}\right)}}{1 + \frac{jt}{12544}\left(\frac{jt}{272.7} - 2\right)}\right) (j-1)$$

Table 5.1 shows the numerical values of the reliability estimation of a 2-out-of-4 system based on the MLE, UMVUE, and BCE methods for various values of time t. It seems that: (i) the system reliability of both UMVUE and BCE methods can produce the same results; and (ii) the difference (in term of percentage) between the two estimates $\widetilde{R}_S(t)$ and $\widehat{R}_S(t)$, that is $(\widetilde{R}_S(t) - \widehat{R}_S(t))$, seems to be getting larger as the time t increases.

5.3.4 Failure-Uncensored Case of Two-Parameter Exponential Distribution

If the lifetime random variable, X, of each component has a two-parameter exponential density function

$$f(x) = \frac{1}{\theta} e^{-\frac{1}{\theta}(x-\mu)} \quad \text{for} \quad x > \mu > 0, \quad \theta > 0 \tag{5.39}$$

then the component reliability, R(t), is

$$R(t) = e^{-\frac{1}{\theta}(t-\mu)} \quad \text{for} \quad t > \mu, \theta > 0 \tag{5.40}$$

From Eq. (5.24), we obtain the reliability function of k-out-of-n system is

$$R_S(t) = \sum_{j=k}^{n} (-1)^{j-k} \binom{j-1}{k-1} \binom{n}{j} e^{-\frac{j(t-\mu)}{\theta}}. \tag{5.41}$$

Suppose m units are subjected to test and the test is terminated after all the units have failed. Let $X_1, X_2, \ldots, X_m$ be the random failure times and suppose that the failure times are two-parameter exponentially distributed with density function as given in Eq. (5.39). Let

$$\bar{x} = \frac{\sum\limits_{i=1}^{m} x_i}{m}; \quad x_{(1)} = \min(x_1, x_2, \ldots, x_m); \quad S_m = \sum_{i=2}^{m} \left(x_{(i)} - x_{(1)}\right). \quad (5.42)$$

We now discuss the reliability estimation of a k-out-of-n system, $R_s(t)$, based on the MLE, UMVUE, and bias-corrected estimator methods.

Maximum Likelihood Estimator of $R_s(t)$

The likelihood function is

$$L(\theta, \mu) = \prod_{i=1}^{m} f(x_i) = \frac{1}{\theta^m} e^{-\frac{1}{\theta} \sum\limits_{i=1}^{m} (x_i - \mu)}$$

The MLE of μ and θ are

$$\hat{\mu} = x_{(1)} \quad \hat{\theta} = \bar{x} - x_{(1)} \quad (5.43)$$

It should be noted that $\hat{\mu}$ and $\hat{\theta}$ are biased. The MLE of component reliability is:

$$\widehat{R}(t) = e^{-\left(\frac{t - x_{(1)}}{\bar{x} - x_{(1)}}\right)} \quad \text{for} \quad t > x_{(1)} \quad (5.44)$$

From the invariance property and Eq. (5.24), the MLE of the reliability function of k out of n system, $R_s(t)$, is given by

$$\widehat{R}_S(t) = \sum_{j=k}^{n} (-1)^{j-k} \binom{j-1}{k-1} \binom{n}{j} e^{-\left(\frac{j(t - x_{(1)})}{\bar{x} - x_{(1)}}\right)} \quad (5.45)$$

UMVUE of $R_s(t)$

The UMVUE of μ and θ can be easily obtained as follows:

$$\tilde{\mu} = \frac{m \, x_{(1)} - \bar{x}}{m - 1}; \quad \tilde{\theta} = \frac{m(\bar{x} - x_{(1)})}{m - 1}. \quad (5.46)$$

It can be shown that $\tilde{\mu}$ and $\tilde{\theta}$ are unbiased. The UMVUE of component reliability is:

$$\tilde{R}(t) = \begin{cases} 1 & \text{if } t \leq \tilde{\mu} \\ \left(1 - \frac{1}{m}\right)\left(1 - \frac{t - x_{(1)}}{S_m}\right)^{m-2} & \tilde{\mu} < t \leq S_m + x_{(1)} \\ 0 & t > S_m + x_{(1)} \end{cases} \quad (5.47)$$

Therefore, the UMVUE of function $R^j(t)$ is given by:

$$\widetilde{R}^j(t) = \left(1 - \frac{j}{m}\right)\left(1 - \frac{j(t - x_{(1)})}{S_m}\right)^{m-2} \qquad \tilde{\mu} < t \le S_m + x_{(1)}$$

Although the MLE has an important invariance property of the maximum like-lihood estimator, the minimum variance unbiased estimator, in general, does not, possess the invariance property of the MLE (Pham 2010). Based on the polynomial combination of UMVUE functions, we can show that the UMVUE of reliability function of k out of n systems, $R_s(t)$, is (Pham and Pham 2010):

$$\widetilde{R}_S(t) = \begin{cases} 1 & \text{if } t \le \tilde{\mu} \\ \sum\limits_{j=k}^{n} (-1)^{j-k}\binom{j-1}{k-1}\binom{n}{j}\left(1 - \frac{j}{m}\right)\left(1 - \frac{j(t-x_{(1)})}{S_m}\right)^{m-2} & \text{if } \tilde{\mu} < t \le S_m + x_{(1)} \\ 0 & \text{if } t > S_m + x_{(1)} \end{cases}$$

$$(5.48)$$

Bias-corrected Estimator Approach—Delta Method

We now estimate the reliability $\hat{R}(t) = e^{-\frac{1}{\theta}(t - \tilde{\mu})}$ using a bias-corrected estimator approach. We obviously can see that $E(\hat{R}(t)) \ne e^{-\frac{1}{\theta}(t-\mu)}$. This implies that $\hat{R}(t)$ is a biased estimate. Denote $\theta = E(X)$. Using the Delta method (Pham 2010), the reliability estimation function of k out of n system using the biased-corrected estimator approach is given by

$$\hat{\underset{\sim}{R}}_S(t) = \sum_{j=k}^{n} (-1)^{j-k}\binom{j-1}{k-1}\binom{n}{j}\frac{e^{-\frac{j(t-\tilde{\mu})}{\tilde{\theta}}}}{1 + \frac{j(t-\tilde{\mu})\left[j(t-\tilde{\mu})-2\tilde{\theta}\right]}{2(m-1)\tilde{\theta}^2} - \frac{j\left[j(t-\tilde{\mu})-\tilde{\theta}\right]}{m(m-1)\tilde{\theta}} + \frac{j^2}{2m(m-1)}}$$

$$(5.49)$$

where

$$\tilde{\mu} = \frac{m\,x_{(1)} - \bar{x}}{m-1}; \quad \tilde{\theta} = \frac{m(\bar{x} - x_{(1)})}{m-1}.$$

5.3.5 Applications

The power usages (in watts) of a computer system using NComputing approach with quad-core, 8 GB of RAM, and a GeForce 9800GX-2 graphics card, performing various simple- to-complex computing applications were recorded (after every two-minutes interval) for computing the value of π including the first 32-million digits.

From the data set as shown in Pham and Pham (2010, Table 5.1, data set 1), the two-parameter exponential distribution will be considered. We now obtain the reliability calculations for a 2 out of 3 computer system based on MLE, UMVUE and bias-corrected estimation approach.

Reliability Computing – the MLE of μ and θ can be obtained as follows:

$$\hat{\mu} = x_{(1)} = 286 \quad \hat{\theta} = \bar{x} - x_{(1)} = 289.6923 - 286 = 3.6923$$

From Eq. (5.45), the reliability estimation function of a 2 out of 3 system using the MLE is

$$\widehat{R}_S(t) = \sum_{j=k}^{n} (-1)^{j-k} \binom{j-1}{k-1} \binom{n}{j} e^{-\left(\frac{j(t-x_{(1)})}{\bar{x}-x_{(1)}}\right)}$$

$$= \sum_{j=2}^{3} (-1)^{j-2} \binom{j-1}{1} \binom{3}{j} e^{-\left(\frac{j(t-286)}{\frac{7532}{26}-286}\right)}$$

$$= 3e^{-\left(\frac{2(t-286)}{\frac{7532}{26}-286}\right)} - 2e^{-\left(\frac{3(t-286)}{\frac{7532}{26}-286}\right)}$$

The reliability estimation function of a 2 out of 3 system using the UMVUE from Eq. (5.48) is

$$\tilde{R}_S(t) = \sum_{j=k}^{n} (-1)^{j-k} \binom{j-1}{k-1} \binom{n}{j} \left(1-\frac{j}{m}\right) \left(1-\frac{j(t-x_{(1)})}{S_m}\right)^{m-2} = \sum_{j=2}^{3} (-1)^{j-2} \binom{j-1}{1} \binom{3}{j} \left(1-\frac{j}{26}\right)$$

$$\left(1-\frac{j(t-x_{(1)})}{26 \sum_{i=2}^{26}(x_{(i)} - x_{(1)})}\right)^{26-2} = 3\left(\frac{12}{13}\right)\left(1-\frac{2(t-286)}{26 \sum_{i=2}^{26}(x_{(i)} - 286)}\right)^{26-2} - \left(\frac{23}{13}\right)\left(1-\frac{3(t-286)}{26 \sum_{i=2}^{26}(x_{(i)} - 286)}\right)^{26-2}$$

The reliability estimation function of 2 out of 3 systems using bias-corrected estimator approach (from Eq. 5.49) is

$$\widetilde{R}_S(t) = \sum_{j=2}^{3} (-1)^{j-2} \binom{j-1}{1} \binom{3}{j} \frac{e^{-\frac{j(t-\tilde{\mu})}{\tilde{\theta}}}}{1 + \frac{j(t-\tilde{\mu})\left[j(t-\tilde{\mu})-2\tilde{\theta}\right]}{2(25)\tilde{\theta}^2} - \frac{j\left[j(t-\tilde{\mu})-\tilde{\theta}\right]}{26(25)\tilde{\theta}} + \frac{j^2}{2*26(25)}}$$

$$= \frac{3e^{-\frac{2(t-\tilde{\mu})}{\tilde{\theta}}}}{1 + \frac{(t-\tilde{\mu})\left[2(t-\tilde{\mu})-2\tilde{\theta}\right]}{25\tilde{\theta}^2} - \frac{\left[2(t-\tilde{\mu})-\tilde{\theta}\right]}{325\tilde{\theta}} + \frac{1}{325}} - \frac{2e^{-\frac{3(t-\tilde{\mu})}{\tilde{\theta}}}}{1 + \frac{3(t-\tilde{\mu})\left[3(t-\tilde{\mu})-2\tilde{\theta}\right]}{50\tilde{\theta}^2} - \frac{3\left[3(t-\tilde{\mu})-\tilde{\theta}\right]}{650\tilde{\theta}} + \frac{9}{1300}}$$

where $\tilde{\mu} = \frac{26 \, x_{(1)} - \bar{x}}{25}$; $\tilde{\theta} = \frac{26 \left(\bar{x} - x_{(1)}\right)}{25}$.

Table 5.2 shows the reliability calculations for various values of t, i.e., energy consumption value, for given data set #1 based on the MLE, UMVUE and bias-corrected estimation approach. Given the data set #1, the MLE, UMVUE, and BCE of the 2 out of 3 system reliability estimation are 0.2661, 0.2713, and 0.2775 when

Table 5.2 Reliability values of a 2 out of 3 system for MLE, UMVUE and Bias-corrected versus t for data set #1

t	MLE	UMVUE	Bias
286	1.0000	1.000	0.989
287	0.858	0.845	0.822
288	0.622	0.621	0.609
289	0.416	0.422	0.421
290	0.266	0.271	0.278
291	0.166	0.168	0.176
292	0.101	0.100	0.108

Fig. 5.1 The 2 out of 3 system reliability function vs. energy consumption t based on MLE, UMVUE and biased-corrected estimators for data set #1 (Pham and Pham 2010)

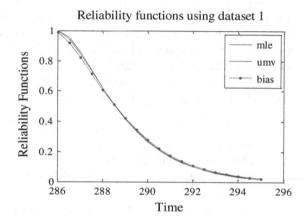

t = 290 wh, respectively. Figure 5.1 shows a 2-out-of-3 system reliability estimation versus the energy consumption. We observe that the UMVUE of $\widetilde{R}_S(t)$ is larger than the MLE $\widehat{R}_S(t)$ at the beginning time t, but getting smaller than the MLE $\widehat{R}_S(t)$ when t is getting larger.

We observe that the actual values of system reliability based on the UMVUE and BCE only differ slightly. In many cases, to obtain the UMVUE of complex system reliability for distributions other than exponential is a lot more difficult. In general, the BCE results such as the bias-corrected estimate (Eq. 5.49) are rather simple and can be used to obtain the reliability estimation function of any complex system.

5.4 Systemability Measures

The traditional reliability definitions and its calculations have commonly been carried out through the failure rate function within a controlled laboratory-test environment. In other words, such reliability functions are applied to the failure testing data and then utilized to make predictions on the reliability of the system used in the field

(Pham 2005). The underlying assumption for such calculation is that the field (or operating) environments and the testing environments are the same.

By defintion, a mathematical reliability function is the probability that a system will be successful in the interval from time 0 to time t, given by

$$R(t) = \int_{t}^{\infty} f(s)ds = e^{-\int_{0}^{t} h(s)ds} \tag{5.50}$$

where $f(s)$ and $h(s)$ are, respectively, the failure time density and failure rate function.

The operating environments are, however, often unknown and yet different due to the uncertainties of environments in the field (Pham and Xie 2003). A new look at how reliability researchers can take account of the randomness of the field environments into mathematical reliability modeling covering system failure in the field is great interest. Pham (2005) introduced a new mathematical function called *systemability*, considering the uncertainty of the operational environments in the function for predicting the reliability of systems.

Notation

$h_i(t)$ ith component hazard rate function
$R_i(t)$ ith component reliability function
λ_i Intensity parameter of Weibull distribution for ith component
$\underline{\lambda}$ $\underline{\lambda} = (\lambda_1, \lambda_2, \lambda_3 \ldots, \lambda_n)$
γ_i Shape parameter of Weibull distribution for ith component
$\underline{\gamma}$ $\underline{\gamma} = (\gamma_1, \gamma_2, \gamma_3 \ldots, \gamma_n)$
η A common environment factor
$G(\eta)$ Cumulative distribution function of η
α Shape parameter of Gamma distribution
β Scale parameter of Gamma distribution

Definition 5.1 (Pham 2005) *Systemability* is defined as the probability that the system will perform its intended function for a specified mission time under the random operating environments.

In a mathematical form, the *systemabililty* function is given by

$$R_s(t) = \int_{\eta} e^{-\eta \int_{0}^{t} h(s)ds} dG(\eta) \tag{5.51}$$

where η is a random variable that represents the system operating environments with a distribution function G.

This new function captures the uncertainty of complex operating environments of systems in terms of the system failure rate. It also would reflect the reliability estimation of the system in the field.

If we assume that η has a gamma distribution with parameters α and β, *i.e.*, $\eta \sim gamma(\alpha, \beta)$ where the pdf of η is given by

$$f_\eta(x) = \frac{\beta^\alpha x^{\alpha-1} e^{-\beta x}}{\Gamma(\alpha)} \quad \text{for} \quad \alpha, \beta > 0; x \geq 0 \tag{5.52}$$

then the systemability function of the system in Eq. (2.30), using the Laplace transform (see Appendix B), is given by

$$R_s(t) = \left[\frac{\beta}{\beta + \int\limits_0^t h(s)ds} \right]^\alpha \tag{5.53}$$

5.4.1 Systemability Calculations

This subsection presents several systemability results and variances of some system configurations such as series, parallel, and k-out-of-n systems (Pham 2005). Consider the following assumptions:

1. A system consists of n independent components where the system is subject to a random operational environment η.
2. ith component lifetime is assumed to follow the Weibull density function, *i.e.* Component hazard rate

$$h_i(t) = \lambda_i \gamma_i t^{\gamma_i - 1} \tag{5.54}$$

Component reliability

$$R_i(t) = e^{-\lambda_i t^{\gamma_i}} \quad t > 0 \tag{5.55}$$

Given common environment factor $\eta \sim gamma(\alpha, \beta)$, the systemability functions for different system structures can be obtained as follows.

Series System Configuration

In a series system, all components must operate successfully if the system is to function. The conditional reliability function of series systems subject to an actual operational random environment η is given by

$$R_{Series}(t|\eta, \underline{\lambda}, \underline{\gamma}) = e^{\left(-\eta \sum\limits_{i=1}^n \lambda_i t^{\gamma_i}\right)} \tag{5.56}$$

The series systemability is given as follows

$$R_{Series}(t|\underline{\lambda}, \underline{\gamma}) = \int_{\eta} \exp\left(-\eta \sum_{i=1}^{n} \lambda_i t^{\gamma_i}\right) dG(\eta) = \left[\frac{\beta}{\beta + \sum_{i=1}^{n} \lambda_i t^{\gamma_i}}\right]^{\alpha} \tag{5.57}$$

The variance of a general function $R(t)$ is given by

$$Var[R(t)] = E[R^2(t)] - (E[R(t)])^2 \tag{5.58}$$

Given $\eta \sim gamma(\alpha, \beta)$, the variance of systemability for any system structure can be easily obtained. Therefore, the variance of series systemability is given by

$$Var\left[R_{Series}(t|\underline{\lambda}, \underline{\gamma})\right] = \int_{\eta} \exp\left(-\eta\left(2\sum_{i=1}^{n} \lambda_i t^{\gamma_i}\right)\right) dG(\eta)$$
$$- \left(\int_{\eta} \exp\left(-\eta \sum_{i=1}^{n} \lambda_i t^{\gamma_i}\right) dG(\eta)\right)^2 \tag{5.59}$$

or

$$Var\left[R_{Series}(t|\underline{\lambda}, \gamma)\right] = \left[\frac{\beta}{\beta + 2\sum_{i=1}^{n} \lambda_i t^{\gamma_i}}\right]^{\alpha} - \left[\frac{\beta}{\beta + \sum_{i=1}^{n} \lambda_i t^{\gamma_i}}\right]^{2\alpha} \tag{5.60}$$

Parallel System Configuration

A parallel system is a system that is not considered to have failed unless all components have failed. The conditional reliability function of parallel systems subject to the uncertainty operational environment η is given by Pham (2005)

$$R_{Parallel}(t|\eta, \lambda, \gamma) = \exp(-\eta\lambda_i t^{\gamma_i}) - \sum_{\substack{i_1,i_2=1 \\ i_1 \neq i_2}}^{n} \exp\left(-\eta\left(\lambda_{i_1} t^{\gamma_{i_1}} + \lambda_{i_2} t^{\gamma_{i_2}}\right)\right)$$

$$+ \sum_{\substack{i_1,i_2,i_2=1 \\ i_1 \neq i_2 \neq i_3}}^{n} \exp\left(-\eta\left(\lambda_{i_1} t^{\gamma_{i_1}} + \lambda_{i_2} t^{\gamma_{i_2}} + \lambda_{i_3} t^{\gamma_{i_3}}\right)\right) -$$

$$\cdots\cdots$$

$$+ (-1)^{n-1} \exp\left(-\eta \sum_{i=1}^{n} \lambda_i t^{\gamma_i}\right) \tag{5.61}$$

Hence, the parallel systemability is given by

$$R_{parallel}(t|\underline{\lambda}, \underline{\gamma}) = \sum_{i=1}^{n} \left[\frac{\beta}{\beta + \lambda_i t^{\gamma_i}} \right]^{\alpha} - \sum_{\substack{i_1, i_2 = 1 \\ i_1 \neq i_2}}^{n} \left[\frac{\beta}{\beta + \lambda_{i_1} t^{\gamma_{i_1}} + \lambda_{i_2} t^{\gamma_{i_2}}} \right]^{\alpha}$$

$$+ \sum_{\substack{i_1, i_2, i_2 = 1 \\ i_1 \neq i_2 \neq i_3}}^{n} \left[\frac{\beta}{\beta + \lambda_{i_1} t^{\gamma_{i_1}} + \lambda_{i_2} t^{\gamma_{i_2}} + \lambda_{i_3} t^{\gamma_{i_3}}} \right]^{\alpha} -$$

$$\cdots$$

$$+ (-1)^{n-1} \left[\frac{\beta}{\beta + \sum_{i=1}^{n} \lambda_i t^{\gamma_i}} \right]^{\alpha} \tag{5.62}$$

or

$$R_{parallel}(t|\underline{\lambda}, \underline{\gamma}) = \sum_{k=1}^{n} (-1)^{k-1} \sum_{\substack{i_1, i_2 \ldots, i_k = 1 \\ i_1 \neq i_2 \ldots \neq i_k}}^{n} \left[\frac{\beta}{\beta + \sum_{j=i_1, \ldots i_k} \lambda_j t^{\gamma_j}} \right]^{\alpha} \tag{5.63}$$

To simplify the calculation of a general n-component parallel system, we only consider here a parallel system consisting of two components. It is easy to see that the second-order moments of the systemability function can be written as

$$E\left[R_{Parallel}^2(t|\underline{\lambda}, \underline{\gamma}) \right] = \int_{\eta} (e^{-2\eta \lambda_1 t^{\gamma_1}} + e^{-2\eta \lambda_2 t^{\gamma_2}} + e^{-2\eta(\lambda_1 t^{\gamma_1} + \lambda_2 t^{\gamma_2})}$$

$$+ e^{-\eta(\lambda_1 t^{\gamma_1} + \lambda_2 t^{\gamma_2})} - e^{-\eta(2\lambda_1 t^{\gamma_1} + \lambda_2 t^{\gamma_2})} - e^{-\eta(\lambda_1 t^{\gamma_1} + 2\lambda_2 t^{\gamma_2})}) \, dG(\eta)$$

The variance of series systemability of a two-component parallel system is given by

$$Var\left[R_{Parallel}(t|\underline{\lambda}, \underline{\gamma}) \right] = \left[\frac{\beta}{\beta + 2\lambda_1 t^{\gamma_1}} \right]^{\alpha} + \left[\frac{\beta}{\beta + 2\lambda_2 t^{\gamma_2}} \right]^{\alpha}$$

$$+ \left[\frac{\beta}{\beta + 2\lambda_1 t^{\gamma_1} + 2\lambda_2 t^{\gamma_2}} \right]^{\alpha} + \left[\frac{\beta}{\beta + \lambda_1 t^{\gamma_1} + \lambda_2 t^{\gamma_2}} \right]^{\alpha}$$

$$- \left[\frac{\beta}{\beta + 2\lambda_1 t^{\gamma_1} + \lambda_2 t^{\gamma_2}} \right]^{\alpha} - \left[\frac{\beta}{\beta + \lambda_1 t^{\gamma_1} + 2\lambda_2 t^{\gamma_2}} \right]^{\alpha}$$

$$- \left[\left[\frac{\beta}{\beta + \lambda_1 t^{\gamma_1}} \right]^{\alpha} + \left[\frac{\beta}{\beta + \lambda_2 t^{\gamma_2}} \right]^{\alpha} - \left[\frac{\beta}{\beta + \lambda_1 t^{\gamma_1} + \lambda_2 t^{\gamma_2}} \right]^{\alpha} \right]^2 \tag{5.64}$$

K-out-of-n System Configuration

In a k-out-of-n configuration, the system will operate if at least k out of n components
are operating. To simplify the complexity of the systemability function, we assume
that all the components in the k-out-of-n systems are identical. Therefore, for a given
common environment η, the conditional reliability function of a component is given
by

$$R(t|\eta, \lambda, \gamma) = e^{-\eta\lambda t^\gamma} \tag{5.65}$$

The conditional reliability function of k-out-of-n systems subject to the uncertainty operational environment η can be obtained as follows:

$$R_{k-out-of-n}(t|\eta, \lambda, \gamma) = \sum_{j=k}^{n} \binom{n}{j} e^{-\eta j \lambda t^\gamma} (1 - e^{-\eta\lambda t^\gamma})^{(n-j)} \tag{5.66}$$

Note that

$$(1 - e^{-\eta\lambda t^\gamma})^{(n-j)} = \sum_{l=0}^{n-j} \binom{n-j}{l} (-e^{-\eta\lambda t^\gamma})^l$$

The conditional reliability function of k-out-of-n systems, from Eq. (5.66), can
be rewritten as

$$R_{K-out-of-N}(t|\eta, \lambda, \gamma) = \sum_{j=k}^{n} \binom{n}{j} \sum_{l=0}^{n-j} \binom{n-j}{l} (-1)^l e^{-\eta(j+l)\lambda t^\gamma}$$

Then if $\eta \sim gamma(\alpha, \beta)$ then the k-out-of-n systemability is given by

$$R_{(T_{1_1},\ldots,T_n)}(t|\lambda, \gamma) = \sum_{j=k}^{n} \binom{n}{j} \sum_{l=0}^{n-j} \binom{n-j}{l} (-1)^l \left[\frac{\beta}{\beta + \lambda(j+l)t^\gamma} \right]^\alpha$$

It can be easily shown that

$$R_{k-out-of-n}^2(t|\eta, \lambda, \gamma) = \sum_{i=k}^{n} \binom{n}{i} \sum_{j=k}^{n} \binom{n}{j} e^{-\eta(i+j)\lambda t^\gamma} (1 - e^{-\eta\lambda t^\gamma})^{(2n-i-j)} \tag{5.67}$$

Since

$$(1 - e^{-\eta\lambda t^\gamma})^{(2n-i-j)} = \sum_{l=0}^{2n-i-j} \binom{2n-i-j}{l} (-e^{-\eta\lambda t^\gamma})^l$$

we can rewrite Eq. (5.67), after several simplifications, as follows

$$R^2_{k-out-of-n}(t|\eta,\lambda,\gamma) = \sum_{i=k}^{n}\binom{n}{i}\sum_{j=k}^{n}\binom{n}{j}(-1)^l\sum_{l=0}^{2n-i-j}\binom{2n-i-j}{l}e^{-\eta(i+j+l)\lambda t^\gamma} \quad (5.68)$$

Therefore, the variance of k-out-of-n system systemability function is given by

$$Var(R_{k/n}(t|\lambda,\gamma) = \int_\eta R^2_{k/n}(t|\eta,\lambda,\gamma)dG(\eta) - \left[\int_\eta R_{k/n}(t|\eta,\lambda,\gamma)dG(\eta)\right]^2$$

$$= \sum_{i=k}^{n}\binom{n}{i}\sum_{j=k}^{n}\binom{n}{j}\sum_{l=0}^{2n-i-j}\binom{2n-i-j}{l}(-1)^l\left(\frac{\beta}{\beta+(i+j+l)\lambda t^\gamma}\right)^2$$

$$- \left(\sum_{j=k}^{n}\binom{n}{j}\sum_{l=0}^{n-j}\binom{n-j}{l}(-1)^l\left(\frac{\beta}{\beta+(j+l)\lambda t^\gamma}\right)^2\right)^2 \quad (5.69)$$

Example 5.3 Consider a k-out-of-n system where $\lambda = 0.0001$, $\gamma = 1.5$, $n = 5$, and $\eta \sim gamma(\alpha,\beta)$. Calculate the systemability of various k-out-of-n system configurations.

Solution: The systemability of generalized k-out-of-5 system configurations is given as follows:

$$R_{k-out-of-n}(t|\lambda,\gamma) = \sum_{j=k}^{5}\binom{5}{j}\sum_{l=0}^{5-j}\binom{5-j}{l}(-1)^l\left[\frac{\beta}{\beta+\lambda(j+l)t^\gamma}\right]^\alpha \quad (5.70)$$

Figure 5.2 shows the reliability function (conventional reliability function) and systemability function (Eq. 5.70) of a series system (here $k = 5$) for $\alpha = 2$, $\beta = 3$. Figure 5.3 shows the reliability and systemability functions of a parallel system (here

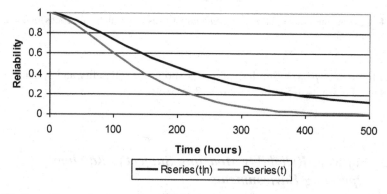

Fig. 5.2 Comparisons of series system reliability versus systemability functions for $\alpha = 2$ and $\beta = 3$ (Pham 2005)

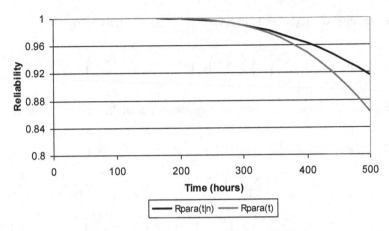

Fig. 5.3 Comparisons of parallel system reliability versus systemability function for $\alpha = 2$ and $\beta = 3$ (Pham 2005)

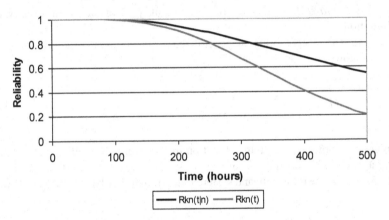

Fig. 5.4 Comparisons of k-out-of-n system reliability versus systemability functions for $\alpha = 2$ and $\beta = 3$ (Pham 2005)

$k = 1$) for $\alpha = 2, \beta = 3$. Similarly, Fig. 5.4 shows the reliability and systemability functions of a 3-out-of-5 system for $\alpha = 2, \beta = 3$.

5.4.2   Software Reliability Modeling Subject to Random Operating Environments

Many existing software reliability models (Li and Pham 2017a, 2017b, 2019; Pham 1996, 2003, 2006a, 2014, 2014a, 2019a; Pham et al. 2014b, Zhu and Pham

2019a, b; Sharma et al. 2019; Song et al. 2018, 2019a, b) have been studied in the past three decades and carried out through the fault intensity rate function and the mean value functions within a controlled testing environment. These models can be used to estimate reliability measures such as the number of remaining errors, failure intensity, and reliability of software. In general most of existing models are applied to the software failure data assuming that the operating environments and the developing environments are about the same. This is often not the case because the operating environments are usually unknown due to the uncertainty of environments in the real world applications (Sgarbossa et al. 2015; Zhu and Pham 2020). In this section, we discuss a software reliability model addressing the uncertainty of the operating environments based on nonhomogeneous Poisson process (NHPP) models. Several model selection criteria are also discussed.

Let η be a random variable that represents the uncertainty of software error detection rate in the operating environments with the probability density function g.

Notation

$m(t)$ expected number of software failures detected by time t
N e   expected number of faults that exist in the software before testing
$b(t)$ time dependent fault detection rate per fault per unit of time

A generalized mean value function $m(t)$ with the uncertainty of operating environments can be formulated in the form of the following differential equation (Pham 2014):

$$\frac{dm(t)}{dt} = \eta b(t)[N - m(t)] \tag{5.71}$$

The solution for the mean value function $m(t)$ of Eq. (5.71), where the initial condition $m(0) = 0$, is given by (Pham 2014):

$$m(t) = \int_{\eta} N \left(1 - e^{-\eta \int_0^t b(x)dx} \right) dg(\eta) \tag{5.72}$$

Assuming that the random variable η has a probability density function g with two parameters $\alpha \geq 0$ and $\beta \geq 0$ so that the mean value function from Eq. (5.72) can be obtained as the general form below:

$$m(t) = N \left(1 - \frac{\beta}{\beta + \int_0^t b(s)ds} \right)^{\alpha} \tag{5.73}$$

Consider the fault detection rate function per fault per unit of time, $b(t)$, is as follows:

$$b(t) = \frac{c}{1 + ae^{-bt}} \quad \text{for} \quad a \geq 0, b \geq 0, c > 0.$$

Substitute the function $b(t)$ above into Eq. (5.73), we can obtain the expected number of software failures detected by time t subject to the uncertainty of the environments, $m(t)$:

$$m(t) = N\left(1 - \frac{\beta}{\beta + \left(\frac{c}{b}\right)\ln\left(\frac{a+e^{bt}}{1+a}\right)}\right)^{\alpha}$$

Table 5.3 presents some existing models in the software reliability engineering literature.

Example 5.4 Consider a system test data as given in Table 5.4, referred to as Phase 2 data set (Pham 2006). In this data set the number of faults detected in each week of testing is found and the cumulative number of faults since the start of testing is recorded for each week up to 21 weeks.

Table 5.3 A summary of existing software reliability models (Pham 2018)

Model	m(t)
Goel-Okumoto (G-O)	$m(t) = a(1 - e^{-bt})$
Delayed S-shaped	$m(t) = a(1 - (1 + bt)e^{-bt})$
Inflection S-shaped	$m(t) = \frac{a(1-e^{-bt})}{1+\beta e^{-bt}}$
Yamada Imperfect debugging 1	$m(t) = a[1 - e^{-bt}][1 - \frac{\alpha}{b}] + \alpha\,a\,t$
Pham Inflexion	$m(t) = N\left(1 - \frac{1}{\left(\frac{\beta+e^{bt}}{1+\beta}\right)^{\frac{\alpha}{b}}}\right)$
PNZ model	$m(t) = \frac{a}{1+\beta e^{-bt}}\left([1 - e^{-bt}][1 - \frac{\alpha}{b}] + \alpha t\right)$
Pham-Zhang model	$m(t) = \frac{1}{1+\beta e^{-bt}}\left((c + a)(1 - e^{-bt}) - \frac{ab}{b-\alpha}(e^{-\alpha t} - e^{-bt})\right)$
Dependent-parameter model	$m(t) = \alpha(1 + \gamma t)(\gamma t + e^{-\gamma t} - 1)$
Vtub-shaped fault-detection rate model	$m(t) = N\left(1 - \left(\frac{\beta}{\beta+a^{t^b}-1}\right)^{\alpha}\right)$
Logistic fault detection model	$m(t) = N\left(1 - \frac{\beta}{\beta+\left(\frac{c}{b}\right)\ln\left(\frac{a+e^{bt}}{1+a}\right)}\right)^{\alpha}$

Table 5.4 Phase 2 system test data (Pham 2006)

Week index	Exposure time (cum. system test hours)	Fault	Cum. fault
1	416	3	3
2	832	1	4
3	1248	0	4
4	1664	3	7
5	2080	2	9
6	2496	0	9
7	2912	1	10
8	3328	3	13
9	3744	4	17
10	4160	2	19
11	4576	4	23
12	4992	2	25
13	5408	5	30
14	5824	2	32
15	6240	4	36
16	6656	1	37
17	7072	2	39
18	7488	0	39
19	7904	0	39
20	8320	3	42
21	8736	1	43

Table 5.5 summarizes the results of the model parameter estimates using the least square estimation (LSE) technique of all the 10 models from Table 5.3 based on the data set as shown in Table 5.4 and three model selection criteria such as mean squared error (MSE), predictive-ratio risk (PRR) and predictive power (PP). The logistic fault detection model (model 10) in Table 5.5 seems to provides the best fit based on those criteria (MSE, PRR, PP).

5.5 Life Testing Cost Model

Some highly reliable electronic products require long testing periods before useful failure data can be obtained. This is an unrealistic situation in practice since manufacturers often need to deliver their products to the market as soon as possible. It is desirable to find ways of speeding up testing. Accelerated testing can be accomplished by reducing the time required for testing by placing many units on the test and terminating the test after a pre-assigned number of failures occur. The appropriate

Table 5.5 Model parameter estimation and model criteria

Model name	LSEs	MSE	PRR	PP
1. G -O Model	$\hat{a} = 98295$ $\hat{b} = 5.2\ 10^{-8}$	6.61	0.69	1.10
2. Delayed S-shaped	$\hat{a} = 62.3$ $\hat{b} = 2.85\ 10^{-4}$	3.27	44.27	1.43
3. Inflection S-shaped	$\hat{a} = 46.6$ $\hat{b} = 5.78\ 10^{-4}$ $\hat{\beta} = 12.20$	1.87	5.94	0.90
4. Yamada imperfect debugging model	$\hat{a} = 1.5$ $\hat{b} = 1.1\ 10^{-3}$ $\hat{\alpha} = 3.8\ 10^{-3}$	4.98	4.30	0.81
5. Pham Inflexion	$N = 45.8270$ $a = 0.2961$ $b = 0.2170$ $\beta = 13.6298$	1.5108	3.1388	0.6800
6. PNZ model	$\hat{a} = 45.99$ $\hat{b} = 6.0\ 10^{-4}$ $\hat{\alpha} = 0$ $\hat{\beta} = 13.24$	1.99	6.83	0.96
7. Pham-Zhang model	$\hat{a} = 0.06$ $\hat{b} = 6.0\ 10^{-4}$ $\hat{\alpha} = 1.0\ 10^{-4}$ $\hat{\beta} = 13.2$ $\hat{c} = 45.9$	2.12	6.79	0.95
8. Dependent parameter model	$\hat{\alpha} = 3.0\ 10^{-6}$ $\hat{\gamma} = 0.49$	43.69	601.34	4.53
9. Vtub-shaped fault-detection rate model (Pham Vtub model)	$\widehat{N} = 43.25$ $\hat{a} = 2.662$ $\hat{b} = 0.196$ $\hat{\alpha} = 4.040$ $\hat{\beta} = 35.090$	1.80	2.06	0.77

(continued)

Table 5.5 (continued)

Model name	LSEs	MSE	PRR	PP
10. Logistic fault detection model	$\widehat{N} = 50.56$	0.9033	0.1163	0.1356
	$\hat{a} = 10000$			
	$\hat{b} = 0.7364$			
	$\hat{c} = 0.2038$			
	$\hat{\alpha} = 0.2554$			
	$\hat{\beta} = 1.778$			

choice of the pre-assigned number of failures depends on economic considerations and involves balancing the cost of increasing the waiting time with the gains due to decreasing the risk of making an error in the products (Pham 1992b). The advantage of this strategy is that testing decisions can be reached in less time than if we wait for all items on test to fail. More specifically, to increase the testing information and decrease the risk, we can increase the sample size used to reach a decision timely and reduce the expected waiting time of the test. Obviously, it will increase the cost due to placing more sample on test. This section we discuss a cost model that can be used to determine the optimum sample size on test such that the expected total cost in the non-replacement procedure is minimized.

Assume that:

(i) n units are drawn at random from the exponential probability density function as follows:

$$f(x) = \frac{1}{\mu}e^{-\frac{x}{\mu}} \quad \text{for} \quad x > 0, \mu > 0$$

(ii) The observations become available in order so that $x_{1,n} \leq x_{2,n} \leq \ldots \leq x_{r,n} \leq \ldots \leq x_{n,n}$
where $x_{i,n}$ is meant the ith smallest observation in a sample of n ordered observations.

(iii) The experiment is terminated as soon as the rth failure occurs $(r \leq n)$.

Let

n	be the number of units to be put on test in the non-replacement procedure
c_1	be the cost of waiting per unit time until the test is completed
c_2	be the cost of placing a unit on test.
c_3	be the one-time set-up cost for the experiment
c_4	be the cost due to the uncertainty of the test results based on the variance measure
$X_{r,n}$	be a random variable that represents the waiting time to get the rth failure in a sample of size n

$E(X_{r,n})$ be the expected waiting time to observe first r failures from a sample size of n

Then the expected waiting time for the rth failure, $E(X_{r,n})$, is

$$E\left(X_{r,n}\right) = \mu \sum_{i=1}^{r} \tfrac{1}{(n-i+1)} \tag{5.74}$$

The variance of the random variable of $X_{i,n}$ can be written as

$$V\left(X_{r,n}\right) = \mu^2 \sum_{i=1}^{r} \tfrac{1}{(n-i+1)^2}$$

We can define the expected total cost function, $C(n)$, as follows:

$$C(n) = c_1 E\left(X_{r,n}\right) + c_2 n + c_3 + c_4 V\left(X_{r,n}\right) \tag{5.75}$$

In other words, the expected total cost is the expected cost of waiting time plus the cost of all number of units placed on test, plus the set-up cost, and the cost due to the uncertainty of the test results based on the variance measure. Clearly. the increasing number of units on test N will on the one hand reduce the expected waiting time and the variance of the test results, but will, on the other hand, increase the cost due to placing more units on test. Therefore, we now wish to determine the optimum number of units to place on test.

For given values of c_1, c_2, c_3, c_4, μ and r, we now show that there exists a unique value n^* that minimizes the expected total cost function $C(n)$.

Theorem 5.3 For given values of c_1, c_2, c_3, c_4, μ and r, there exists a unique value n^* that minimizes the expected total cost function $C(n)$ and

$$n^* = \inf\left\{n \geq r : f(n) < \frac{c_2}{\mu r}\right\} \tag{5.76}$$

where

$$f(n) = \left(\frac{1}{(n+1)(n-r+1)}\right)\left(c_1 + c_4\mu\left(\frac{2n-r+2}{(n+1)(n-r+1)}\right)\right) \tag{5.77}$$

If there exists a value n_0 such that

$$f(n_0) = \frac{c_2}{\mu r}, \text{ that is, } \Delta C(n_0) = C(n_0 + 1) - C(n_0) = 0,$$

then both n_0 and $(n_0 + 1)$ minimize the expected total cost function $C(n)$.
Proof: Define $\Delta C(n) = C(n + 1) - C(n)$. We obtain

$$C(n) = c_1 E(X_{r,n}) + c_2 n + c_3 + c_4 V(X_{r,n})$$

$$= c_1 \mu \sum_{i=1}^{r} \frac{1}{(n-i+1)} + c_2 n + c_3 + c_4 \mu^2 \sum_{i=1}^{r} \frac{1}{(n-i+1)^2}$$

and

$$C(n+1) = c_1 E(X_{r,n+1}) + c_2(n+1) + c_3 + c_4 V(X_{r,n+1})$$

$$= c_1 \mu \sum_{i=1}^{r} \frac{1}{(n-i+2)} + c_2(n+1) + c_3 + c_4 \mu^2 \sum_{i=1}^{r} \frac{1}{(n-i+2)^2}$$

Then

$$\Delta C(n) = C(n+1) - C(n)$$

$$= \left[c_1 \mu \sum_{i=1}^{r} \frac{1}{(n-i+2)} + c_2(n+1) + c_3 + c_4 \mu^2 \sum_{i=1}^{r} \frac{1}{(n-i+2)^2} \right]$$

$$- \left[c_1 \mu \sum_{i=1}^{r} \frac{1}{(n-i+1)} + c_2(n) + c_3 + c_4 \mu^2 \sum_{i=1}^{r} \frac{1}{(n-i+1)^2} \right]$$

After simplifications, we obtain

$$\Delta C(n) = C(n+1) - C(n)$$

$$= c_1 \mu \left[\frac{1}{n+1} - \frac{1}{n-r+1} \right] + c_2 + c_4 \mu^2 \left[\frac{1}{(n+1)^2} - \frac{1}{(n-r+1)^2} \right]$$

$$= c_2 - \mu r f(n)$$

where $f(n)$ is given in Eq. (5.77). One has

$$\Delta C(n) \leq 0 \quad \text{if and only if} \quad f(n) \geq \frac{c_2}{\mu r}$$

It can be easily shown that the function $f(n)$ is decreasing in n, then there exists a value n such that

$$f(n) < \frac{c_2}{\mu r}.$$

Let n^* denote the smallest value such n. Thus,

$$n^* = \inf \left\{ n \geq r : f(n) < \frac{c_2}{\mu r} \right\}$$

Note that if there exists a value n_0 such that

$$f(n_0) = \frac{c_2}{\mu r}, \text{ that is, } \Delta C(n_0) = C(n_0 + 1) - C(n_0) = 0,$$

then both n_0 and $(n0 +1)$ minimize the expected total cost function $C(n)$.

Example 5.5 Given: $c_1 = \$25/\text{hour}, c_2 = \$500$ per unit, $c_3 = 100, c_4 = 15, \mu = 1500$ and $r = 10$. Using the Theorem 1, obtain the optimal value n that minimizes the expected total cost function, $C(n)$ where $C(n)$ is given in Eq. (5.74). We calculate the optimal value of n, say n^*, and $C(n^*)$. From Eq. (5.77), we have

$$f(n) = \left(\frac{1}{(n+1)(n-10+1)}\right)\left(25 + (15)(1500)\left(\frac{2n-10+2}{(n+1)(n-10+1)}\right)\right)$$

$$= \left(\frac{1}{(n+1)(n-9)}\right)\left(25 + (22500)\left(\frac{2n-8}{(n+1)(n-9)}\right)\right)$$

and

$$\frac{c_2}{\mu r} = \frac{500}{(1500)(10)} = 0.03333$$

The number of units on test n, a function $f(n)$ and the expected total cost $C(n)$ are listed in Table below. Since

$$f(117) = 0.033272 < \frac{c_2}{\mu r}$$

therefore, the optimal value of n is $n^* = 117$ and the corresponding expected total cost is 88654.42. That is, $C(n^* = 117) = \$ 88,654.42$

N	f(n)	C(n)
115	0.035071	
116	0.0341551	88,666.75
117	**0.033272**	**88,654.42**
118	0.032418	88,655.35
119	0.031594	88,669.08
120	0.030798	88,695.16

5.6 Stress-Strength Interval-System Reliability

The interval-system is defined as a system with a series of chance events that occur in a given interval of time (Pham 2020b). A k-out-of-n interval-system is a system with a series of n events in a given interval of time which successes (or functions) if and only if at least k of the events succeed (function). This section discusses the reliability estimates of k-out-of-n interval-system based on stress-strength events where X (stress) and Y (strength) are independent two-parameter exponential random variables using the UMVUE and MLE methods. A numerical application in human heart conditions is discussed to illustrate the model results.

Let X be the random variable that represents the stress placed on the system by the operating environment and Y represents the strength of the system. A system is able to perform its intended function if its strength is greater than the stress imposed upon it. Reliability of the system is defined as the probability that the system is strong enough to overcome the stress, that is $R = P(Y > X)$, where X and Y are independent random variables with the following two-parameter exponential pdf

$$f_X(x) = \frac{1}{\sigma_x} e^{-\frac{(x-\mu_x)}{\sigma_x}} \quad \text{for} \quad x > \mu_x \geq 0, \quad \sigma_x > 0 \tag{5.78}$$

and

$$f_Y(y) = \frac{1}{\sigma_y} e^{-\frac{(y-\mu_y)}{\sigma_y}} \quad \text{for} \quad y > \mu_y \geq 0, \quad \sigma_y > 0 \tag{5.79}$$

respectively.

5.6.1 UMVUE and MLE of Reliability Function

We assume that σ_x and σ_y are known. We plan to report the modeling results when both σ_x and σ_y are unknown in a near future The stress–strength reliability can be obtained as follows.

Reliability Function

When $\mu_y \geq \mu_x$: The stress-strength R is given by

$$P(Y > X) = P_X P_{Y|X}\left(Y > X | \mu_x < X < \mu_y\right) + P_X P_{Y|X}\left(Y > X | X > \mu_y\right)$$

$$= \frac{1}{\sigma_x} \int_{\mu_x}^{\mu_y} e^{-\left(\frac{x-\mu_x}{\sigma_x}\right)} dx + \frac{1}{\sigma_y} \int_{\mu_y}^{\infty} e^{-\left(\frac{x-\mu_y}{\sigma_y}\right)} \cdot e^{-\left(\frac{x-\mu_x}{\sigma_x}\right)} dx$$

$$= 1 - \frac{\sigma_x e^{\left(\frac{\mu_x-\mu_y}{\sigma_x}\right)}}{\sigma_x + \sigma_y}. \tag{5.80}$$

When $\mu_y \leq \mu_x$: The stress-strength R is given by

$$P(Y > X) = E_X P_{Y|X}(Y > X | X)$$

$$= \frac{1}{\sigma_x} \int_{\mu_y}^{\infty} e^{-\left(\frac{x-\mu_y}{\sigma_y}\right)} e^{-\left(\frac{x-\mu_x}{\sigma_x}\right)} dx$$

$$= \frac{\sigma_y \, e^{\left(\frac{\mu_y-\mu_x}{\sigma_y}\right)}}{\sigma_x + \sigma_y}. \tag{5.81}$$

From Eqs. (5.80) and (5.81), the stress-strength reliability function R can be written as,

$$R = P(Y > X) = \begin{cases} 1 - \dfrac{\sigma_x \, e^{\left(\frac{\mu_x-\mu_y}{\sigma_x}\right)}}{\sigma_x+\sigma_y} & \text{for } \mu_y \geq \mu_x \\[2mm] \dfrac{\sigma_y \, e^{\left(\frac{\mu_y-\mu_x}{\sigma_y}\right)}}{\sigma_x+\sigma_y} & \text{for } \mu_y \leq \mu_x \end{cases} \tag{5.82}$$

Assume independent random samples $X_1, X_2, \ldots, X_m$ and $Y_1, Y_2, \ldots, Y_a$ are drawn from the stress pdf $f_X(x)$ and strength pdf $f_Y(y)$ given in Eq. (5.78) and Eq. (5.79), respectively. Let $X_{(1)}, X_{(2)}, \ldots, X_{(m)}$ and $Y_{(1)}, Y_{(2)}, \ldots, Y_{(a)}$ be the corresponding ordered statistics. It can be shown that $X_{(1)}$ and $Y_{(1)}$ are complete sufficient statistic for μ_x and μ_y, respectively, when σ_x and σ_y are known.

UMVUE of R^k

We now wish to obtain the statistical function T_k, the UMVUE of $R^k = (P(Y > X))^k$ where σ_x and σ_y are known. From Eq. (5.82), we have

When $\mu_y \geq \mu_x$

$$R^k \equiv (P(Y > X))^k = \left(1 - \frac{\sigma_x}{\sigma_x + \sigma_y} e^{-\left(\frac{\mu_y - \mu_x}{\sigma_x} \right)} \right)^k$$

$$= \sum_{i=0}^{k} (-1)^i \binom{k}{i} \left(\frac{\sigma_x}{\sigma_x + \sigma_y} \right)^{-\frac{(\mu_y - \mu_x)i}{\sigma_x}}.$$

When $\mu_y \leq \mu_x$

$$R^k \equiv (P(Y > X))^k = \left(\frac{\sigma_y}{\sigma_x + \sigma_y} e^{\left(\frac{\mu_y - \mu_x}{\sigma_y} \right)} \right)^k$$

$$= \frac{\sigma_y^k}{(\sigma_x + \sigma_y)^k} e^{\frac{(\mu_y - \mu_x)k}{\sigma_y}}.$$

Thus, R^k can be written as

$$R^k = \begin{cases} \sum_{i=0}^{k} (-1)^i \binom{k}{i} \left(\frac{\sigma_x}{\sigma_x + \sigma_y} \right)^i e^{-\frac{(\mu_y - \mu_x)i}{\sigma_x}} & \text{for } \mu_y \geq \mu_x \\ \frac{\sigma_y^k}{(\sigma_x + \sigma_y)^k} e^{\frac{(\mu_y - \mu_x)k}{\sigma_y}} & \text{for } \mu_y \leq \mu_x \end{cases} \tag{5.83}$$

Let $W = Y_{(1)} - X_{(1)}$. An unbiased estimator of R^k is given by

$$Z_k = \begin{cases} 1 \text{ if } X_i \geq Y_i \\ 0 \text{ otherwise} \end{cases} \quad \text{for } i = 1, 2, \ldots, k$$

where $k < \min\{m, a\}$.

In general, the method of finding the UMVUE is to search for any unbiased statistic $T(X_1, X_2, \ldots, X_k)$ and a complete sufficient statistic $\hat{\theta}$ if one exists. Then the UMVUE is given by

$$E\left[T(X_1, X_2, \ldots, X_k) | \hat{\theta} \right].$$

From Lehmann-Scheffe theorem (Lehmann and Casella (1988), the UMVUE of R^k is

$$E\left(Z_k | X_{(1)}, Y_{(1)} \right) = P\left(X_1 < Y_1, X_2 < Y_2, \ldots, X_k < Y_k | X_{(1)}, Y_{(1)} \right)$$

$$= P\big(X_1 - X_{(1)} < Y_1 - Y_{(1)} + W, X_2 - X_{(1)} < Y_2 - Y_{(1)} + W, X_k - X_{(1)} < Y_k - Y_{(1)} + W < Y_k | W\big)$$

which will be a function of W, σ_x and σ_y since the distribution of $X_i - X_{(1)}$ and $Y_i - Y_{(1)}$ for $i = 1, 2, \ldots, k$ do not involve μ_x and μ_y, respectively. Therefore, based on Lehmann-Scheffe theorem, an unbiased estimator of R^k based on W is the UMVUE of R^k. Note that, the pdf of W where $W = Y_{(1)} - X_{(1)}$ is given by

$$
f_W(w) = \begin{cases} \dfrac{ma}{(a\sigma_x + m\sigma_y)} e^{-\frac{(\mu_y - \mu_x - w)}{\sigma_x}} & \text{for } w < (\mu_y - \mu_x) \\[3mm] \dfrac{ma}{(a\sigma_x + m\sigma_y)} e^{-\frac{a(w - (\mu_y - \mu_x))}{\sigma_y}} & \text{for } w \geq (\mu_y - \mu_x) \end{cases}
$$

Theorem 5.4 Pham (2020b): The statistic T_k is the UMVUE of R^k where σ_x and σ_y are known, $k < \min\{m, a\}$, and.

$$
T_k = \begin{cases} \dfrac{(a-k)(k\sigma_x + m\sigma_y)\sigma_y^{k-1}}{ma(\sigma_x + \sigma_y)^k} e^{\frac{(Y_{(1)} - X_{(1)})k}{\sigma_y}} \, I\big(Y_{(1)} < X_{(1)}\big) \\[4mm] + \sum_{i=0}^{k} (-1)^i \dbinom{k}{i} \dfrac{(m-i)(a\sigma_x + i\sigma_y)\sigma_x^{i-1}}{ma(\sigma_x + \sigma_y)^i} e^{-\frac{i(Y_{(1)} - X_{(1)})}{\sigma_x}} \, I\big(Y_{(1)} \geq X_{(1)}\big) \end{cases} \tag{5.84}
$$

The Proof of Theorem 4 (see Problem 1).

MLE of R^k

It can be shown that the MLE of μ_x and μ_y where σ_x and σ_y are known from the pdf $f_X(x)$ and $f_Y(y)$, respectively, are

$$\hat{\mu}_x = X_{(1)} \quad \text{and} \quad \hat{\mu}_y = Y_{(1)} \tag{5.85}$$

From Eq. (5.82), we have

$$R^j(t) = \dfrac{\sigma_y^j}{(\sigma_x + \sigma_y)^j} e^{\frac{(\mu_y - \mu_x)j}{\sigma_y}} \, I(\mu_y \leq \mu_x) + \sum_{i=0}^{j} (-1)^i \dbinom{j}{i} \left(\dfrac{\sigma_x}{\sigma_x + \sigma_y}\right)^i e^{-\frac{(\mu_y - \mu_x)i}{\sigma_x}} \, I(\mu_y \geq \mu_x). \tag{5.86}$$

Thus, the MLE of R^j is

$$
\widehat{R}^j(t) = \dfrac{\sigma_y^j}{(\sigma_x + \sigma_y)^j} e^{\frac{(Y_{(1)} - X_{(1)})j}{\sigma_y}} \, I\big(Y_{(1)} \leq X_{(1)}\big)
$$

$$
+ \sum_{i=0}^{j} (-1)^i \dbinom{j}{i} \left(\dfrac{\sigma_x}{\sigma_x + \sigma_y}\right)^i e^{-\frac{(Y_{(1)} - X_{(1)})i}{\sigma_x}} \, I\big(Y_{(1)} \geq X_{(1)}\big). \tag{5.87}
$$

5.6.2 Interval-System Reliability Estimation

This section we discuss the UMVUE and MLE of k-out-of-n interval-system reliability based on stress-strength events where X (stress) and Y (strength) are independent two-parameter exponential random variables. A system is able to perform its intended function if its strength is greater than the stress imposed upon it. Reliability of the system, $R = P(Y > X)$, is the probability that the strengths of the unit are greater than the stresses.

The k-out-of-n interval-system is a system with a series of n events in a given interval of time which successes (or functions) if and only if at least k out of n independent events success (function).

The reliability of the k-out-of-n interval-system is the probability that at least k out of n events in a given interval of time success, and is given by

$$R_s(t) = \sum_{i=k}^{n} \binom{n}{i} (R(t))^i (1 - R(t))^{n-i}. \tag{5.88}$$

The reliability function of the k-out-of-n interval-system (as from Eq. 5.88) can be rewritten as (Pham (2010) and Pham and Pham (2010)):

$$R_S(t) = \sum_{j=k}^{n} (-1)^{j-k} \binom{j-1}{k-1} \binom{n}{j} R^j(t) \tag{5.89}$$

where $R(t) = P(Y > X)$.

UMVUE of $R_s(t)$

Let

$$\widetilde{R}_S(t) = \sum_{j=k}^{n} (-1)^{j-k} \binom{j-1}{k-1} \binom{n}{j} \widetilde{R}^j(t) \tag{5.90}$$

where (from Eq. (5.84))

$$\widetilde{R}^j(t) = \begin{cases} \dfrac{(a-j)(j\sigma_x+m\sigma_y)\sigma_y^{j-1}}{ma(\sigma_x+\sigma_y)^k} \, e^{\frac{(Y_{(1)}-X_{(1)})j}{\sigma_y}} \, \mathrm{I}\big(Y_{(1)} < X_{(1)}\big) \\ + \displaystyle\sum_{i=0}^{j} (-1)^i \binom{j}{i} \dfrac{(m-i)(a\sigma_x+i\sigma_y)\sigma_x^{i-1}}{ma(\sigma_x+\sigma_y)^i} \, e^{-\frac{i(Y_{(1)}-X_{(1)})}{\sigma_x}} \, \mathrm{I}\big(Y_{(1)} \geq X_{(1)}\big). \end{cases} \tag{5.91}$$

Theorem 5.5 The statistic $\widetilde{R}_S(t)$ is the UMVUE of system reliability $R_S(t)$ where σ_x and σ_y are known and $k < \min\{m,a\}$.

Proof:

Based on the polynomial combination of UMVUE functions, we can show that the UMVUE of reliability function of k out of n interval-system, $R_s(t)$, is

$$\widetilde{R}_S(t) = \sum_{i=k}^{n} (-1)^{j-k} \binom{j-1}{k-1} \binom{n}{j} \widetilde{R}^j(t)$$

where $\widetilde{R}^j(t)$ is given in Eq. (5.84).

MLE of $R_s(t)$

Hence, the MLE of the system reliability $R_S(t)$, where σ_x and σ_y are known, is given by

$$\widehat{R}_S(t) = \sum_{j=k}^{n} (-1)^{j-k} \binom{j-1}{k-1} \binom{n}{j} \widehat{R}^j(t) \tag{5.92}$$

where (also from Eq. (5.87))

$$\widehat{R}^j(t) = \frac{\sigma_y^j}{(\sigma_x + \sigma_y)^j} e^{\frac{(Y_{(1)} - X_{(1)})j}{\sigma_y}} I(Y_{(1)} < X_{(1)}) + \sum_{i=0}^{j} (-1)^i \binom{j}{i} \left(\frac{\sigma_x}{\sigma_x + \sigma_y}\right)^i e^{-\frac{(Y_{(1)} - X_{(1)})i}{\sigma_x}} I(Y_{(1)} \geq X_{(1)}).$$

5.6.3   Applications

One in every four deaths in the United States occurs as a result of heart disease (CDC NCHS 2015). Monitoring your heart rate can help prevent heart complications. The heart rate is one of the important indicators of health in the human body. It measures the number of times the heart beats per minute. Stress tests are carried out to measure the heart's ability to respond to external/emotion stress in a controlled environment. The stresses induced by exercise on a treadmill monitoring the average heart rate in 12 15-s periods were found as follows (Pham 2020b):

15.1, 12.3, 19.8, 13.0, 12.5, 13.3, 18.6, 17.2, 18.7, 22.4, 14.6, and 15.1 (per 15 - second period)..

The values of normal heart rate of an individual on tests found by monitoring 15 15-s periods (multiply by four to obtain BPM) were:

19.2, 24.2, 21.1, 16.7, 17.6, 14.3, 22.4, 20.8, 14.6, 17.7, 22.7, 15.9, 19.2, 15.3, and 25.2 (per 15 - second period)

Given $m = 12$, $a = 15$, $k = 2$, $n = 3$, $\sigma_x = 3.5$, and $\sigma_y = 3.7$. Here $X_{(1)} = 12.3$, $Y_{(1)} = 14.3$. We now calculate the reliability of 2-out-of-3 interval-system in a given interval of 45 s.

Calculate UMVUE

Since $Y_{(1)} = 14.30 > X_{(1)} = 12.30$, from Eq. (5.91),

$$
\begin{aligned}
\widetilde{R}^j(t) &= \sum_{i=0}^{j} (-1)^i \binom{j}{i} \frac{(m-i)\left(a\sigma_x + i\sigma_y\right)\sigma_x^{i-1}}{ma\left(\sigma_x + \sigma_y\right)^i} e^{-\frac{i(Y_{11})-X_{(1)})}{\sigma_x}} \\
&= \sum_{i=0}^{j} (-1)^i \binom{j}{i} \frac{(12-i)(15(3.5)+i(3.7))(3.5)^{i-1}}{12(15)(3.5+3.7)^i} e^{-\frac{i(14.3-12.3)}{3.5}} \\
&= \sum_{i=0}^{j} (-1)^i \binom{j}{i} \frac{(12-i)(52.5+3.7i)(3.5)^{i-1}}{180(7.2)^i} e^{-0.5714i}
\end{aligned}
$$

So, we obtain

$$
\widetilde{R}^j(t) = \begin{cases} 0.5328917 \text{ for } j = 2 \\ 0.3880224 \text{ for } j = 3 \end{cases}
$$

Thus, the reliability of 2-out-of-3 interval-system in a given interval of 45 s is

$$
\begin{aligned}
\widetilde{R}_S(t) &= \sum_{j=2}^{3} (-1)^{j-2} \binom{j-1}{2-1} \binom{3}{j} \widetilde{R}^j(t) \\
&= 3\widetilde{R}^2(t) - 2\widetilde{R}^3(t) \\
&= 3(0.5328917) - 2(0.3880224) \\
&= 0.8226303
\end{aligned}
$$

Calculate MLE

From Eq. (5.92), the MLE of 2-out-of-3 interval-system reliability $R_S(t)$ is given by

$$
\widehat{R}_S(t) = \sum_{j=2}^{3} (-1)^{j-2} \binom{j-1}{2-1} \binom{3}{j} \widehat{R}^j(t) = 3\widehat{R}^2(t) - 2\widehat{R}^3(t)
$$

Since $Y_{(1)} = 14.30 > X_{(1)} = 12.30$ and from Eq. (5.87), we obtain

$$
\widehat{R}^j(t) = \sum_{i=0}^{j} (-1)^i \binom{j}{i} \left(\frac{3.5}{3.5+3.7}\right)^i e^{-\frac{(14.3-12.3)i}{3.5}}
$$

$$= \sum_{i=0}^{j} (-1)^i \binom{j}{i} (0.486111)^i \, e^{-0.5714286i}.$$

then

$$\widehat{R}^j(t) = \begin{cases} 0.5263275 \text{ for } j = 2 \\ 0.3818423 \text{ for } j = 3 \end{cases}$$

Hence,

$$\widehat{R}_S(t) = 3\widehat{R}^2(t) - 2\widehat{R}^3(t)$$
$$= 3(0.5263275) - 2(0.3818423)$$
$$= 0.8152979$$

Thus, the UMVUE and MLE reliability of 2-out-of-3 interval system $R_S(t)$ are 0.8226303 and 0.8152979, respectively. From the data on the stress and strength tests measuring the heart rate conditions, the reliability estimates of stress-strength 2-out-of-3 interval-system based on the UMVUE and MLE are $\widetilde{R}_S(t) = 0.8226303$ and $\widehat{R}_S(t) = 0.8152979$, respectively.

5.7 Problems

(Theorem 1) Show that the statistic T_k is the UMVUE of R^k where σ_x and σ_y are known, $k < \min\{m, a\}$, and

$$T_k = \begin{cases} \dfrac{(a-k)(k\sigma_x + m\sigma_y)\sigma_y^{k-1}}{ma(\sigma_x + \sigma_y)^k} e^{\frac{(Y_{(1)} - X_{(1)})k}{\sigma_y}} \, \mathrm{I}\big(Y_{(1)} < X_{(1)}\big) \\ + \displaystyle\sum_{i=0}^{k} (-1)^i \binom{k}{i} \dfrac{(m-i)(a\sigma_x + i\sigma_y)\sigma_x^{i-1}}{ma(\sigma_x + \sigma_y)^i} e^{-\frac{i(Y_{(1)} - X_{(1)})}{\sigma_x}} \, \mathrm{I}\big(Y_{(1)} \geq X_{(1)}\big). \end{cases}$$

References

Basu AP (1964) Estimates of reliability for some distributions useful in life testing. Technometrics 215–219

Kwang Yoon Song KY, Chang H, Pham H (2018) Optimal release time and sensitivity analsyis using a new NHPP software reliability model with probability of fault removal subject to operating environments. Appl Sci 8(5), 714–722

Laurent AG (1963) Conditional distribution of order statistics and distribution of the reduced *i*th order statistic of the exponential model. Ann Math Stat 34:652–657

Lehmann EL, Casella G (1998) Theory of point estimation, 2nd ed., Springer

Li Q, Pham H (2017a) NHPP software reliability model considering the uncertainty of operating environments with imperfect debugging and testing coverage. Appl Math Model 51(11), 68–83 (2017)

Li Q, Pham H (2017b) A testing-coverage software reliability model considering fault removal efficiency and error generation. Plos One. https://doi.org/10.1371/journal.pone.0181524

Li Q, Pham H (2019) A generalized software reliability growth model with consideration of the uncertainty of operating environments. IEEE Access 7, 84253–84267

Pham H (1992b) Optimal design of life testing for ULSI circuit manufacturing. IEEE Trans Semicond Manuf 5(1), 68–70

Pham H (1996) A software cost model with imperfect debugging, random life cycle and penalty cost. Int J Syst Sci 5:455–463

Pham H (2003) Handbook of reliability engineering. Springer, London

Pham H (2006a) System software reliability. Springer

Pham H (2010) On the estimation of reliability of k-out-of-n systems. Int J Syst Assur Eng Manag 1(1), 32–35 (2010)

Pham H (2005) A new generalized systemability model. Int J Perform Eng 1(2), 145–156 (2005)

Pham H (2014) A new software reliability model with Vtub-shaped fault-detection rate and the uncertainty of operating environments. Optimization 63(10), 1481–1490

Pham H (2014a) Loglog fault-detection rate and testing coverage software reliability models subject to random environments. Vietnam J Comput Sci 1(1)

Pham H (2018) A logistic fault-dependent detection software reliability model. J Univers Comput Sci 24(12), 1717–1730

Pham H (2019a) A distribution function and its applications in software reliability. Int J Perform Eng 15(5), 1306–1313

Pham H (2020b) On stress-strength interval-system reliability with applications in heart conditions. Int J Math, Eng Manag Sci 5(11), 84253–84267

Pham H, Xie M (2002) A generalized surveillance model with applications to systems safety. IEEE Trans Syst, Man Cybern, Part C 32:485–492

Pham H, Pham H, Jr. (2010) Improving energy and power efficiency using Ncomputing and approaches for predicting reliability of complex computing systems. Int J Autom Comput 7(2), 153–159 (2010)

Pham H, Pham DH, Pham H Jr (2014) A new mathematical logistic model and its applications. Int J Inf Manage Sci 25:79–99

Rutemiller HC (1966) Point estimation of reliability of a system comprised of K elements from the same exponential distribution. Am Stat Assoc J 1029–1032

Sgarbossa F, Persona A, Pham H (2015) Using systemability function for periodic replacement policy in real environments. Qual Reliab Eng Int 31(4):617–633

Sharma M, Pham H, Singh VB (2019) Modeling and analysis of leftover issues and release time planning in multi-release open source software using entropy based measure. Int J Comput Syst Sci Eng 34(1)

Song KY, Chang I-H, Pham H (2019a) A testing coverage model based on nhpp software reliability considering the software operating environment and the sensitivity analysis. Mathematics 7(5). https://doi.org/10.3390/math7050450

Song KY, Chang I-H, Pham H (2019b) NHPP software reliability model with inflection factor of the fault detection rate considering the uncertainty of software operating environments and predictive analysis. Symmetry 11(4). https://doi.org/10.3390/sym11040521

Zhu M, Pham H (2019) A software reliability model incorporating martingale process with gamma-distributed environmental factors. Ann Oper Res. https://doi.org/10.1007/s10479-018-2951-7

Zhu M, Pham H (2019) A novel system reliability modeling of hardware, software, and interactions of hardware and software. Mathematics 7(11)

Zhu M, Pham H (2020) A generalized multiple environmental factors software reliability model with stochastic fault detection process. Ann Oper Res

Chapter 6
Stochastic Processes

6.1 Introduction

Stochastic processes are used for the description of a systems operation over time. There are two main types of stochastic processes: continuous and discrete. The complex continuous process is a process describing a system transition from state to state. The simplest process that will be discussed here is a Markov process. Given the current state of the process, its future behavior does not depend on the past. This chapter describes the concepts of stochastic processes including Markov process, Poisson process, renewal process, quasi-renewal process, and nonhomogeneous Poisson process, and their applications in reliability and availability for degraded systems with repairable components.

6.2 Markov Processes

In this section, we will discuss discrete stochastic processes. As an introduction to the Markov process, let us examine the following example.

Example 6.1 Consider a parallel system consisting of two components (see Fig. 6.1). From a reliability point of view, the states of the system can be described by.

State 1: Full operation (both components operating).
State 2: One component operating - one component failed.
State 3: Both components have failed.

Define

$$P_i(t) = P[X(t) = i] = P[\text{system is in state } i \text{ at time } t]$$

and

H. Pham, *Statistical Reliability Engineering*, Springer Series in Reliability Engineering, https://doi.org/10.1007/978-3-030-76904-8_6

Fig. 6.1 A two-component
parallel system

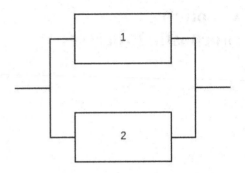

$P_i(t + dt) = P[X(t + dt) = i] = P[\text{system is in state } i \text{ at time } t + dt].$

Define a random variable $X(t)$ which can assume the values 1, 2, or 3 corresponding to the above-mentioned states. Since $X(t)$ is a random variable, one can discuss $P[X(t) = 1]$, $P[X(t) = 2]$ or conditional probability, $P[X(t_1) = 2 \mid X(t_0) = 1]$. Again, $X(t)$ is defined as a function of time t, the last stated conditional probability, $P[X(t_1) = 2 \mid X(t_0) = 1]$, can be interpreted as the probability of being in state 2 at time t_1, given that the system was in state 1 at time t_0. In this example, the "state space" is discrete, i.e., 1, 2, 3, etc., and the parameter space (time) is continuous. The simple process described above is called a stochastic process, i.e., a process which develops in time (or space) in accordance with some probabilistic (stochastic) laws. There are many types of stochastic processes. In this section, the emphasis will be on Markov processes which are a special type of stochastic process.

Definition 6.1 Let $t_0 < t_1 < \cdots < t_n$. If.

$$P[X(t_n) = A_n \mid X(t_{n-1}) = A_{n-1}, X(t_{n-2}) = A_{n-2}, \ldots, X(t_0) = A_0]$$
$$= P[X(t_n) = A_n \mid X(t_{n-1}) = A_{n-1}] \tag{6.1}$$

then the process is called a Markov process. Given the present state of the process, its future behavior does not depend on past information of the process.

The essential characteristic of a Markov process is that it is a process that has no memory; its future is determined by the present and not the past. If, in addition to having no memory, the process is such that it depends only on the difference $(t + dt) - t = dt$ and not the value of t, i.e., $P[X(t + dt) = j \mid X(t) = i]$ is independent of t, then the process is Markov with stationary transition probabilities or homogeneous in time. This is the same property noted in exponential event times, and referring back to the graphical representation of $X(t)$, the times between state changes would in fact be exponential if the process has stationary transition probabilities.

Thus, a Markov process which is time homogeneous can be described as a process where events have exponential occurrence times. The random variable of the process is $X(t)$, the state variable rather than the time to failure as in the exponential failure density. To see the types of processes that can be described, a review of the exponential distribution and its properties will be made. Recall that, if $X_1, X_2, \ldots, X_n$, are independent random variables, each with exponential density and a mean equal to $1/\lambda_i$ then min $\{ X_1, X_2, \ldots, X_n \}$ has an exponential density with mean $\left(\sum \lambda_i \right)^{-1}$. The significance of the property is as follows:

1. The failure behavior of the simultaneous operation of components can be characterized by an exponential density with a mean equal to the reciprocal of the sum of the failure rates.
2. The joint failure/repair behavior of a system where components are operating and/or undergoing repair can be characterized by an exponential density with a mean equal to the reciprocal of the sum of the failure and repair rates.
3. The failure/repair behavior of a system such as 2 above, but further complicated by active and dormant operating states and sensing and switching, can be characterized by an exponential density.

The above property means that almost all reliability and availability models can be characterized by a time homogeneous Markov process if the various failure times and repair times are exponential. The notation for the Markov process is $\{X(t), t > 0\}$, where $X(t)$ is discrete (state space) and t *is* continuous (parameter space). By convention, this type of Markov process is called a continuous parameter Markov chain.

From a reliability/availability viewpoint, there are two types of Markov processes. These are defined as follows:

1. *Absorbing Process*: Contains what is called an "absorbing state" which is a state from which the system can never leave once it has entered, e.g., a failure which aborts a flight or a mission.
2. *Ergodic Process*: Contains no absorbing states such that $X(t)$ can move around indefinitely, e.g., the operation of a ground power plant where failure only temporarily disrupts the operation.

Pham (2000) presents a summary of the processes to be considered broken down by absorbing and ergodic categories. Both reliability and availability can be described in terms of the probability of the process or system being in defined "up" states, e.g., states 1 and 2 in the initial example. Likewise, the mean time between failures (MTBF) can be described as the total time in the "up" states before proceeding to the absorbing state or failure state.

Define the incremental transition probability as

$$P_{ij}(dt) = P[X(t + dt) = j | X(t) = i]$$

This is the probability that the process (random variable $X(t)$) will go to state j during the increment t to $(t + dt)$, given that it was in state i at time t. Since we are dealing with time homogeneous Markov processes, i.e., exponential failure and repair times, the incremental transition probabilities can be derived from an analysis of the exponential hazard function. In Sect. 2.1, it was shown that the hazard function for the exponential with mean $1/\lambda$ was just λ. This means that the limiting (as $dt \to 0$) conditional probability of an event occurrence between t and $t + dt$, given that an event had not occurred at time t, *is* just λ, i.e.,

$$h(t) = \lim_{dt \to 0} \frac{P[t < X < t + dt | X > t]}{dt} = \lambda$$

The equivalent statement for the random variable $X(t)$ is

$$h(t)dt = P[X(t + dt) = j | X(t) = i] = \lambda dt$$

Now, $h(t)dt$ is in fact the incremental transition probability, thus the $P_{ij}(dt)$ can be stated in terms of the basic failure and/or repair rates. Define.

P_i (t):   the probability that the system is in state i at time t
r_{ij} (t):   transition rate from state i to state j

In general, the differential equations can be written as follows:

$$\frac{\partial P_i(t)}{\partial t} = -\sum_j r_{ij}(t) P_i(t) + \sum_j r_{ji}(t) P_j(t). \tag{6.2}$$

Solving the above different equations, one can obtain the time-dependent probability of each state.

Example 6.2 Returning to Example 6.1, a state transition can be easily constructed showing the incremental transition probabilities for process between all possible states:

State 1: Both components operating.
State 2: One component up - one component down.
State 3: Both components down (absorbing state).

The loops (see Fig. 6.2) indicate the probability of remaining in the present state during the dt increment

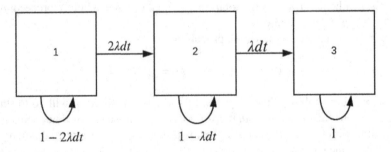

Fig. 6.2 State transition diagram for a two-component system

$$P_{11}(dt) = 1 - 2\lambda dt \quad P_{12}(dt) = 2\lambda dt \quad P_{13}(dt) = 0$$
$$P_{21}(dt) = 0 \quad\quad\quad P_{22}(dt) = 1 - \lambda dt \quad P_{23}(dt) = \lambda dt$$
$$P_{31}(dt) = 0 \quad\quad\quad P_{32}(dt) = 0 \quad\quad\quad P_{33}(dt) = 1$$

"up" states before proceeding to the absorbing state or failure state.

The zeros on P_{ij}, $i > j$, denote that the process cannot go backwards, i.e., this is not a repair process. The zero on P_{13} denotes that in a process of this type, the probability of more than one event (e.g., failure, repair, etc.) in the incremental time period dt approaches zero as dt approaches zero.

Except for the initial conditions of the process, i.e., the state in which the process starts, the process is completely specified by the incremental transition probabilities. The reason for the latter is that the assumption of exponential event (failure or repair) times allows the process to be characterized at any time t since it depends only on what happens between t and $(t + dt)$. The incremental transition probabilities can be arranged into a matrix in a way which depicts all possible statewide movements. Thus, for the parallel configurations,

$$[P_{ij}(dt)] = \begin{bmatrix} 1 - 2\lambda dt & 2\lambda dt & 0 \\ 0 & 1 - \lambda dt & \lambda dt \\ 0 & 0 & 1 \end{bmatrix}$$

for $i, j = 1, 2$, or 3. The matrix $[P_{ij}(dt)]$ is called the incremental, one-step transition matrix. It is a stochastic matrix, i.e., the rows sum to 1.0. As mentioned earlier, this matrix along with the initial conditions completely describes the process.

Now, $[P_{ij}(dt)]$ gives the probabilities for either remaining or moving to all the various states during the interval t to $t + dt$, hence,

$$P_1(t + dt) = (1 - 2\lambda dt)P_1(t)$$
$$P_2(t + dt) = 2\lambda dt P_1(t)(1 - \lambda dt)P_2(t)$$
$$P_3(t + dt) = \lambda dt P_2(t) + P_3(t) \tag{6.3}$$

By algebraic manipulation, we have

$$\frac{[P_1(t + dt) - P_1(t)]}{dt} = -2\lambda P_1(t)$$
$$\frac{[P_2(t + dt) - P_2(t)]}{dt} = 2\lambda P_1(t) - \lambda P_2(t)$$
$$\frac{[P_3(t + dt) - P_3(t)]}{dt} = \lambda P_2(t)$$

Taking limits of both sides as $dt \to 0$, we obtain (also see Fig. 6.3)

$$P_1'(t) = -2\lambda P_1(t)$$

Fig. 6.3 Markov transition
rate diagram for a
two-component parallel
system

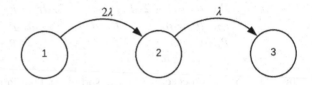

$$P_2'(t) = 2\lambda P_1(t) - 2\lambda P_2(t)$$
$$P_3'(t) = \lambda P_2(t) \tag{6.4}$$

The above system of linear first-order differential equations can be easily solved
for $P_1(t)$ and $P_2(t)$, and therefore, the reliability of the configuration can be obtained:

$$R(t) = \sum_{i=1}^{2} P_i(t)$$

Actually, there is no need to solve all three equations, but only the first two as $P_3(t)$
does not appear and also $P_3(t) = 1 - P_1(t) - P_2(t)$. The system of linear, first-order
differential equations can be solved by various means including both manual and
machine methods. For purposes here, the manual methods employing the Laplace
transform (see Appendix B) will be used.

$$L[P_i(t)] = \int_0^\infty e^{-st} P_i(t)dt = f_i(s) \tag{6.5}$$

$$L[P_i'(t)] = \int_0^\infty e^{-st} P_i'(t)dt = s\,f_i(s) - P_i(0)$$

The use of the Laplace transform will allow transformation of the system of linear,
first-order differential equations into a system of linear algebraic equations which
can easily be solved, and by means of the inverse transforms, solutions of $P_i(t)$ can
be determined.

Returning to the example, the initial condition of the parallel configuration is
assumed to be "full-up" such that

$$P_1(t=0) = 1, \; P_2(t=0) = 0, \; P_3(t=0) = 0$$

transforming the equations for $P'_1(t)$ and $P'_2(t)$ gives

$$sf_1(s) - P_1(t)|_{t=0} = -2\lambda f_1(s)$$
$$sf_2(s) - P_2(t)|_{t=0} = 2\lambda f_1(s) - \lambda f_2(s)$$

Evaluating $P_1(t)$ and $P_2(t)$ at $t = 0$ gives

$$sf_1(s) - 1 = -2\lambda f_1(s)$$
$$sf_2(s) - 0 = 2\lambda f_1(s) - \lambda f_2(s)$$

from which we obtain

$$(s + 2\lambda) f_1(s) = 1$$
$$- 2\lambda f_1(s) + (s + \lambda) f_2(s) = 0$$

Solving the above equations for $f_1(s)$ and $f_2(s)$, we have

$$f_1(s) = \frac{1}{(s + 2\lambda)}$$
$$f_2(s) = \frac{2\lambda}{[(s + 2\lambda)(s + \lambda)]}$$

From Appendix B of the inverse Laplace transforms,

$$P_1(t) = e^{-2\lambda t}$$
$$P_2(t) = 2e^{-\lambda t} - 2e^{-2\lambda t}$$
$$R(t) = P_1(t) + P_2(t) = 2e^{-\lambda t} - e^{-2\lambda t} \tag{6.6}$$

The example given above is that of a simple absorbing process where we are concerned about reliability If repair capability in the form of a repair rate μ were added to the model, the methodology would remain the same with only the final result changing.

Example 6.3 Continued from Example 6.2 with a repair rate μ added to the parallel configuration (see Fig. 6.4), the incremental transition matrix would be

$$[P_{ij}(dt)] = \begin{bmatrix} 1 - 2\lambda dt & 2\lambda dt & 0 \\ \mu dt & 1 - (\lambda + \mu)dt & \lambda dt \\ 0 & 0 & 1 \end{bmatrix}$$

The differential equations would become

Fig. 6.4 Markov transition rate diagram for a two-component repairable system

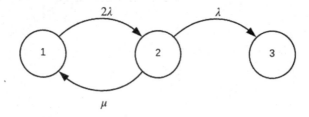

$$P_1'(t) = -2\lambda P_1(t) + \mu P_2(t)$$
$$P_2'(t) = 2\lambda P_1(t) - (\lambda + \mu) P_2(t) \tag{6.7}$$

and the transformed equations would become

$$(s + 2\lambda) f_1(s) - \mu f_2(s) = 1$$
$$- 2\lambda f_1(s) + (s + \lambda + \mu) f_2(s) = 0$$

Hence, we obtain

$$f_1(s) = \frac{(s + \lambda + \mu)}{(s - s_1)(s - s_2)}$$

$$f_2(s) = \frac{2\lambda}{(s - s_1)(s - s_2)}$$

where

$$s_1 = \frac{-(3\lambda + \mu) + \sqrt{(3\lambda + \mu)2 - 8\lambda^2}}{2}$$

$$s_2 = \frac{-(3\lambda + \mu) - \sqrt{(3\lambda + \mu)2 - 8\lambda^2}}{2} \tag{6.8}$$

Using the Laplace transform (see Appendix B), we obtain

$$P_1(t) = \frac{(s_1 + \lambda + \mu)e^{-s_1 t}}{(s_1 - s_2)} + \frac{(s_2 + \lambda + \mu)e^{-s_2 t}}{(s_2 - s_1)}$$

$$P_2(t) = \frac{2\lambda e^{-s_1 t}}{(s_1 - s_2)} + \frac{2\lambda e^{-s_2 t}}{(s_2 - s_1)}$$

Reliability function $R(t)$, is defined as the probability that the system continues to function throughout the interval $(0,t)$. Thus, the reliability of two-component in a parallel system is given by

$$R(t) = P_1(t) + P_2(t)$$
$$= \frac{(s_1 + 3\lambda + \mu)e^{-s_1 t} - (s_2 + 3\lambda + \mu)e^{-s_2 t}}{(s_1 - s_2)} \tag{6.9}$$

6.2.1 Three-Non-Identical Unit Load-Sharing Parallel Systems

Let $f_{if}(t)$, $f_{ih}(t)$, and $f_{ie}(t)$ be the pdf for time to failure of unit i, for $i = 1$, 2, and 3 under full load and half load and equal-load condition (this occurs when all three units are working), respectively. Also let $R_{if}(t)$, $R_{ih}(t)$, and $R_{ie}(t)$ be the reliability of unit i under full load and half load condition, respectively. The following events would be considered for the three-unit load-sharing parallel system to work:

Event 1: All the three units are working until the end of mission time t;
Event 2: All three units are working til time t_1; at time t_1 one of the three units fails. The remaining two units are working til the end of the mission.
Event 3: All three units are working til time t_1; at time t_1 one of the three units fails. Then at time t_2 the second unit fails, and the remaining unit is working until the end of mission t.

Example 6.4 Consider a three-unit shared load parallel system where.

λ_0 is the constant failure rate of a unit when all the three units are operational;
λ_h is the constant failure rate of each of the two surviving units, each of which shares half of the total load; and.
λ_f is the constant failure rate of a unit at full load.

For a shared-load parallel system to fail, all the units in the system must fail.
We now derive the reliability of a 3-unit shared-load parallel system using the Markov method. The following events would be considered for the three-unit load-sharing system to work:

Event 1: All the three units are working until the end of the mission time t where each unit shares one-third of the total load.
Event 2: All the three units are working until time t_1 (each shares one-third of the total load). At time t_1, one of the units (say unit 1) fails, and the other two units (say units 2 and 3) remain to work until the mission time t. Here, once a unit fails at time t_1, the remaining two working units would take half each of the total load and have a constant rate λ_h. As for all identical units, there are 3 possibilities under this situation.
Event 3: All the three units are working until time t_1 (each shares one-third of the total load) when one (say unit 1) of the three units fails. At time t_2, $(t_2 > t_1)$ one more unit fails (say unit 2) and the remaining unit works until the end of the mission time t. Under this event, there are 6 possibilities that the probability of two units failing before time t and only one unit remains to work until time t.

Define State i represents that i components are working. Let $P_i(t)$ denote the probability that the system is in state i at time t for i = 0, 1, 2, 3. Figure 6.5 below shows the Markov diagram of the system.
The Markov equations can be obtained as follows:

Fig. 6.5 Markov diagram
for a three-unit shared load
parallel system

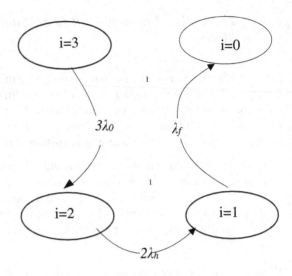

$$
\begin{cases}
\frac{dP_3(t)}{dt} = -3\lambda_0 P_3(t) \\
\frac{dP_2(t)}{dt} = 3\lambda_0 P_3(t) - 2\lambda_h P_2(t) \\
\frac{dP_1(t)}{dt} = 2\lambda_h P_2(t) - \lambda_f P_1(t) \\
\frac{dP_0(t)}{dt} = \lambda_f P_1(t) \\
P_3(0) = 1 \\
P_j(0) = 0,\ j \neq 3 \\
P_0(t) + P_1(t) + P_2(t) + P_3(t) = 1
\end{cases}
$$

Solving the above differential equations using the Laplace transform method, we can easily obtain the following results:

$$
P_3(t) = e^{-3\lambda_0 t}
$$

$$
P_2(t) = \frac{3\lambda_0}{2\lambda_h - 3\lambda_0} \left(e^{-3\lambda_0 t} - e^{-2\lambda_h t} \right)
$$

$$
P_1(t) = \frac{6\lambda_0\lambda_h}{(2\lambda_h - 3\lambda_0)} \left[\frac{e^{-3\lambda_0 t}}{(\lambda_f - 3\lambda_0)} - \frac{e^{-2\lambda_h t}}{(\lambda_f - 2\lambda_h)} + \frac{(2\lambda_h - 3\lambda_0)e^{-\lambda_f t}}{(\lambda_f - 3\lambda_0)(\lambda_f - 2\lambda_h)} \right]
$$

Hence, the reliability of a three-unit shared-load parallel system is

$$
R(t) = P_3(t) + P_2(t) + P_1(t)
$$

$$
= e^{-3\lambda_0 t} + \frac{3\lambda_0}{2\lambda_h - 3\lambda_0} \left(e^{-3\lambda_0 t} - e^{-2\lambda_h t} \right)
$$

$$+ \frac{6\lambda_0\lambda_h}{(2\lambda_h - 3\lambda_0)} \left[\frac{e^{-3\lambda_0 t}}{(\lambda_f - 3\lambda_0)} - \frac{e^{-2\lambda_h t}}{(\lambda_f - 2\lambda_h)} + \frac{(2\lambda_h - 3\lambda_0)e^{-\lambda_f t}}{(\lambda_f - 3\lambda_0)(\lambda_f - 2\lambda_h)} \right]$$

$$(6.10)$$

which is the same as Eq. (6.60) in Chapt. 4.

6.2.2 System Mean Time Between Failures

Another parameter of interest in absorbing Markov processes is the mean time between failures (MTBF) (Pham et al. 1997). Recalling the previous Example 6.3 of a parallel configuration with repair, the differential equations $P_1'(t)$ and $P_2'(t)$ describing the process were (see Eq. 6.7):

$$P_1'(t) = -2\lambda P_1(t) + \mu P_2(t)$$
$$P_2'(t) = 2\lambda P_1(t) - (\lambda + \mu)P_2(t).$$

$$(6.11)$$

Integrating both sides of the above equations yields

$$\int_0^\infty P_1'(t)dt = -2\lambda \int_0^\infty P_1(t)dt + \mu \int_0^\infty P_2(t)dt$$

$$\int_0^\infty P_2'(t)dt = 2\lambda \int_0^\infty P_1(t)dt - (\lambda + \mu) \int_0^\infty P_2(t)dt$$

From Chap. 1,

$$\int_0^\infty R(t)dt = MTTF$$

$$(6.12)$$

Similarly,

$$\int_0^\infty P_1(t)dt = \text{mean time spent in state 1, and}$$

$$\int_0^\infty P_2(t)dt = \text{mean time spent in state 2}$$

Designating these mean times as T_1 and T_2, respectively, we have

$$P_1(t)dt|_0^\infty = -2\lambda T_1 + \mu T_2$$
$$P_2(t)dt|_0^\infty = 2\lambda T_1 - (\lambda + \mu)T_2$$

But $P_1(t) = 0$ as $t \to \infty$ and $P_1(t) = 1$ for $t = 0$. Likewise, $P_2(t) = 0$ as $t \to \infty$ and $P_2(t) = 0$ for $t = 0$. Thus,

$$-1 = -2\lambda T_1 + \mu T_2$$
$$0 = 2\lambda T_1 - (\lambda + \mu)T_2$$

or, equivalently,

$$\begin{bmatrix} -1 \\ 0 \end{bmatrix} = \begin{bmatrix} -2\lambda & \mu \\ 2\lambda & -(\lambda + \mu) \end{bmatrix} \begin{bmatrix} T_1 \\ T_2 \end{bmatrix}$$

Therefore,

$$T_1 = \frac{(\lambda+\mu)}{2\lambda^2} \quad T_2 = \frac{1}{\lambda}$$
$$MTTF = T_1 + T_2 = \frac{(\lambda+\mu)}{2\lambda^2} + \frac{1}{\lambda} = \frac{(3\lambda+\mu)}{2\lambda^2} \tag{6.13}$$

The MTBF for non-maintenance processes is developed exactly the same way as just shown. What remains under absorbing processes is the case for availability for maintained systems. The difference between reliability and availability for absorbing processes is somewhat subtle. A good example is that of a communica-tion system where, if such a system failed temporarily, the mission would continue, but, if it failed permanently, the mission would be aborted.

Example 6.5 Consider the following cold-standby configuration consisting of two units: one main unit and one spare unit (see Fig. 6.6):

State 1: Main unit operating—spare OK.
State 2: Main unit out—restoration underway.
State 3: Spare unit installed and operating.
State 4: Permanent failure (no spare available).

Fig. 6.6 A cold-standby system

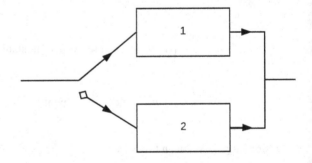

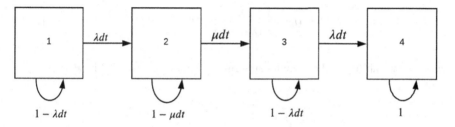

Fig. 6.7 State transition diagram for the cold-standby system

From Fig. 6.7, the incremental transition matrix is given by

$$[P_{ij}(dt)] = \begin{bmatrix} 1 - \lambda dt & \lambda dt & 0 & 0 \\ 0 & 1 - \mu dt & \mu dt & 0 \\ 0 & 0 & 1 - \lambda dt & \lambda dt \\ 0 & 0 & 0 & 1 \end{bmatrix}$$

We obtain

$$P_1'(t) = -\lambda P_1(t)$$
$$P_2'(t) = \lambda P_1(t) - \mu P_2(t)$$
$$P_3'(t) = \mu P_2(t) - \lambda P_3(t)$$

Using the Laplace transform, we obtain

$$s f_1(s) - 1 = -\lambda f_1(s)$$
$$s f_2(s) = \lambda f_1(s) - \mu f_2(s)$$
$$s f_3(s) = \mu f_2(s) - \lambda f_3(s)$$

After simplifications,

$$f_1(s) = \frac{1}{(s + \lambda)}$$
$$f_2(s) = \frac{\lambda}{[(s + \lambda)(s + \mu)]}$$
$$f_3(s) = \frac{\lambda \mu}{[(s + \lambda)^2(s + \mu)]}$$

Therefore, the probability of full-up performance, $P_1(t)$, is given by

$$P_1(t) = e^{-\lambda t} \tag{6.14}$$

Similarly, the probability of the system being down and under repair, $P_2(t)$, is

$$P_2(t) = \left[\frac{\lambda}{(\lambda - \mu)}\right](e^{-\mu t} - e^{-\lambda t}) \tag{6.15}$$

and the probability of the system being full-up but no spare available, $P_3(t)$, is

$$P_3(t) = \left[\frac{\lambda\mu}{(\lambda - \mu)^2}\right][e^{-\mu t} - e^{-\lambda t} - (\lambda - \mu)te^{-\lambda t}] \tag{6.16}$$

Hence, the point availability, $A(t)$, is given by

$$A(t) = P_1(t) + P_3(t) \tag{6.17}$$

If average or interval availability is required, this is achieved by

$$\left(\frac{1}{t}\right)\int_0^T A(t)dt = \left(\frac{1}{t}\right)\int_0^T [P_1(t) + P_3(t)]dt \tag{6.18}$$

where T is the interval of concern.

With the above example, cases of the absorbing process (both maintained and non-maintained) have been covered insofar as "manual" methods are concerned. In general, the methodology for treatment of absorbing Markov processes can be "packaged" in a fairly simplified form by utilizing matrix notation. Thus, for example, if the incremental transition matrix is defined as follows:

$$[P_{ij}(dt)] = \begin{bmatrix} 1 - 2\lambda dt & 2\lambda dt & 0 \\ \mu dt & 1 - (\lambda + \mu)dt & \lambda dt \\ 0 & 0 & 1 \end{bmatrix}$$

then if the dts are dropped and the last row and the last column are deleted, the remainder is designated as the matrix T:

$$[T] = \begin{bmatrix} 1 - 2\lambda & 2\lambda \\ \mu & 1 - (\lambda + \mu) \end{bmatrix}$$

Define $[Q] = [T]' - [I]$, where $[T]'$ is the transposition of $[T]$ and $[I]$ is the unity matrix:

$$[Q] = \begin{bmatrix} 1 - 2\lambda & \mu \\ 2\lambda & 1 - (\lambda + \mu) \end{bmatrix} - \begin{bmatrix} 1 & 0 \\ 0 & 1 \end{bmatrix}$$

$$= \begin{bmatrix} -2\lambda & \mu \\ 2\lambda & -(\lambda + \mu) \end{bmatrix}$$

Further define $[P(t)]$ and $[P'(t)]$ as column vectors such that

$$[P_1(t)] = \begin{bmatrix} P_1(t) \\ P_2(t) \end{bmatrix}, \quad [P'(t)] = \begin{bmatrix} P_1'(t) \\ P_2'(t) \end{bmatrix}$$

then

$$[P'(t)] = [Q][P(t)]$$

At the above point, solution of the system of differential equations will produce solutions to $P_1(t)$ and $P_2(t)$. If the MTBF is desired, integration of both sides of the system produces

$$\begin{bmatrix} -1 \\ 0 \end{bmatrix} = [Q]\begin{bmatrix} T_1 \\ T_2 \end{bmatrix}$$

$$\begin{bmatrix} -1 \\ 0 \end{bmatrix} = \begin{bmatrix} -2\lambda & \mu \\ 2\lambda & -(\lambda + \mu) \end{bmatrix}\begin{bmatrix} T_1 \\ T_2 \end{bmatrix} \quad \text{or}$$

$$[Q]^{-1}\begin{bmatrix} 1 \\ 0 \end{bmatrix} = \begin{bmatrix} T_1 \\ T_2 \end{bmatrix}$$

where $[Q]^{-1}$ is the inverse of $[Q]$ and the MTBF is given by

$$\text{MTBF} = T_1 + T_2 = \frac{3\lambda + \mu}{(2\lambda)^2}$$

In the more general MTBF case,

$$[Q]^{-1}\begin{bmatrix} -1 \\ 0 \\ \cdot \\ \cdot \\ \cdot \\ 0 \end{bmatrix} = \begin{bmatrix} T_1 \\ T_2 \\ \cdot \\ \cdot \\ \cdot \\ T_{n-1} \end{bmatrix} \quad \text{where } \sum_{i=1}^{n-1} T_i = \text{MTBF}$$

and $(n - 1)$ is the number of non-absorbing states.

For the reliability/availability case, utilizing the Laplace transform, the system of linear, first-order differential equations is transformed to

$$s\begin{bmatrix} f_1(s) \\ f_2(s) \end{bmatrix} - \begin{bmatrix} 1 \\ 0 \end{bmatrix} = [Q]\begin{bmatrix} f_1(s) \\ f_2(s) \end{bmatrix}$$

$$[sI - Q] \begin{bmatrix} f_1(s) \\ f_2(s) \end{bmatrix} = \begin{bmatrix} 1 \\ 0 \end{bmatrix}$$

$$\begin{bmatrix} f_1(s) \\ f_2(s) \end{bmatrix} = [sI - Q]^{-1} \begin{bmatrix} 1 \\ 0 \end{bmatrix}$$

$$L^{-1} \begin{bmatrix} f_1(s) \\ f_2(s) \end{bmatrix} = L^{-1} \left\{ [sI - Q]^{-1} \begin{bmatrix} 1 \\ 0 \end{bmatrix} \right\}$$

$$\begin{bmatrix} p_1(s) \\ p_2(s) \end{bmatrix} = L^{-1} \left\{ [sI - Q]^{-1} \begin{bmatrix} 1 \\ 0 \end{bmatrix} \right\}$$

Generalization of the latter to the case of $(n - 1)$ non-absorbing states is straightforward.

Ergodic processes, as opposed to absorbing processes, do not have any absorbing states, and hence, movement between states can go on indefinitely For the latter reason, availability (point, steady-state, or interval) is the only meaningful measure. As an example for ergodic processes, a ground-based power unit configured in parallel will be selected.

Example 6.6 Consider a parallel system consisting of two identical units each with exponential failure and repair times with constant rates λ and μ, respectively (see Fig. 6.8). Assume a two-repairmen capability if required (both units down), then.

State 1: Full-up (both units operating).
State 2: One unit down and under repair (other unit up).
State 3: Both units down and under repair.

It should be noted that, as in the case of failure events, two or more repairs cannot be made in the dt interval.

$$[P_{ij}(dt)] = \begin{bmatrix} 1 - 2\lambda dt & 2\lambda dt & 0 \\ \mu dt & 1 - (\lambda + \mu)dt & \lambda dt \\ 0 & 2\mu dt & 1 - 2\mu dt \end{bmatrix}$$

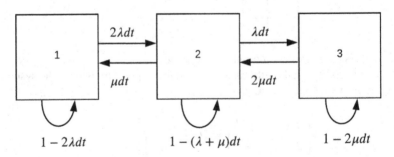

Fig. 6.8 State transition diagram with repair for a parallel system

Case I: Point Availability—Ergodic Process. For an ergodic process, as $t \rightarrow \infty$ the availability settles down to a constant level. Point availability gives a measure of things before the "settling down" and reflects the initial conditions on the process. Solution of the point availability is similar to the case for absorbing processes except that the last row and column of the transition matrix must be retained and entered into the system of equations. For example, the system of differential equations becomes

$$
\begin{bmatrix} P_1'(t) \\ P_2'(t) \\ P_3'(t) \end{bmatrix} = \begin{bmatrix} -2\lambda & \mu & 0 \\ 2\lambda & -(\lambda + \mu) & 2\mu \\ 0 & \lambda & -2\mu \end{bmatrix} \begin{bmatrix} P_1(t) \\ P_2(t) \\ P_3(t) \end{bmatrix}
$$

Similar to the absorbing case, the method of the Laplace transform can be used to solve for $P_1(t)$, $P_2(t)$, and $P_3(t)$, with the point availability, $A(t)$, given by

$$
A(t) = P_1(t) + P_2(t) \tag{6.19}
$$

Case II: Interval Availability—Ergodic Process. This is the same as the absorbing case with integration over time period T of interest. The interval availability, $A(T)$, is

$$
A(T) = \frac{1}{T} \int_0^T A(t)dt \tag{6.20}
$$

Case III: Steady State Availability—Ergodic Process. Here the process is examined as $t \rightarrow \infty$ with complete "washout" of the initial conditions. Letting $t \rightarrow \infty$ the system of differential equations can be transformed to linear algebraic equations. Thus,

$$
\lim_{t \to \infty} \begin{bmatrix} P_1'(t) \\ P_2'(t) \\ P_3'(t) \end{bmatrix} = \lim_{t \to \infty} \begin{bmatrix} -2\lambda & \mu & 0 \\ 2\lambda & -(\lambda + \mu) & 2\mu \\ 0 & \lambda & -2\mu \end{bmatrix} \begin{bmatrix} P_1(t) \\ P_2(t) \\ P_3(t) \end{bmatrix}
$$

As $t \rightarrow \infty$, $P_i(t) \rightarrow$ constant and $P_i'(t) \rightarrow 0$. This leads to an unsolvable sys-tem, namely

$$
\begin{bmatrix} 0 \\ 0 \\ 0 \end{bmatrix} = \begin{bmatrix} -2\lambda & \mu & 0 \\ 2\lambda & -(\lambda + \mu) & 2\mu \\ 0 & \lambda & -2\mu \end{bmatrix} \begin{bmatrix} P_1(t) \\ P_2(t) \\ P_3(t) \end{bmatrix}
$$

To avoid the above difficulty, an additional equation is introduced:

$$\sum_{i=1}^{3} P_i(t) = 1$$

With the introduction of the new equation, one of the original equations is deleted and a new system is formed:

$$\begin{bmatrix} 1 \\ 0 \\ 0 \end{bmatrix} = \begin{bmatrix} 1 & 1 & 1 \\ -2\lambda & \mu & 0 \\ 2\lambda & -(\lambda + \mu) & 2\mu \end{bmatrix} \begin{bmatrix} P_1(t) \\ P_2(t) \\ P_3(t) \end{bmatrix}$$

or, equivalently,

$$\begin{bmatrix} P_1(t) \\ P_2(t) \\ P_3(t) \end{bmatrix} = \begin{bmatrix} 1 & 1 & 1 \\ -2\lambda & \mu & 0 \\ 2\lambda & -(\lambda + \mu) & 2\mu \end{bmatrix}^{-1} \begin{bmatrix} 1 \\ 0 \\ 0 \end{bmatrix}$$

We now obtain the following results:

$$P_1(t) = \frac{\mu^2}{(\mu + \lambda)^2}$$

$$P_2(t) = \frac{2\lambda\mu}{(\mu + \lambda)^2}$$

and

$$P_3(t) = 1 - P_1(t) - P_2(t) = \frac{\lambda^2}{(\mu + \lambda)^2}$$

Therefore, the steady state availability, $A(\infty)$, is given by

$$A_3(\infty) = P_1(t) + P_2(t) = \frac{\mu(\mu + 2\lambda)}{(\mu + \lambda)^2}. \tag{6.21}$$

Note that Markov methods can also be employed where failure or repair times are not exponential, but can be represented as the sum of exponential times with identical means (Erlang distribution or Gamma distribution with integer valued shape parameters). Basically, the method involves the introduction of "dummy" states which are of no particular interest in themselves, but serve the purpose of changing the hazard function from constant to increasing.

Fig. 6.9 System state diagram

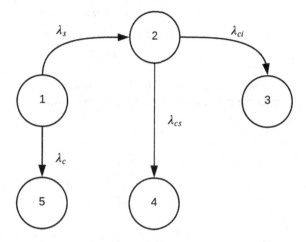

Example 6.7 We now discuss two Markov models (Cases 1 and 2 below) which allow integration of control systems of nuclear power plant reliabiltiy and safey analysis. A basic system transition diagram for both models is presented in Fig. 6.9. In both models, it is assumed that the control system is composed of a control rod and an associated safety system. The following assumptions are applied in this example.

(i)   All failures are statistically independent.
(ii)   Each unit has a constant rate.
(iii) The control system fails when the control rod fails.

The following notations are assocaited with the system shown in Fig. 6.9.

i ith state of the system: $i = 1$ (control and its associated safety system operating normally); $i = 2$ (control operating normally, safety system failed), $i = 3$ (control failed with an accident), $i = 4$ (control failed safely); $i = 5$ (control failed but its associated safety system operating normally).

$P_i(t)$ probability that the control system is in state i at time t, $i = 1, 2, ..., 5$

λ_i ith constant failure rate: $i = s$ (state 1 to state 2), $i = ci$ (state 2 to state 3), $i = cs$ (state 2 to state 4), $i = c$ (state 1 to state 5).

$P_i(s)$ Laplace transform of the probability that the control system is in state i; $i = 1, 2, ..., 5$.

s Laplace transform variable.

Case 1: The system represented by Model 1 is shown in Fig. 6.9. Using the Markov approach, the system of differential equations (associated with Fig. 6.9) is given below:

$$P_1'(t) = -(\lambda_s + \lambda_c)P_1(t)$$
$$P_2'(t) = \lambda_s P_1(t) - (\lambda_{ci} + \lambda_{cs})P_2(t)$$

$$P_3'(t) = \lambda_{ci} P_2(t)$$
$$P_4'(t) = \lambda_{cs} P_2(t)$$
$$P_5'(t) = \lambda_c P_1(t)$$

Assume that at time $t = 0$, $P_1(0) = 1$, and $P_2(0) = P_3(0) = P_4(0) = P_5(0) = 0$. Solving the above system of equations, we obtain

$$P_1(t) = e^{-At}$$

$$P_2(t) = \frac{\lambda_s}{B}\left(e^{-Ct} - e^{-At}\right)$$

$$P_3(t) = \frac{\lambda_s \lambda_{ci}}{AC}\left(1 - \frac{Ae^{-Ct} - Ce^{-At}}{B}\right)$$

$$P_4(t) = \frac{\lambda_s \lambda_{cs}}{AC}\left(1 - \frac{Ae^{-Ct} - Ce^{-At}}{B}\right)$$

$$P_5(t) = \frac{\lambda_c}{A}\left(1 - e^{-At}\right)$$

where

$$A = \lambda_s + \lambda_c; \quad B = \lambda_s + \lambda_c - \lambda_{cs} - \lambda_{ci}; \quad C = \lambda_{ci} + \lambda_{cs}.$$

The reliability of both the control and its safety system working normally, R_{cs}, is given by

$$R_{cs}(t) = P_1(t) = e^{-At}.$$

The reliability of the control system working normally with or without the safety system functioning successfully is

$$R_{ss}(t) = P_1(t) + P_2(t) = e^{-At} + \frac{\lambda_s}{B}\left(e^{-Ct} - e^{-At}\right).$$

The mean time to failure (MTTF) of the control with the safety system up is

$$MTTF_{cs} = \int_0^\infty R_{cs}(t)dt = \frac{1}{A}.$$

Similarly, the MTTF of the control with the safety system up or down is

$$MTTF_{ss} = \int_0^\infty R_{ss}(t)dt = \frac{1}{A} + \frac{\lambda_s}{B}\left(\frac{1}{C} - \frac{1}{A}\right).$$

Case 2: This model is the same as *Case 1* except that a repair is allowed when the safety system fails with a constant rate μ. The system of differential equations for this model is as follows:

$$P_1'(t) = \mu P_2(t) - A P_1(t)$$
$$P_2'(t) = \lambda_s P_1(t) - (\lambda_{ci} + \lambda_{cs} + \mu) P_2(t)$$
$$P_3'(t) = \lambda_{ci} P_2(t)$$
$$P_4'(t) = \lambda_{cs} P_2(t)$$
$$P_5'(t) = \lambda_c P_1(t)$$

We assume that at time t = 0, $P_1(0) = 1$, and $P_2(0) = P_3(0) = P_4(0) = P_5(0) = 0$. Solving the above system of equations, we obtain

$$P_1(t) = e^{-At} + \mu \lambda_s \left[\frac{e^{-At}}{(r_1 + A)(r_2 + A)} + \frac{e^{r_1 t}}{(r_1 + A)(r_1 - r_2)} + \frac{e^{r_2 t}}{(r_2 + A)(r_2 - r_1)} \right]$$

$$P_2(t) = \lambda_s \frac{e^{r_1 t} - e^{r_2 t}}{(r_1 - r_2)}$$

$$P_3(t) = \frac{\lambda_s \lambda_{ci}}{r_1 r_2} \left(\frac{r_1 e^{r_2 t} - r_2 e^{-r_1 t}}{r_2 - r_1} + 1 \right)$$

$$P_4(t) = \frac{\lambda_s \lambda_{cs}}{r_1 r_2} \left(\frac{r_1 e^{r_2 t} - r_2 e^{-r_1 t}}{r_2 - r_1} + 1 \right)$$

$$P_5(t) = \frac{\lambda_c}{A} \left(1 - e^{-At} \right) + \mu \lambda_s \lambda_c$$

$$\left[\frac{1}{r_1 r_2 A} - \frac{e^{-At}}{A(r_1 + A)(r_2 + A)} + \frac{e^{r_1 t}}{r_1 (r_1 + A)(r_1 - r_2)} + \frac{e^{r_2 t}}{r_2 (r_2 + A)(r_2 - r_1)} \right]$$

where

$$r_1, r_2 = \frac{-a \pm \sqrt{a^2 - 4b}}{2},$$

$$a = A + C + \mu, \quad b = \lambda_{ci} \lambda_s + \lambda_{cs} \lambda_s + (\lambda_{ci} + \lambda_{cs} + \mu) \lambda_c.$$

The reliability of both the control and its associated safety system working normally with the safety repairable system is.

$$R_{cs}(t) = e^{-At} + \mu \lambda_s \left[\frac{e^{-At}}{(r_1 + A)(r_2 + A)} + \frac{e^{r_1 t}}{(r_1 + A)(r_1 - r_2)} + \frac{e^{r_2 t}}{(r_2 + A)(r_2 - r_1)} \right].$$

The reliability of the control operating normal with or without the safety system operating (but having safety system repair) is

$$R_{ss}(t) = e^{-At} + \frac{\lambda_s \left(e^{r_1 t} - e^{r_2 t} \right)}{(r_1 - r_2)}$$
$$+ \mu \lambda_s \left[\frac{e^{-At}}{(r_1 + A)(r_2 + A)} + \frac{e^{r_1 t}}{(r_1 + A)(r_1 - r_2)} + \frac{e^{r_2 t}}{(r_2 + A)(r_2 - r_1)} \right].$$

The MTTF of the control with the safety system operating is

$$MTTF_{cs} = \int_0^\infty R_{cs}(t)dt = \frac{1}{A}\left(1 + \frac{\mu \lambda_s}{b} \right).$$

We can see that the repair process has helped to improve the system's MTTF. Similarly, the MTTF of the control with the safety system up or down but with accessible repair is given by

$$MTTF_{ss} = \int_0^\infty R_{ss}(t)dt = \frac{1}{A}\left(1 + \frac{\mu \lambda_s}{b} \right) + \frac{\lambda_s}{A}.$$

Example 6.8 A system is composed of eight identical active power supplies, at least seven of the eight are required for the system to function. In other words, when two of the eight power supplies fail, the system fails. When all eight power supplies are operating, each has a constant failure rate λ_a per hour. If one power supply fails, each remaining power supply has a failure rate λ_b per hour where $\lambda_a \leq \lambda_b$ We assume that a failed power supply can be repaired with a constant rate μ per hour. The system reliability function, $R(t)$, is defined as the probability that the system continues to function throughout the interval $(0, t)$. Here we wish to determine the system mean time to failure (MTTF).

Define.

State 0: All 8 units are working.
State 1: 7 units are working.
State 2: More than one unit failed and system does not work.

The initial condition: $P_0(0) = 1$, $P_1(0) = P_2(0) = 0$.
The Markov modeling of differential equations (see Fig. 6.10) can be written as follows:

$$P_0'(t) = -8\lambda_a P_0(t) + \mu P_1(t)$$
$$P_1'(t) = 8\lambda_a P_0(t) - (7\lambda_b + \mu)P_1(t)$$
$$P_2'(t) = 7\lambda_b P_1(t) \tag{6.22}$$

Using the Laplace transform, we obtain

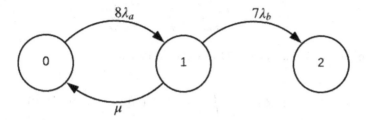

Fig. 6.10 Markov transition rate diagram for a 7-out-8 dependent system

$$\begin{cases} sF_0(s) - P_0(0) = -8\lambda_a F_0(s) + \mu F_1(s) \\ sF_1(s) - P_1(0) = 8\lambda_a F_0(s) - (7\lambda_b + \mu)F_1(s) \\ sF_2(s) - P_2(0) = 7\lambda_b F_1(s) \end{cases} \qquad (6.23)$$

When s = 0:

$$F_i(0) = \int_0^\infty P_i(t)dt.$$

Thus, the system reliability function and system MTTF, respectively, are

$$R(t) = P_0(t) + P_1(t). \qquad (6.24)$$

and

$$MTTF = \int_0^\infty R(t)dt = \int_0^\infty [P_0(t) + P_1(t)]dt = \sum_{i=1}^2 F_i(0). \qquad (6.25)$$

From Eq. (6.23), when s = 0, we have

$$\begin{cases} -1 = -8\lambda_a F_0(0) + \mu F_1(0) \\ 0 = 8\lambda_a F_0(0) - (7\lambda_b + \mu)F_1(0) \end{cases} \qquad (6.26)$$

From Eq. (6.26), after some arrangements, we can obtain

$$7\lambda_b F_1(0) = 1 \quad \Rightarrow \quad F_1(0) = \frac{1}{7\lambda_b}$$

and

$$F_0(0) = \frac{7\lambda_b + \mu}{8\lambda_a} F_1(0)$$

$$= \frac{7\lambda_b + \mu}{8\lambda_a} \frac{1}{7\lambda_b} = \frac{7\lambda_b + \mu}{56\lambda_a\lambda_b}.$$

From Eq. (6.25), the system MTTF can be obtained

$$MTTF = \int_0^\infty R(t)dt = \int_0^\infty [P_0(t) + P_1(t)]dt = F_0(0) + F_1(0)$$

$$= \frac{7\lambda_b + \mu}{56\lambda_a\lambda_b} + \frac{1}{7\lambda_b} = \frac{\mu + 8\lambda_a + 7\lambda_b}{56\lambda_a \lambda_b}.$$

Given $\lambda_a = 3 \times 10^{-3} = 0.003$, $\lambda_b = 5 \times 10^{-2} = 0.05$, and $\mu = 0.8$, then the system mean time to failure is given by:

$$MTTF = \frac{\mu + 8\lambda_a + 7\lambda_b}{56 \lambda_a \lambda_b}$$

$$= \frac{0.8 + 8(0.003) + 7(0.05)}{56(0.003)(0.05)} = \frac{1.174}{0.0084} = 139.762 \text{ h.}$$

Example 6.9 A system consists of two independent components operating in parallel (see Fig. 6.1) with a single repair facility where repair may be completed for a failed component before the other component has failed. Both the components are assumed to be functioning at time $t = 0$. When both components have failed, the system is considered to have failed and no recovery is possible. Assuming component i has the constant failure rate λ_i and repair rate μ_i for $i = 1$ and 2. The system reliability function, $R(t)$, is defined as the probability that the system continues to function throughout the interval $(0, t)$.

(a)   Derive the system reliability function and system mean time to failure (MTTF) and calculate the MTTF.
(b)   Assume that both components have the same failure rate λ and repair rate μ. That is, $\lambda_1 = \lambda_2 = \lambda$ and $\mu_1 = \mu_2 = \mu$. Calculate the reliability function and system MTTF when $\lambda = 0.003$ per hour, and $\mu = 0.1$ per hour, and t = 25 h.

Define.

State 1: both components are working.
State 2: component 1 failed, component 2 is working.
State 3: component 2 failed, component 1 is working.
State 4: Both components 1 and 2 failed.

The initial conditions: $P_1(0) = 1$, $P_2(0) = P_3(0) = P_4(0) = 0$. From Fig. 6.11, the Markov modeling of differential equations can be written as follows:

Fig. 6.11 A degraded
system rate diagram

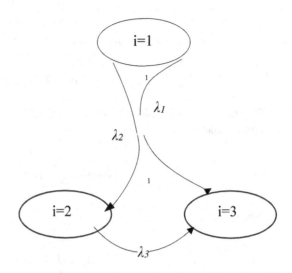

$$\begin{cases} \frac{dP_1(t)}{dt} = -(\lambda_1 + \lambda_2)P_1(t) + \mu_1 P_2(t) + \mu_2 P_3(t) \\ \frac{dP_2(t)}{dt} = \lambda_1 P_1(t) - (\lambda_2 + \mu_1)P_2(t) \\ \frac{dP_3(t)}{dt} = \lambda_2 P_1(t) - (\lambda_1 + \mu_2)P_3(t) \\ \frac{dP_4(t)}{dt} = \lambda_2 P_2(t) + \lambda_1 P_3(t) \\ P_1(0) = 1, \quad P_j(0) = 0, j \neq 1. \end{cases} \qquad (6.27)$$

Let $\ell\{P_i(t)\} = F_i(s)$. Then $\ell\left\{\frac{\partial P_i(t)}{\partial t}\right\} = sF_i(s) - F_i(0)$. Using the Laplace transform, we obtain

$$\begin{cases} sF_1(s) - 1 = -(\lambda_1 + \lambda_2)F_1(s) + \mu_1 F_2(s) + \mu_2 F_3(s) \\ sF_2(s) = \lambda_1 F_1(s) - (\lambda_2 + \mu_1)F_2(s) \\ sF_3(s) = \lambda_2 F_1(s) - (\lambda_1 + \mu_2)F_3(s) \\ sF_4(s) = \lambda_2 F_2(s) + \lambda_1 F_3(s) \end{cases} \qquad (6.28)$$

From Eq. (6.28), we obtain

$$F_1(s) = \frac{(s + a_2)(s + a_3)}{s^3 + b_1 s^2 + c_1 s + c_2}$$

$$F_2(s) = \frac{\lambda_1}{s + a_2} F_1(s)$$

$$F_3(s) = \frac{\lambda_2}{s + a_3} F_1(s)$$

where

$$a_1 = \lambda_1 + \lambda_2; \quad a_2 = \lambda_2 + \mu_1; \quad a_3 = \lambda_1 + \mu_2;$$
$$a_4 = \lambda_1\mu_1; \quad a_5 = \lambda_2\mu_2; \quad b_1 = a_1 + a_2 + a_3;$$
$$b_2 = a_1 a_2 + a_1 a_3 + a_2 a_3; \quad b_3 = a_1 a_2 a_3;$$
$$c_1 = b_2 - a_4 - a_5; \quad c_2 = b_3 - a_3 a_4 - a_2 a_5.$$

Take the inverse of Laplace transform, that is $P_i(t) = \ell^{-1}\{F_i(s)\}$, then the system reliability function is

$$R(t) = \sum_{i=1}^{3} P_i(t). \tag{6.29}$$

When s = 0:

$$F_i(0) = \int_0^\infty P_i(t).$$

Thus, the system MTTF is

$$MTTF = \int_0^\infty R(t)dt = \int_0^\infty [P_1(t) + P_2(t) + P_3(t)]dt = \sum_{i=1}^{3} F_i(0).$$

Substitute s = 0 into Eq. (6.28), we have

$$\begin{cases} -1 = -(\lambda_1 + \lambda_2)F_1(0) + \mu_1 F_2(0) + \mu_2 F_3(0) \\ 0 = \lambda_1 F_1(0) - (\lambda_2 + \mu_1)F_2(0) \\ 0 = \lambda_2 F_1(0) - (\lambda_1 + \mu_2)F_3(0) \\ 0 = \lambda_2 F_2(0) + \lambda_1 F_3(0) \end{cases}$$

Solving for $F_i(0)$, we obtain

$$F_1(0) = \frac{a_2 a_3}{a_1 a_2 a_3 - a_3 a_4 - a_2 a_5}$$

$$F_2(0) = \frac{a_2 a_3 \lambda_1}{a_1 a_2^2 a_3 - a_2 a_3 a_4 - a_2^2 a_5}$$

$$F_3(0) = \frac{a_2 a_3 \lambda_2}{a_1 a_2 a_3^2 - a_2 a_3 a_5 - a_3^2 a_4}. \tag{6.30}$$

Thus, the system MTTF is

$$MTTF = \sum_{i=1}^{3} F_i(0)$$

$$= \frac{a_2 a_3}{a_1 a_2 a_3 - a_3 a_4 - a_2 a_5} + \frac{a_2 a_3 \lambda_1}{a_1 a_2^2 a_3 - a_2 a_3 a_4 - a_2^2 a_5}$$

$$+ \frac{a_2 a_3 \lambda_2}{a_1 a_2 a_3^2 - a_2 a_3 a_5 - a_3^2 a_4}. \tag{6.31}$$

When $\lambda_1 = \lambda_2 = \lambda$ and $\mu_1 = \mu_2 = \mu$, from Eq. (6.29) and (6.31), we can show that the system reliability and the MTTF are given as follows:

$$R(t) = \frac{2\lambda^2}{\alpha_1 - \alpha_2} \left(\frac{e^{-\alpha_2 t}}{\alpha_2} - \frac{e^{-\alpha_1 t}}{\alpha_1} \right) \tag{6.32}$$

where

$$\alpha_1, \alpha_2 = \frac{(3\lambda + \mu) \pm \sqrt{\lambda^2 + 6\lambda\mu + \mu^2}}{2}$$

and

$$MTTF = \frac{3}{2\lambda} + \frac{\mu}{2\lambda^2} \tag{6.33}$$

respectively.

(b) Calculate the reliability function and system MTTF when $\lambda_1 = \lambda_2 = \lambda = 0.003$ per hour, and $\mu_1 = \mu_2 = \mu = 0.1$ per hour, and $t = 25$ h.

Substitute $\lambda = 0.003$ and $\mu = 0.1$ into Eq. (6.32), we obtain

$$\alpha_1 = 0.1088346, \quad \alpha_2 = 0.0001654$$

thus, the system reliability at the mission time $t = 25$ h is

$$R(t = 25) = 0.99722$$

Similarly, from Eq. (6.33), we obtain the system MTTF is 6055.56 h.

6.2.3 Degraded Systems

In real life, there are many systems may continue to function in a degraded system state (Pham et al. 1996, 1997; Li and Pham a, b). Such systems may perform its function but not at the same full operational state. Define the states of a system, where the transition rate diagram is shown in Fig. 6.11, are as follows:

State 1: operational state.

State 2: degraded state.
State 3: failed state.

We denote the probability of being in state i at time t as $P_i(t)$.

From the rate diagram (Fig. 6.11) we can obtain the following differential equations:

$$
\begin{cases}
\frac{dP_1(t)}{dt} = -(\lambda_1 + \lambda_2)P_1(t) \\
\frac{dP_2(t)}{dt} = \lambda_2 P_1(t) - \lambda_3 P_2(t) \\
\frac{dP_3(t)}{dt} = \lambda_1 P_1(t) + \lambda_3 P_2(t) \\
P_1(0) = 1, \quad P_j(0) = 0, \ j \neq 1
\end{cases}
\tag{6.34}
$$

From Eq. (6.34), we can obtain the solution

$$
P_1(t) = e^{-(\lambda_1 + \lambda_2)t}
\tag{6.35}
$$

We can also show, from Eq. (6.34), that

$$
P_2(t) = \frac{\lambda_2}{(\lambda_1 + \lambda_2 - \lambda_3)}\left(e^{-\lambda_3 t} - e^{-(\lambda_1 + \lambda_2)t}\right)
\tag{6.36}
$$

Finally,

$$
P_3(t) = 1 - P_1(t) - P_2(t).
$$

The system reliability is given by

$$
R(t) = P_1(t) + P_2(t)
$$
$$
= e^{-(\lambda_1 + \lambda_2)t} + \frac{\lambda_2}{(\lambda_1 + \lambda_2 - \lambda_3)}\left(e^{-\lambda_3 t} - e^{-(\lambda_1 + \lambda_2)t}\right)
\tag{6.37}
$$

The system mean time to a complete failure is

$$
MTTF = \int_0^\infty R(t)dt
$$
$$
= \frac{1}{\lambda_1 + \lambda_2} + \frac{\lambda_2}{(\lambda_1 + \lambda_2 - \lambda_3)}\left(\frac{1}{\lambda_3} - \frac{1}{\lambda_1 + \lambda_2}\right).
\tag{6.38}
$$

Example 6.10 A computer system used in a data computing center experiences degrade state and complete failure state as follows:

$\lambda_1 = 0.0003$ per hour, $\lambda_2 = 0.00005$ per hour, and $\lambda_3 = 0.008$ per hour.

Here for example, when the system is in degraded state, it will fail at a constant rate 0.008 per hour. From Eqs. (6.35–6.36), we obtain the following results. Table

Time (h)	$P_1(t)$	$P_2(t)$	R(t)
10	0.996506	0.000480	0.996986
20	0.993024	0.000921	0.993945
50	0.982652	0.002041	0.984693
100	0.965605	0.003374	0.968979

Table 6.1 Reliability and MTTF of degraded computer system

6.1 shows the reliability results for various mission times.

$$P_1(t) = e^{-(0.0003+0.00005)t}$$

$$P_2(t) = \frac{0.00005}{(0.0003 + 0.00005 - 0.008)} \left(e^{-0.008t} - e^{-(0.0003+0.00005)t}\right)$$

From Eq. (6.38), the system MTTF is given by

$$MTTF = \frac{1}{\lambda_1 + \lambda_2} + \frac{\lambda_2}{(\lambda_1 + \lambda_2 - \lambda_3)} \left(\frac{1}{\lambda_3} - \frac{1}{\lambda_1 + \lambda_2}\right) = 2875.$$

6.2.4 k-*Out-Of*-n *Systems with Degradation*

In some environments the components may not fail fully but can degrade and there may exist multiple states of degradation. In such cases, the efficiency especially the performance of the system may decrease (Pham et al. 1996; Yu et al. 2018). This section discusses the reliability of the k-out-of-n systems considering that:

(1) The system consists of n independent and identically distributed (i.i.d.) non-repairable components;
(2) Each component can have d stages of degradation; degradation stage (d + 1) is a failed state and stage (d + 2) is a catastrophic failure state;
(3) The system functions when at least k out of n components function;
(4) The components may fail catastrophically and can reach the failed state directly from a good state as well as from a degraded state;
(5) A component can survive either until its last degradation or until a catastrophic failure at any stage;
(6) All transition rates (i.e., catastrophic and degradation) are constant; and
(7) The degradation rate and catastrophic failure rate of a component depends on the state of the component.

Let λ_i be a transition (degradation) rate of the component from state i to state (i + 1) for i = 1,2, ..., d. Let μ_i be a transition rate of the component from state i to state (d + 2), i.e. catastrophic failure stage. A component may fail catastrophically

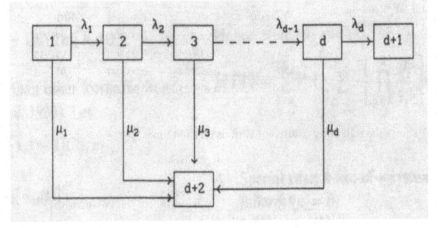

Fig. 6.12 A component flow diagram (Pham 1996)

while it is in any degraded state. A block diagram of such a component is shown in Fig. 6.12.

In general, the system consists of n i.i.d. non-repairable components and at least k components are required for the system to function. Each component starts in a good state ("state 1"). The component continues to perform its function within the system in all of its d level of degraded states of operation. The component no longer performs it function when it reaches its last degradation state at $(d + 1)$ at which point it has degraded to a failed state or when it has failed catastrophically, i.e. state $(d + 2)$, from any of its operational states of degradation. The rate at which the components degrade to a lower state of degradation or fail catastrophically increases as the components degrades from one state to a lower state of degradation. Components that reached the failed state either by degradation or by catastrophic failure cannot longer perform their function and cannot be repaired. In other words, once a component has reached a failed (either degradation or catastrophic) state, it cannot be restored to a good state or any degradation state.

The successful operation of the entire system is expressed as a combination of component success and failure events. We can formulate the component reliability function using the Markov approach. Denote $P_i(t)$ as the probability that a component is in state i at time t. From Fig. 6.12, we can easily obtain the following differential equations using the Markov approach:

$$\frac{dP_1(t)}{dt} = -(\lambda_1 + \mu_1)P_1(t)$$

$$\frac{dP_i(t)}{dt} = \lambda_{i-1}P_{i-1}(t) - (\lambda_i + \mu_i)P_i(t) \quad \text{for } i = 2, 3, \dots, d$$

$$\frac{dP_{d+1}(t)}{dt} = \lambda_d P_d(t)$$

$$\frac{dP_{d+2}(t)}{dt} = \sum_{j=1}^{d} \mu_j P_j(t). \tag{6.39}$$

Solving the above system of differential equations, we obtain the state probability as follows:

$$P_m(t) = \prod_{k=1}^{m} \lambda_{k-1} \left(\sum_{i=1}^{m} \frac{e^{-(\lambda_1+\mu_2)t}}{\prod_{\substack{j=1 \\ j \neq i}}^{m} (\lambda_j + \mu_j - \lambda_i - \mu_i)} \right) \tag{6.40}$$

for $m = 1, 2, ..., d$ and $\lambda_0 = 1$. Thus, the component reliability is

$$R_c(t) = \sum_{i=1}^{d} B_i e^{-(\lambda_i+\mu_i)t} \tag{6.41}$$

where

$$B_i = \sum_{m=i}^{d} \frac{\prod_{k=1}^{m} \lambda_{k-1}}{\prod_{\substack{j=1 \\ j \neq i}}^{m} (\lambda_j + \mu_j - \lambda_i - \mu_i)}. \tag{6.42}$$

The mean time to failure of the component (MTTF$_C$) is given by

$$
\begin{aligned}
MTTF_C &= \int_0^\infty R_C(t)dt \\
&= \int_0^\infty \sum_{i=1}^{d} \left(B_i e^{-(\lambda_i+\mu_i)t} \right) dt \\
&= \sum_{i=1}^{d} \left(\frac{B_i}{\lambda_i + \mu_i} \right).
\end{aligned} \tag{6.43}
$$

The k-out-of-n system reliability is

$$R_S(t) = \sum_{i=k}^{n} \binom{n}{i} [R_C(t)]^i [1 - R_C(t)]^{n-i} \tag{6.44}$$

where $R_C(t)$ is in Eq. (6.41). After several algebra simplifications, we obtain the system reliability (Pham et al., 1996):

$$R_s(t) = \sum_{i=k}^{n} i! \, A_i \sum_{\substack{\sum_{j=1}^{d} i_j = i}} \left[\prod_{j=1}^{d} \left(\frac{B_j^{i_j}}{i_j!} \right) \right] e^{-\sum_{j=1}^{d} i_j (\lambda_j + \mu_j) t} \tag{6.45}$$

where $A_i = (-1)^{i-k} \binom{i-1}{k-1} \binom{n}{i}$. The MTTF of the system is

$$MTTF_S = \int_0^{\infty} R_s(t) dt$$

where $R_s(t)$ is given in Eq. (6.45). Therefore, system MTTF is

$$MTTF_S = \sum_{i=k}^{n} i! \, A_i \sum_{\substack{\sum_{j=1}^{d} i_j = i}} \left(\frac{\prod_{j=1}^{d} \left(\frac{B_j^{i_j}}{i_j!} \right)}{\sum_{j=1}^{d} i_j (\lambda_j + \mu_j)} \right). \tag{6.46}$$

For components without catastrophic failure. When there is no catastrophic failure, the catastrophic failure rate μ_i in Eq. (6.45) becomes zero. From Eq. (6.41), we obtain

$$R_C(t) = \sum_{i=1}^{d} B_i e^{-\lambda_i t} \tag{6.47}$$

where

$$B_i = \prod_{\substack{j=1 \\ j \neq i}}^{d} \frac{\lambda_j}{\lambda_j - \lambda_i} \qquad i = 1, 2, \ldots, d$$

Similarly, the component MTTF is

$$MTTF_C = \sum_{i=1}^{d} \frac{B_i}{\lambda_i} = \sum_{i=1}^{d} \frac{1}{\lambda_i} \tag{6.48}$$

The system reliability and MTTF for this special case are computed from the general forms of the Eqs. (6.45) and (6.46), respectively:

$$R_s(t) = \sum_{i=k}^{n} i! \, A_i \sum_{\substack{\sum_{j=1}^{d} i_j = i}} \left[\prod_{j=1}^{d} \left(\frac{B_j^{i_j}}{i_j!} \right) \right] e^{-\sum_{j=1}^{d} i_j \lambda_j t} \tag{6.49}$$

where $A_i = (-1)^{i-k} \binom{i-1}{k-1} \binom{n}{i}$ and

$$MTTF_S = \sum_{i=k}^{n} i! \, A_i \sum_{\substack{\sum_{j=1}^{d} i_j = i}} \left[\frac{\prod_{j=1}^{d} \left(\frac{B_j^{i_j}}{i_j!} \right)}{\sum_{j=1}^{d} i_j \lambda_j} \right] \tag{6.50}$$

Example 6.11 Consider a 2-out-of-5 system where components consist of two stages of degradation ($d = 2$) with the following values:

$$\lambda_1 = 0.015/h, \quad \mu_1 = 0.0001/h, \quad \lambda_2 = 0.020/h, \text{ and } \mu_2 = 0.0002/h$$

Given $n = 5$, $k = 2$, $\lambda_1 = 0.015$, $\lambda_2 = 0.02$, $\mu_1 = 0.0001$, $\mu_2 = 0.0002$, $d = 2$. There are two cases as follows.

Case 1: From Eq. (6.42), components can fail by degradation and by catastrophic events:

$$B_1 = \lambda_0 + \frac{\lambda_1 \lambda_2}{\lambda_2 + \mu_2 - \lambda_1 - \mu_1}$$

$$= 1 + \frac{(0.015)(1)}{0.02 + 0.0002 - 0.015 - 0.0001} = 3.94$$

$$B_2 = \frac{\lambda_0 \lambda_1}{\lambda_1 + \mu_1 - \lambda_2 - \mu_2} = \frac{(1)(0.015)}{0.015 + 0.0001 - 0.020 - 0.0002} = -2.94$$

$$MTTF_C = \frac{B_1}{\lambda_1 + \mu_1} + \frac{B_2}{\lambda_2 + \mu_2} = 115.4$$

$$R_C(t) = \sum_{i=1}^{2} B_i e^{-(\lambda_i + \mu_i)t} = B_1 e^{-(\lambda_1 + \mu_1)t} + B_2 e^{-(\lambda_2 + \mu_2)t}$$

For $t = 1$, then

$$R_C(t = 1) = 3.94\, e^{-(0.015+0.0001)(1)} + (-2.94)\, e^{-(0.02+0.0002)(1)} = 0.9998$$

Similarly,

$$R_S(t) = \sum_{i=2}^{5} \binom{5}{i} [R_C(t)]^i [1 - R_C(t)]^{5-i}$$

For $t = 1$, then $R_s(t = 1) \approx 1$ and MTTF $= \int_0^\infty R_S(t)dt = 144.5$ h. Tables 6.2 and 6.3 present the tabulated reliability of the 2-out-of-5 system with and without catastrophic failures for varying mission time t, respectively.

Case 2: Components can only fail by degradation (no catastrophic failures ($\mu_1 = \mu_2 = 0$)):

$$B_1 = \frac{\lambda_2}{\lambda_2 - \lambda_1} = \frac{0.02}{0.02 - 0.015} = 4$$

$$B_2 = \frac{\lambda_1}{\lambda_1 - \lambda_2} = \frac{0.015}{0.015 - 0.02} = -3$$

Then we can easily obtain as follows:

$$R_C(t = 1) = 0.9999; \quad R_C(t = 5) = 0.9965$$

Table 6.2 Reliability of 2-out-of-5 system with catastrophic failures for varying time t

Time	Component reliability	System reliability
1	0.9998	1.0000
5	0.9959	1.0000
10	0.9856	1.0000
20	0.9502	1.0000
25	0.9269	0.9999
50	0.7812	0.9905
75	0.6235	0.9298
100	0.4805	0.7872
150	0.2671	0.4032
200	0.1406	0.1477
250	0.0715	0.0443

Table 6.3 Reliability of 2-out-of-5 system without catastrophic failures for varying time t

Time	Component reliability	System reliability
1	0.9999	1.0000
5	0.9965	1.0000
10	0.9866	1.0000
20	0.9523	1.0000
25	0.9296	0.9999
50	0.7858	0.9913
75	0.6292	0.9335
100	0.4865	0.7952
150	0.2722	0.4140
200	0.1442	0.1542
250	0.0739	0.0469

$$\text{MTTF}_C = \frac{1}{\lambda_1} + \frac{1}{\lambda_2} = \frac{1}{0.015} + \frac{1}{0.02} = 116.67 \text{ h}$$

The system MTTF is: MTTF $= 145.9$ h. The catastrophic failure process has decreased both the system reliability and component reliability.

6.2.5 Degraded Systems with Partial Repairs

In some environments, systems might not always fail fully, but can degrade and there can be multiple stages of degradation. In such cases, the efficiency of the system may decrease. After a certain stage of degradation the efficiency of the system may decrease to an unacceptable limit and can be considered as a total failure (Pham et al. 1997). In addition, the system can fail partially from any stage and can be repaired. The repair action cannot bring the system to the good stage but can make it operational and the failure rate of the system will remain the same as before the failure. This section discusses a model for predicting the reliability and availability of multistage degraded systems with partial repairs based on the results by Pham et al. (Pham et al. 1997).

Initially, the system is considered to be in its good state. After some time, it can either go to the first degraded state upon degradation or can go to a failed state upon a partial failure. If the system fails partially, the repair action starts immediately, and after repair the system will be restored to the good state, will be kept in operation and the process will be repeated. However, if the system reaches the first degraded state (state 3), it can either go to the second degraded state (state 5) upon degradation or can go to the failed state upon a partial failure with increased transition rates. If the system fails partially at this stage, after repair the system will be restored back to the first degraded state, and will be kept in operation. Figure 6.13 shows the

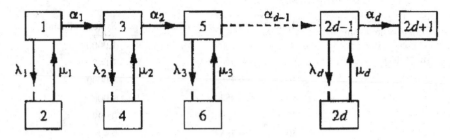

Fig. 6.13 A system state diagram (Pham 1997)

system flow diagram transition process where: State (i): State of the system; State (1): Good; State (2i-1): Degraded; State (2i): Failed (Partially); and State (2d + 1) Failed (Completely).

Assumptions

1. System can have d-stages of degradation (dth stage is complete failure state).
2. The system might fail partially from a good state as well as from any degraded state.
3. System can be restored back from a partially failed state to its original state just before the failure.
4. All transition rates are constant (i.e., degradation, partial failure, and partial repair rates).
5. The degradation as well as repair rates of the system depend upon the state of the system (i.e., degradation level).

Figure 1: System flow diagram.
Notation.

d: number of operational states.
State (2i):   partially failed states; $i = 1, 2, \ldots, d$
State 1:   good state.
State (2i − 1):   degraded operational states; $i = 2, 3, \ldots, d$
α_i: transition (degradation) rate from state $(2i − 1)$ to $(2i + 1)$.
λ_i: transition (partial failure) rate from state $(2i − 1)$ to $(2i)$.
μ_i: transition (partial repair) rate from state $(2i)$ to $(2i − 1)$.

Using the Markov approach, we can obtain the following equations:

$$\frac{dP_1(t)}{dt} = -(\alpha_1 + \lambda_1)P_1(t) + \mu_1 P_2(t)$$

$$\frac{dP_{(2i-1)}(t)}{dt} = -(\alpha_i + \lambda_1)P_{(2i-1)}(t) + \mu_i P_{(2i)}(t) + \alpha_{i-1} P_{(2i-3)}(t) \quad \text{for } i = 2, 3, \ldots, d$$

$$\frac{dP_{(2i)}(t)}{dt} = -\mu_i P_{(2i)}(t) + \lambda_i P_{(2i-1)}(t) \quad \text{for } i = 1, 2, \ldots, d$$

$$\frac{dP_{(2d+1)}(t)}{dt} = \alpha_d P_{(2d-1)}(t). \tag{6.51}$$

Taking Laplace transformations of each of these equations and simplifications, we obtain the following equations:

$$P_{(2i-1)}(t) = \sum_{k=1}^{i} \left(A_{ik} e^{-\beta_k t} + B_{ik} e^{-\gamma_i t} \right)$$

$$P_{(2i)}(t) = \sum_{k=1}^{i} \left(C_{ik} e^{-\beta_k t} + D_{ik} e^{-\gamma_k t} \right) \tag{6.52}$$

where

$$\beta_i = \frac{(\alpha_i + \lambda_i + \mu_i) + \sqrt{(\alpha_i + \lambda_i + \mu_i)^2 - 4\alpha_i\mu_i}}{2}$$

$$\gamma_i = \frac{(\alpha_i + \lambda_i + \mu_i) - \sqrt{(\alpha_i + \lambda_1 + \mu_i)^2 - 4\alpha_i\mu_i}}{2} \quad \text{for } i = 1, 2, \ldots, d$$

and

$$A_{ik} = \begin{cases} \frac{(\mu_1 - \beta_1)}{(\gamma_1 - \beta_1)} & \text{for } i = 1, k = 1 \\[2ex] \left(\prod_{m=1}^{i-1} \alpha_m\right)\left(\prod_{m=1}^{i} \mu_m - \beta_k\right)\left(\prod_{\substack{m=1 \\ m \neq k}}^{i} \frac{1}{(\beta_m - \beta_k)}\right)\left(\prod_{m=1}^{i} \frac{1}{(\gamma_m - \beta_k)}\right) & \text{for } i = 2, \ldots, d \end{cases}$$

$$B_{ik} = \begin{cases} \frac{(\mu_1 - \gamma_1)}{(\beta_1 - \gamma_1)} & \text{for } i = 1, k = 1 \\[2ex] \left(\prod_{m=1}^{i-1} \alpha_m\right)\left(\prod_{m=1}^{i} \mu_m - \gamma_k\right)\left(\prod_{\substack{m=1 \\ m \neq k}}^{i} \frac{1}{(\gamma_m - \gamma_k)}\right)\left(\prod_{m=1}^{i} \frac{1}{(\beta_m - \gamma_k)}\right) & \text{for } i = 2, \ldots, d \end{cases}$$

$$C_{ik} = \begin{cases} \frac{\lambda_1}{(\gamma_1 - \beta_1)} & \text{for } i = 1, k = 1 \\[2ex] \lambda_i \left(\prod_{m=1}^{i-1} \alpha_m\right)\left(\prod_{m=1}^{i} \mu_m - \beta_k\right)\left(\prod_{\substack{m=1 \\ m = k}}^{i} \frac{1}{(\beta_m - \beta_k)}\right)\left(\prod_{m=1}^{i} \frac{1}{(\gamma_m - \beta_k)}\right) & \text{for } i = 2, \ldots, d \end{cases}$$

$$D_{ik} = \begin{cases} \frac{\lambda_1}{(\beta_1 - \gamma_1)} & \text{for } i = 1, k = 1 \\[2ex] \lambda_i \left(\prod_{m=1}^{i-1} \alpha_m\right)\left(\prod_{m=1}^{i} \mu_m - \gamma_k\right)\left(\prod_{\substack{m=1 \\ m = 1}}^{i} \frac{1}{(\gamma_m - \gamma_k)}\right)\left(\prod_{m=1}^{i} \frac{1}{(\beta_m - \gamma_k)}\right) & \text{for } i = 2, \ldots, d \end{cases}$$

$$\tag{6.53}$$

The availability A(t) of the system (i.e., the probability that system will be found in an operational (either good or degraded) state at time t) is given by:

$$A(t) = \sum_{i=1}^{d} P_{(2i-1)}(t) = \sum_{i=1}^{d}\sum_{k=1}^{i} \left(A_{ik}e^{-\beta_i t} + B_{ik}e^{-\gamma_i t}\right). \tag{6.54}$$

The system unavailability due to partial failures is

$$D(t) = \sum_{i=1}^{d} P_{(2i)}(t) = \sum_{i=1}^{d}\sum_{k=1}^{i} \left(C_{ik}e^{-\beta_k t} + D_{ik}e^{-\gamma_k t}\right). \tag{6.55}$$

Thus, the probability that the system fails completely before time t is:

$$F(t) = 1 - A(t) - D(t)$$
$$= 1 - \sum_{i=1}^{d}\sum_{k=1}^{i} \left[(A_{ik} + C_{ik})e^{-\beta_k t} + (B_{ik} + D_{ik})e^{-\gamma_k t}\right]. \tag{6.56}$$

After simplifications, we obtain

$$F(t) = 1 - \sum_{i=1}^{d} \left(X_i e^{-\beta_i t} + Y_i e^{-\gamma_i t}\right) \tag{6.57}$$

where

$$X_i = \frac{1}{\beta_i}\left(\prod_{m=1}^{d} \frac{\alpha_m(\mu_m - \beta_i)}{(\gamma_m - \beta_i)}\right)\left(\prod_{\substack{m=1 \\ m \neq i}}^{d} \frac{1}{(\beta_m - \beta_i)}\right)$$

$$Y_i = \frac{1}{\gamma_i}\left(\prod_{m=1}^{d} \frac{\alpha_m(\mu_m - \gamma_i)}{(\beta_m - \gamma_i)}\right)\left(\prod_{\substack{m=1 \\ m \neq i}}^{d} \frac{1}{(\gamma_m - \gamma_i)}\right). \tag{6.58}$$

If the repair time tends to infinity (or repair rate is zero), then the total operational time becomes the time to first failure. Therefore, the system reliability $R(t)$ can be obtained from $A(t)$ by substituting zeros for all repair rates. Thus, we obtain

$$R(t) = \sum_{i=1}^{d} L_i e^{-(\alpha_i + \lambda_i)t} \tag{6.59}$$

where

$$L_i = \sum_{m=i}^{d} \left(\frac{\prod_{k=1}^{m} \alpha_{k-1}}{\prod_{\substack{j=1 \\ j \neq i}}^{m} (\alpha_j + \lambda_j - \alpha_i - \lambda_i)} \right) \quad \text{for } i = 1, 2, \ldots, d \text{ and } \alpha_0 = 1. \tag{6.60}$$

The mean time to first failure of the system (MTTF) is given by

$$MTTF = \int_0^\infty R(t)dt = \int_0^\infty \left(\sum_{i=1}^{d} L_i e^{-(\alpha_i + \lambda_i)t} \right) dt$$

$$= \sum_{i=1}^{d} \frac{L_i}{\alpha_i + \lambda_i}. \tag{6.61}$$

Example 6.12 Consider a multistage repairable system with $d = 2$ (stages of degradation) and degradation rates: $\alpha_1 = 0.001$, $\alpha_2 = 0.002$; with partial failure rates: $\lambda_1 = 0.01$, $\lambda_2 = 0.05$, and repairing rates: $\mu_1 = 0.02$, and $\mu_2 = 0.01$. Calculate the system availability and reliability using Eqs. (6.54) and (6.59) (Fig. 6.14).

From Eqs. (6.54), (6.55), and (6.59), we obtain the reliability results as shown in Table 6.4. The system mean time to first failure (MTTF) is 92.7 (units of time).

Fig. 6.14 System flow diagram with d = 2

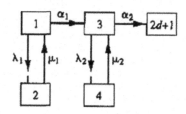

Table 6.4 Reliability
measures for various time t

t	A(t)	D(t)	R(t)
10	0.9120	0.0860	0.9032
20	0.8845	0.1489	0.8135
30	0.7929	0.1947	0.7313
40	0.7534	0.2280	0.6567
50	0.7232	0.2521	0.5892
100	0.6471	0.2992	0.3409

6.3 Counting Processes

Among discrete stochastic processes, counting processes in reliability engineering are widely used to describe the appearance of events in time, e.g., failures, number of perfect repairs, etc. The simplest counting process is a Poisson process. The Poisson process plays a special role to many applications in reliability (Pham 2000). A classic example of such an application is the decay of uranium. Radioactive particles from nuclear material strike a certain target in accordance with a Poisson process of some fixed intensity. A well-known counting process is the so-called renewal process. This process is described as a sequence of events, the intervals between which are independent and identically distributed random variables. In reliability theory, this type of mathematical model is used to describe the number of occurrences of an event in the time interval. In this section we also discuss the quasi-renewal process and the non-homogeneous Poisson process.

A non-negative, integer-valued stochastic process, $N(t)$, is called a counting process if $N(t)$ represents the total number of occurrences of the event in the time interval [0, t] and satisfies these two properties:

1. If $t_1 < t_2$, then $N(t_1) \leq N(t_2)$
2. If $t_1 < t_2$, then $N(t_2) - N(t_1)$ is the number of occurrences of the event in the interval $[t_1, t_2]$.

For example, if $N(t)$ equals the number of persons who have entered a restaurant at or prior to time t, then $N(t)$ is a counting process in which an event occurs whenever a person enters the restaurant.

6.3.1 Poisson Processes

One of the most important counting processes is the Poisson process.

Definition 6.2 A counting process, $N(t)$, is said to be a Poisson process with intensity λ if.

1. The failure process, $N(t)$, has stationary independent increments
2. The number of failures in any time interval of length s has a Poisson distribution with mean λs, that is,

$$P\{N(t+s) - N(t) = n\} = \frac{e^{-\lambda s}(\lambda s)^n}{n!} \quad n = 0, 1, 2, \ldots \tag{6.62}$$

3. The initial condition is $N(0) = 0$

This model is also called a homogeneous Poisson process indicating that the failure rate λ does not depend on time t. In other words, the number of failures occurring during the time interval $(t, t + s]$ does not depend on the current time t but only the length of time interval s. A counting process is said to possess independent increments if the number of events in disjoint time intervals are independent.

For a stochastic process with independent increments, the auto-covariance function is

$$Cov[X(t_1), X(t_2)] = \begin{cases} Var[N(t_1 + s) - N(t_2)] & \text{for } 0 < t_2 - t_1 < s \\ 0 & \text{otherwise} \end{cases}$$

where

$$X(t) = N(t + s) - N(t).$$

If $X(t)$ is Poisson distributed, then the variance of the Poisson distribution is

$$Cov[X(t_1), X(t_2)] = \begin{cases} \lambda[s - (t_2 - t_1)] & \text{for } 0 < t_2 - t_1 < s \\ 0 & \text{otherwise} \end{cases}$$

This result shows that the Poisson increment process is covariance stationary. We now present several properties of the Poisson process.

Property 6.1 The sum of independent Poisson processes, $N_1(t), N_2(t), \ldots, N_k(t)$, with mean values $\lambda_1 t, \lambda_2 t, \ldots, \lambda_k t$ respectively, is also a Poisson process with mean $\left(\sum_{i=1}^{k} \lambda_i\right) t$. In other words, the sum of the independent Poisson processes is also a Poisson process with a mean that is equal to the sum of the individual Poisson process' mean.

Property 6.2 The difference of two independent Poisson processes, $N_1(t)$, and $N_2(t)$, with mean $\lambda_1 t$ and $\lambda_2 t$, respectively, is not a Poisson process. Instead, it has the probability mass function.

$$P[N_1(t) - N_2(t) = k] = e^{-(\lambda_1 + \lambda_2)t} \left(\frac{\lambda_1}{\lambda_2}\right)^{\frac{k}{2}} I_k(2\sqrt{\lambda_1 \lambda_2 t}), \qquad (6.63)$$

where $I_k(.)$ is a modified Bessel function of order k.

Proof Define $N(t) = N_1(t) - N_2(t)$. We have

$$P[N(t) = k] = \sum_{i=0}^{\infty} P[N_1(t) = k + i]\, P[N_2(t) = i].$$

Since $N_i(t)$ for $i = 1, 2$ is a Poisson process with mean $\lambda_i t$, therefore,

$$P[N(t) = k] = \sum_{i=0}^{\infty} \frac{e^{-\lambda_1 t}(\lambda_1 t)^{k+i}}{(k+i)!} \frac{e^{-\lambda_2 t}(\lambda_2 t)^{i}}{i!}$$

$$= e^{-(\lambda_1 + \lambda_2)t} \left(\frac{\lambda_1}{\lambda_2}\right)^{\frac{k}{2}} \sum_{i=0}^{\infty} \frac{\left(\sqrt{\lambda_1 \lambda_2 t}\right)^{2i+k}}{i!(k+i)!}$$

$$= e^{-(\lambda_1 + \lambda_2)t} \left(\frac{\lambda_1}{\lambda_2}\right)^{\frac{k}{2}} I_k(2\sqrt{\lambda_1 \lambda_2 t}).$$

Property 6.3 If the Poisson process, $N(t)$, with mean λt, is filtered such that every occurrence of the event is not completely counted, then the process has a constant probability p of being counted. The result of this process is a Poisson process with mean $\lambda p t$.

Property 6.4 Let $N(t)$ be a Poisson process and Y_i a family of independent and identically distributed random variables which are also independent of $N(t)$. A stochastic process $X(t)$ *is* said to be a compound Poisson process if it can be represented as.

$$X(t) = \sum_{i=1}^{N(t)} Y_i.$$

6.3.2 Renewal Processes

A renewal process is a more general case of the Poisson process in which the inter-arrival times of the process or the time between failures do not necessarily follow the exponential distribution. For convenience, we will call the occurrence of an event a renewal, the inter-arrival time the renewal period, and the waiting time or repair time the renewal time.

Definition 6.3 A counting process $N(t)$ that represents the total number of occurrences of an event in the time interval $(0, t]$ is called a renewal process, if the time between failures are independent and identically distributed random variables.

The probability that there are exactly n failures occurring by time t can be written as

$$P\{N(t) = n\} = P\{N(t) \geq n\} - P\{N(t) > n\} \qquad (6.64)$$

Note that the times between the failures are $T_1, T_2, \ldots, T_n$ so the failures occurring at time W_k are

$$W_k = \sum_{i=1}^{k} T_i$$

and

$$T_k = W_k - W_{k-1}$$

Thus,

$$
\begin{aligned}
P\{N(t) = n\} &= P\{N(t) \geq n\} - P\{N(t) > n\} \\
&= P\{W_n \leq t\} - P\{W_{n+1} \leq t\} \\
&= F_n(t) - F_{n+1}(t)
\end{aligned}
\qquad (6.65)
$$

where $F_n(t)$ is the cumulative distribution function for the time of the nth failure and $n = 0, 1, 2, \ldots$.

Example 6.13 Consider a software testing model for which the time to find an error during the testing phase has an exponential distribution with a failure rate of X. It can be shown that the time of the nth failure follows the gamma distribution with parameters k and n with probability density function. From Eq. (6.65) we obtain

$$
\begin{aligned}
P\{N(t) = n\} &= P\{N(t) \leq n\} - P\{N(t) \leq n - 1\} \\
&= \sum_{k=0}^{n} \frac{(\lambda t)^k}{k!} e^{-\lambda t} - \sum_{k=0}^{n-1} \frac{(\lambda t)^k}{k!} e^{-\lambda t} \\
&= \frac{(\lambda t)^n}{n!} e^{-\lambda t} \quad \text{for } n = 0, 1, 2, \ldots.
\end{aligned}
\qquad (6.66)
$$

Several important properties of the renewal function are given below.

Property 6.5 The mean value function of the renewal process, denoted by $M(t)$, is equal to the sum of the distribution function of all renewal times, that is,

$$M(t) = E[N(t)] = \sum_{n=1}^{\infty} F_n(t)$$

Proof The renewal function can be obtained as

$$M(t) = E[N(t)]$$

$$= \sum_{n=1}^{\infty} n P\{N(t) = n\}$$

$$= \sum_{n=1}^{\infty} n[F_n(t) - F_{n+1}(t)]$$

$$= \sum_{n=1}^{\infty} F_n(t). \tag{6.67}$$

The mean value function, $M(t)$, of the renewal process is also called the renewal function. In other words, the mean value function represents the expected number of renewals in $[0, t]$.

Property 6.6 The renewal function, $M(t)$, satisfies the following equation:

$$M(t) = F(t) + \int_0^t M(t-s) dF(s) \tag{6.68}$$

where $F(t)$ is the distribution function of the inter-arrival time or the renewal period. The proof is left as an exercise for the reader (see Problem 7).

In general, let $y(t)$ be an unknown function to be evaluated and $x(t)$ be any non-negative and integrable function associated with the renewal process. Assume that $F(t)$ is the distribution function of the renewal period. We can then obtain the following result.

Property 6.7 Let the renewal equation be.

$$y(t) = x(t) + \int_0^t y(t-s) dF(s) \tag{6.69}$$

then its solution is given by

$$y(t) = x(t) + \int_0^t x(t-s)dM(s)$$

where $M(t)$ is the mean value function of the renewal process.

The proof of the above property can be easily derived using the Laplace transform. It is also noted that the integral equation given in Property 6.6 is a special case of Property 6.7.

Example 6.14 Let $x(t) = a$. Thus, from Property 6.7, the solution $y(t)$ is given by

$$y(t) = x(t) + \int_0^t x(t-s)dM(s)$$

$$= a + \int_0^t a\,dM(s)$$

$$= a(1 + E[N(t)]).$$

6.3.3 Quasi-Renewal Processes

In this section, a general renewal process, namely, the quasi-renewal process, is discussed. Let $\{N(t), t > 0\}$ be a counting process and let X_n be the time between the $(n-1)$th and the nth event of this process, $n \geq 1$.

Definition 6.4 (Wang and Pham 1996): If the sequence of non-negative random variables $\{X_1, X_2, ...\}$ is independent and.

$$X_i = aX_{i-1} \tag{6.70}$$

for $i \geq 2$ where $\alpha > 0$ is a constant, then the counting process $\{N(t), t \geq 0\}$ is said to be a quasi-renewal process with parameter and the first inter-arrival time X_1.

When $\alpha = 1$, this process becomes the ordinary renewal process as discussed in Sect. 2.6.2. This quasi-renewal process can be used to model reliability growth processes in software testing phases and hardware burn-in stages for $\alpha > 1$, and in hardware maintenance processes when $\alpha \leq 1$.

Assume that the probability density function, cumulative distribution function, survival function, and failure rate of random variable X_1 are $f_1(x)$, $F_1(x)$, $s_1(x)$, and $r_1(x)$, respectively. Then the pdf, cdf, survival function, failure rate of X_n for $n = 1$, 2, 3, ... is respectively given below (Wang and Pham 1996):

$$f_n(x) = \frac{1}{\alpha^{n-1}} f_1\left(\frac{1}{\alpha^{n-1}} x\right)$$

$$F_n(x) = F_1\left(\frac{1}{\alpha^{n-1}} x\right)$$

$$s_n(x) = s_1\left(\frac{1}{\alpha^{n-1}} x\right)$$

$$f_n(x) = \frac{1}{\alpha^{n-1}} r_1\left(\frac{1}{\alpha^{n-1}} x\right). \tag{6.71}$$

Similarly, the mean and variance of X_n is given as

$$E(X_n) = \alpha^{n-1} E(X_1)$$
$$Var(X_n) = \alpha^{2n-2} Var(X_1).$$

Because of the non-negativity of X_1 and the fact that X_1 is not identically 0, we obtain

$$E(X_1) = \mu_1 \neq 0$$

Property 6.8 (Wang and Pham 1996): The shape parameters of X_n are the same for $n = 1, 2, 3, \ldots$ for a quasi-renewal process if X_1 follows the gamma, Weibull, or log normal distribution.

This means that after "renewal", the shape parameters of the inter-arrival time will not change. In software reliability, the assumption that the software debugging process does not change the error-free distribution type seems reasonable. Thus, the error-free times of software during the debugging phase modeled by a quasi-renewal process will have the same shape parameters. In this sense, a quasi-renewal process is suitable to model the software reliability growth. It is worthwhile to note that

$$\lim_{n\to\infty} \frac{E(X_1 + X_2 + \ldots + X_n)}{n} = \lim_{n\to\infty} \frac{\mu_1(1-\alpha^n)}{(1-\alpha)n}$$
$$= 0 \quad \text{if } \alpha < 1$$
$$= \infty \quad \text{if } \alpha > 1$$

Therefore, if the inter-arrival time represents the error-free time of a software system, then the average error-free time approaches infinity when its debugging process is occurring for a long debugging time.

Distribution of N(t).

Consider a quasi-renewal process with parameter α and the first inter-arrival time X_1. Clearly, the total number of renewals, $N(t)$, that has occurred up to time t and the arrival time of the nth renewal, SS_n, has the following relationship:

$$N(t) \geq n \text{ if and only if } SS_n \leq t$$

that is, $N(t)$ is at least n if and only if the nth renewal occurs prior to time t. It is easily seen that

$$SS_n = \sum_{i=1}^{n} X_i = \sum_{i=1}^{n} \alpha^{i-1} X_1 \quad \text{for} \quad n \geq 1 \tag{6.72}$$

Here, $SS_0 = 0$. Thus, we have

$$\begin{aligned} P\{N(t) = n\} &= P\{N(t) \geq n\} - P\{N(t) \geq n+1\} \\ &= P\{SS_n \leq t\} - P\{SS_{n+1} \leq t\} \\ &= G_n(t) - G_{n+1}(t) \end{aligned}$$

where $G_n(t)$ is the convolution of the inter-arrival times $F_1, F_2, F_3, ..., F_n$. In other words,

$$G_n(t) = P\{F_1 + F_2 + ... + F_n \leq t\}$$

If the mean value of $N(t)$ is defined as the renewal function $M(t)$, then,

$$\begin{aligned} M(t) &= E[N(t)] \\ &= \sum_{n=1}^{\infty} P\{N(t) \geq n\} \\ &= \sum_{n=1}^{\infty} P\{SS_n \leq t\} \\ &= \sum_{n=1}^{\infty} G_n(t). \end{aligned} \tag{6.73}$$

The derivative of $M(t)$ is known as the renewal density

$$m(t) = M'(t).$$

In renewal theory, random variables representing the inter-arrival distributions only assume non-negative values, and the Laplace transform of its distribution $F_1(t)$ is defined by

$$\mathfrak{L}\{F_1(s)\} = \int_0^{\infty} e^{-sx} dF_1(x)$$

Therefore,

$$\mathcal{L}F_n(s) = \int\limits_0^\infty e^{-a^{n-1}st}\,dF_1(t) = \mathcal{L}F_1(\alpha^{n-1}s)$$

and

$$\mathcal{L}m_n(s) = \sum_{n=1}^\infty \mathcal{L}G_n(s)$$

$$= \sum_{n=1}^\infty \mathcal{L}F_1(s)\mathcal{L}F_1(\alpha s)\cdots\cdots\mathcal{L}F_1(\alpha^{n-1}s)$$

Since there is a one-to-one correspondence between distribution functions and its Laplace transform, it follows that.

Property 6.9 (Wang and Pham 1996): The first inter-arrival distribution of a quasi-renewal process uniquely determines its renewal function.

If the inter-arrival time represents the error-free time (time to first failure), a quasi-renewal process can be used to model reliability growth for both software and hardware.

Suppose that all faults of software have the same chance of being detected. If the inter-arrival time of a quasi-renewal process represents the error-free time of a software system, then the expected number of software faults in the time interval [0, t] can be defined by the renewal function, $M(t)$, with parameter $\alpha > 1$. Denoted by $M_r(t)$, the number of remaining software faults at time t, it follows that

$$M_r(t) = M(T_c) - M(t),$$

where $M(T_c)$ is the number of faults that will eventually be detected through a software lifecycle T_c.

6.3.4 Non-homogeneous Poisson Processes

The non-homogeneous Poisson process model (NHPP) that represents the number of failures experienced up to time t is a non-homogeneous Poisson process $\{N(t), t \geq 0\}$. The main issue in the NHPP model is to determine an appropriate mean value function to denote the expected number of failures experienced up to a certain time (Pham 2006a).

With different assumptions, the model will end up with different functional forms of the mean value function. Note that in a renewal process, the exponential assumption for the inter-arrival time between failures is relaxed, and in the NHPP, the stationary assumption is relaxed.

The NHPP model is based on the following assumptions:

- The failure process has an independent increment, i.e., the number of failures during the time interval $(t, t + s)$ depends on the current time t and the length of time interval s, and does not depend on the past history of the process.
- The failure rate of the process is given by

$$P\{\text{exactly one failure in}(t, t + \Delta t)\} = P\{N(t + \Delta t) - N(t) = 1\}$$
$$= \lambda(t)\Delta t + o(\Delta t)$$

where $\lambda(t)$ is the intensity function.

- During a small interval Δt, the probability of more than one failure is negligible, that is,

$$P\{\text{two or more failure in}(t, t + \Delta t)\} = o(\Delta t)$$

- The initial condition is $N(0) = 0$.

On the basis of these assumptions, the probability of exactly n failures occurring during the time interval $(0, t)$ for the NHPP is given by

$$\Pr\{N(t) = n\} = \frac{[m(t)]^n}{n!}e^{-m(t)} \quad n = 0, 1, 2, \ldots \quad (6.74)$$

where $m(t) = E[N(t)] = \int_0^t \lambda(s)ds$ and $\lambda(t)$ is the intensity function. It can be easily shown that the mean value function $m(t)$ is non-decreasing.

Reliability Function.

The reliability $R(t)$, defined as the probability that there are no failures in the time interval $(0, t)$, is given by

$$R(t) = P\{N(t) = 0\}$$
$$= e^{-m(t)}$$

In general, the reliability $R(x|t)$, the probability that there are no failures in the interval $(t, t + x)$, is given by

$$R(x|t) = P\{N(t + x) - N(t) = 0\}$$
$$= e^{-[m(t+x)-m(t)]}$$

and its density is given by

$$f(x) = \lambda(t + x)e^{-[m(t+x)-m(t)]}$$

where

$$\lambda(x) = \frac{\partial}{\partial x}[m(x)]$$

The variance of the NHPP can be obtained as follows:

$$Var[N(t)] = \int_0^t \lambda(s)ds$$

and the auto-correlation function is given by

$$Cor[s] = E[N(t)]E[N(t+s) - N(t)] + E[N^2(t)]$$

$$= \int_0^t \lambda(s)ds \int_0^{t+s} \lambda(s)ds + \int_0^t \lambda(s)ds$$

$$= \int_0^t \lambda(s)ds \left[1 + \int_0^{t+s} \lambda(s)ds\right] \tag{6.75}$$

Example 6.15 Assume that the intensity λ is a random variable with the pdf $f(\lambda)$. Then the probability of exactly n failures occurring during the time interval $(0, t)$ is given by

$$P\{N(t) = n\} = \int_0^\infty e^{-\lambda t} \frac{(\lambda t)^n}{n!} f(\lambda)d\lambda.$$

It can be shown that if the pdf $f(\lambda)$ is given as the following gamma density function with parameters k and m,

$$f(\lambda) = \frac{1}{\Gamma(m)} k^m \lambda^{m-1} e^{-k\lambda} \quad \text{for } \lambda \geq 0$$

then

$$P(N(t) = n) = \binom{n+m-1}{n} [p(t)]^m [q(t)]^n \quad n = 0, 1, 2, \ldots \tag{6.76}$$

is also called a negative binomial density function, where

$$p(t) = \frac{k}{t+k} \quad \text{and} \quad q(t) = \frac{t}{t+k} = 1 - p(t). \tag{6.77}$$

Thus,

$$P(N(t) = n) = \binom{n+m-1}{n}\left(\frac{k}{t+k}\right)^m \left(\frac{t}{t+k}\right)^n \quad n = 0, 1, 2, \ldots \tag{6.78}$$

The reader interested in a deeper understanding of advanced probability theory and stochastic processes should note the following highly recommended books:

Devore, J. L., *Probability and Statistics for Engineering and the Sciences,* 3rd edition, Brooks/Cole Pub. Co., Pacific Grove, 1991.

Gnedenko, B. V and I. A. Ushakov, *Probabilistic Reliability Engineering,* Wiley, New York, 1995.

Feller, W., *An Introduction to Probability Theory and Its Applications,* 3rd edition, Wiley, New York, 1994.

6.4 Problems

1. Calculate the reliability and MTTF of k-out-of-$(2k - 1)$ systems when $d = 3$,

   $$\lambda_1 = 0.0025/h, \quad \lambda_2 = 0.005/h, \quad \lambda_3 = 0.01/h \text{ and } \mu_1 = \mu_2 = \mu_3 = 0$$

   where $k = 1,2,3,4$ and 5 for various time t. (Hints: using Eqs. (6.42) and (6.43)).

2. In a nuclear power plant there are five identical and statistically independent channels to monitor the radioactivity of air in the ventilation system with the aim of alerting reactor operators to the need for reactor shutdown when a dangerous level of radioactivity is present. When at least three channels register a dangerous level of radioactivity, the reactor automatically shuts down. Furthermore, each channel contains three identical sensors and when at least two sensors register a dangerous level of radioactivity, the channel registers the dangerous level of radioactivity. The failure rate of each sensor in any channel is 0.001 per day. However, the common-cause failure rate of all sensors in a channel is 0.0005 per day. Obtain the sensor reliability, channel reliability, and the entire system reliability for various time t.

3. A crucial system operates in a good state during an exponentially distributed time with expected value $\frac{1}{\lambda}$ After leaving the good state, the system enters a degradation state. The system can still function properly in the degradation state during a fixed time $a > 0$, but a failure of the system occurs after this time. The system is inspected every T time units where $T > a$. It is replaced by a new one when the inspection reveals that the system is not in the good state.

(a) What is the probability of having a replacement because of a system failure?

(b) What is the expected time between two replacements?

4. A system consists of two independent components operating in parallel with a single repair facility where repair may be completed for a failed component before the other component has failed. Both the components are assumed to be functioning at time $t = 0$. When both components have failed, the system is considered to have failed and no recovery is possible. Assuming component i has the constant failure rate λ_i and repair rate μ_i for $i = 1$ and 2. The system reliability function, $R(t)$, is defined as the probability that the system continues to function throughout the interval $(0, t)$.

(a) Using the Eqs. (6.29) and (6.31) and the Laplace transform, derive the reliability function for the system. Obtain the system mean time to failure (MTTF)

(b) Calculate (a) with $\lambda_1 = 0.003$ per hour, $\lambda_2 = 0.005$ per hour, $\mu_1 = 0.3$ per hour, $\mu_2 = 0.1$ per hour, and $t = 25$ h.

5. A system is composed of 20 identical active power supplies, at least 19 of the power supplies are required for the system to function. In other words, when 2 of the 20 power supplies fail, the system fails. When all 20 power supplies are operating, each has a constant failure rate λ_a per hour. If one power supply fails, each remaining power supply has a failure rate λ_b per hour where $\lambda_a \leq \lambda_b$. We assume that a failed power supply can be repaired with a constant rate μ per hour. The system reliability function, $R(t)$, is defined as the probability that the system continues to function throughout the interval $(0, t)$.

(a) Determine the system mean time to failure (MTTF).

(b) Given $\lambda_a = 0.0005$, $\lambda_b = 0.004$, and $\mu = 0.5$, calculate the system MTTF.

6. A system is composed of 15 identical active power supplies, at least 14 of the power supplies are required for the system to function. In other words, when 2 of the 15 power supplies fail, the system fails. When all 15 power supplies are operating, each has a constant failure rate λ_a per hour. If one power supply fails, each remaining power supply has a failure rate λ_b per hour where $\lambda_a \leq \lambda_b$. We assume that a failed power supply can be repaired with a constant rate μ per hour. The system reliability function, $R(t)$, is defined as the probability that the system continues to function throughout the interval $(0, t)$.

(a) Determine the system mean time to failure (MTTF).

(b) Given $\lambda_a = 0.0003$, $\lambda_b = 0.005$, and $\mu = 0.6$, calculate the system MTTF.

7. Events occur according to an NHPP in which the mean value function is $m(t) = t^3 + 3t^2 + 6t$ $t > 0$.
What is the probability that n events occur between times $t = 10$ and $t = 15$?

8. Show that the renewal function, $M(t)$, can be written as follows:

$$M(t) = F(t) + \int_0^t M(t-s)\,dF(s)$$

where $F(t)$ is the distribution function of the inter-arrival time or the renewal period.

References

Li W, Pham H (2005a) An inspection-maintenance model for systems with multiple competing processes. IEEE Trans Reliab 54(2):318–327

Li W, Pham H (2005b) Reliability modeling of multi-state degraded systems with multiple multi-competing failures and random shocks. IEEE Trans Reliab 54(2):297–303

Pham H (1996) A software cost model with imperfect debugging, random life cycle and penalty cost. Int J Syst Sci 5:455–463

Pham H, Suprasad A, Misra RB (1996) Reliability and MTTF prediction of k-out-of-n complex systems with components subjected to multiple stages of degradation. Int J Syst Sci 27(10):995–1000

Pham H, Suprasad A, Misra RB (1997) Availability and mean life time prediction of multi-stage degraded system with partial repairs. Reliab Eng Syst Saf 56:169–173

Pham H (1997) Reliability analysis of digital communication systems with imperfect voters. Math Comput Model J 1997(26):103–112

Pham H (2000) Software reliability. Springer, Singapore

Pham H (2006) Springer, system software reliability

Wang HZ, Pham H (1996) A quasi renewal process and its application in the imperfect maintenance. Int J Syst Sci 27/10:1055–1062 and 28/12:1329

Yu J, Zheng S, Pham H, Chen T (2018) Reliability modeling of multi-state degraded repairable systems and its applications to automotive systems Qual Reliab Eng Int 34(3):459–474

Chapter 7
Maintenance Models

7.1 Introduction

It has been commonly known that reliability of any system including telecommunication, computers, aircraft, power plants etc., can be improved by applying the redundancy or maintenance approaches. In general, the failure of such systems is usually costly, if not very dangerous. Maintenance, replacement and inspection problems have been extensively studied in the reliability, maintainability and warranty literature (Pham and Wang, 1996; Wang and Pham, 2006). Maintenance involves corrective (unplanned) and preventive (planned). Corrective maintenance (CM) occurs when the system fails. In other words, CM means all actions performed as a result of failure, to restore an item to a specified condition. Some researchers also refer to CM as repair. Preventive maintenance (PM) occurs when the system is operating. In other words, PM means all actions performed in an attempt to retain an item in specified condition from operation by providing systematic inspection, detection, adjustment, and prevention of failures. Maintenance also can be categorized according to the *degree* to which the operating conditions of an item are restored by maintenance as follows (Wang and Pham, 2006):

a. Replacement policy: A system with no repair is replaced before failure with a new one.
b. Preventive maintenance (pm) policy: A system with repair is maintained preventively before failure.
c. Inspection policy: A system is checked to detect its failure.
d. Perfect repair or perfect maintenance: a maintenance action which restores the system operating condition to 'as good as new', i.e., upon perfect maintenance, a system has the same lifetime distribution and failure rate function as a brand new one. Generally, replacement of a failed system by a new one is a perfect repair.
e. Minimal repair or minimal maintenance: a maintenance action which restores the system to the failure rate it had when it just failed. The operating state of the system under minimal repair is also called 'as bad as old' policy in the literature.

© The Author(s), under exclusive license to Springer Nature Switzerland AG 2022
H. Pham, *Statistical Reliability Engineering*, Springer Series in Reliability Engineering,
https://doi.org/10.1007/978-3-030-76904-8_7

f.   Imperfect repair or imperfect maintenance: a maintenance action may not make
     a system 'as good as new' but younger. Usually, it is assumed that imperfect
     maintenance restores the system operating state.

This chapter discusses a brief introduction in maintenance modeling with various
maintenance policies including age replacement, block replacement and multiple
failure degradation processes and random shocks. We also discuss the reliability
and inspection maintenance modeling for degraded systems with competing failure
processes.

Example 7.1 Suppose that a system is restored to "as good as new" periodically at
intervals of time T. So the system renews itself at time T, $2\,T$, ... Define system
reliability under preventive maintenance (PM) as.

$$R_M(t) = \Pr\{\text{failure has not occurred by time t}\} \tag{7.1}$$

In other words, the system survives to time t if and only if it survives every PM
cycle $\{1,2,\ldots,k\}$ and a further time $(t - kT)$.

(a)   For a system with a failure time probability density function

$$f(t) = \lambda^2 t\, e^{-\lambda t} \quad \text{for } t > 0, \lambda > 0 \tag{7.2}$$

obtain the system reliability under PM and system mean time to first failure.

(b)   Calculate the system mean time to first failure if PM is performed every 20 days
      and $\lambda = 0.005$.

Solution: The system survives to time t if and only if it survives every PM cycle
$\{1,2,\ldots,k\}$ and a further time $(t - kT)$. Thus,

$$R_M(t) = [R(T)]^k R(t - kT) \quad \text{for } kT < t < (k + 1)T \text{ and } k = 0, 1, 2, \ldots \tag{7.3}$$

Interval containing t Formula for $R_M(t)$

Interval containing t	Formula for $R_M(t)$
$0 < t < T$	$R(t)$
$T < t < 2T$	$R(T)R(t - T)$
$2T < t < 3T$	$[R(T)]^2 R(t - 2T)$
...	
$kT < t < (k + 1)T$	$[R(T)]^k R(t - kT)$

Thus, system reliability under PM is

$$R_M(t) = [R(T)]^k R(t - kT) \quad \text{for } kT < t < (k + 1)T \text{ and } k = 0, 1, 2, \ldots \tag{7.4}$$

Given the probability density function as from Eq. (7.1),

$$f(t) = \lambda^2 t e^{-\lambda t} \quad \text{for } t > 0, \ \lambda > 0$$

then

$$R(t) = \int_t^\infty f(x)dx = \int_t^\infty \lambda^2 x \, e^{-\lambda x} dx = (1 + \lambda t)e^{-\lambda t}. \tag{7.5}$$

Thus, from Eq. (7.4), we have

$$R_M(t) = [R(T)]^k R(t - kT) \quad \text{for } k = 0, 1, 2, \ldots$$
$$= \left[(1 + \lambda T)e^{-\lambda T}\right]^k \left[(1 + \lambda(t - kT))e^{-\lambda(t-kT)}\right]. \tag{7.6}$$

The system mean time to first failure (MTTFF) is

$$MTTFF = \int_0^\infty R_M(t)dt$$

$$= \int_0^T R(t)dt + \int_T^{2T} R(T)R(t - T)dt + \int_{2T}^{3T} [R(T)]^2 R(t - 2T)dt + \ldots$$

$$= \sum_{k=0}^\infty [R(T)]^k \int_0^T R(u)du$$

$$= \frac{\int_0^T R(u)du}{1 - R(T)}.$$

Thus,

$$MTTFF = \frac{\int_0^T R(u)du}{1 - R(T)}. \tag{7.7}$$

Note that, from Eq. (7.5)

$$\int_0^T R(x)dx = \int_0^T (1 + \lambda x)e^{-\lambda x}dx$$

$$= \frac{2}{\lambda}\left(1 - e^{-\lambda T}\right) - Te^{-\lambda T}.$$

From Eq. (7.7), we have

$$MTTFF = \frac{\int\limits_{0}^{T} R(u)du}{1 - R(T)} = \frac{\frac{2}{\lambda}\left(1 - e^{-\lambda T}\right) - Te^{-\lambda T}}{1 - (1 + \lambda T)e^{-\lambda T}}.$$

(b)   Here $T = 20$ days, and $\lambda = 0.005$ per day.

$$MTTFF = \frac{\frac{2}{0.005}\left(1 - e^{-(0.005)20}\right) - (20)e^{-(0.005)20}}{1 - (1 + (0.005)(20))e^{-(0.005)(20)}} = \frac{19.9683}{0.0047} = 4,267.8 \text{ days.}$$

7.2   Maintenance and Replacement Policies

A failed system is assumed to immediately replace or repair. There is a cost associated with it. One the one hand, designer may want to maintain a system before its failure. On the other hand, it is better not to maintain the system too often because the cost involved each time. Therefore it is important to determine when to perform the maintenance of the system that can minimize the expected total system cost.

Consider a one-unit system where a unit is replaced upon failure. Let.

c_1 the cost of each failed unit which is replaced
$c_2(<c_1)$ the cost of a planned replacement for each non-failed unit
$N_1(t)$ the number of failures with corrective replacements (CM)
$N_2(t)$ the number of replacements of non-failed units during $(0, t]$ interval.

In general, the expected total system cost during $(0, T]$, $E_c(T)$, can be defined as follows:

$$E_c[T] = c_1 E[N_1(T)] + c_2 E[N_2(T)]. \tag{7.8}$$

We now discuss the optimum policies which minimize the expected costs per unit time of each replacement policy such as age replacement and block replacement.

7.2.1   Age Replacement Policy

A unit is replaced at time T or at failure, whichever occurs first. T is also called a planned replacement policy. Let $\{X_k\}_{k=1}^{\infty}$ be the failure times of successive operating units with a density f and distribution F with finite mean μ. Let $Z_k \equiv \min\{X_k, T\}$ represents the intervals between the replacements caused by either failure or planned replacement for $k = 1,2,\ldots$ The probability of Z_k can be written as follows:

$$\Pr(Z_k \leq t) = \begin{cases} F(t) & t < T \\ 1 & t \geq T. \end{cases} \tag{7.9}$$

The mean time of one cycle is

$$E(Z_k) = \int_0^T t dF(t) + TR(T) = \int_0^T R(t) dt. \tag{7.10}$$

The expected total system cost per cycle is

$$E_c(T) = c_1 F(T) + c_2 R(T) \tag{7.11}$$

where $R(T) = 1 - F(T)$.

The expected total cost per unit time for an infinite time span, $C(T)$, is

$$C(T) = \frac{c_1 F(T) + c_2 R(T)}{\int_0^T R(t) dt}. \tag{7.12}$$

Let $r(t) \equiv f(t)/R(t)$ be the failure rate. We wish to find the optimal replacement policy time T^* which minimizes the expected total cost per unit time $C(T)$ in Eq. (7.12).

Theorem 7.1 *Given c_1, c_2, and μ. Assume the failure rate $r(t)$ is a strictly increasing function and $A = \frac{c_1}{\mu(c_1 - c_2)}$. The optimal replacement policy time T^* that minimizes the expected total system cost per unit time $C(T)$ can be obtained as follows:*
If $r(\infty) > A$ then there exists a finite value

$$T^* = G^{-1}\left(\frac{c_2}{c_1 - c_2}\right) \tag{7.13}$$

where $G(T) = r(T) \int_0^T R(t) dt - F(T)$ and the resulting expected total system cost per unit time C(T) is

$$C(T^*) = (c_1 - c_2) r(T^*).$$

(ii)   If $r(\infty) \leq A$ then the optimum replacement time T is at: $T^* = \infty$. This implies that a unit should not be replaced unless it fails.

The above results can be obtained by differentiating the expected total cost function per unit time $C(T)$ from Eq. (7.12) with respect to T and setting it equal to 0. We have

$$\frac{\partial C(T)}{\partial T} = (c_1 - c_2)\left(r(T) \int_0^T R(t) dt - F(T)\right) - c_2 \equiv 0 \tag{7.14}$$

or, equivalently, $G(T) = \frac{c_2}{c_1 - c_2}$. Since $r(T)$ is strictly increasing and $G(0) = 0$, we can easily show that the function $G(T)$ is strictly increasing in T.

If $r(\infty) > A$ then $G(\infty) > \frac{c_2}{(c_1 - c_2)}$. This shows that there exists a finite value T^* where T^* is given in Eq. (7.13) and it minimizes $C(T)$.

If $r(\infty) \le A$ then $G(\infty) \le \frac{c_2}{(c_1 - c_2)}$. This shows that the optimum replacement time is $T^* = \infty$. This implies that a unit will not be replaced until it fails.

Example 7.2 Under an age replacement policy, the system is replaced at time T or at failure whichever occurs first. The costs of a failed unit and a planned replacement unit are respectively c_1 and c_2 with $c_1 \ge c_2$. For systems with a failure time probability density function.

$$f(t) = \lambda^2 t e^{-\lambda t} \quad \text{for } t > 0, \ \lambda > 0$$

obtain the optimal planned replacement time T^* that minimizes the expected total cost per cycle per unit time. Given $c_1 = 10$ and $c_2 = 1$, what is the optimal planned replacement time T^* that minimizes the expected total cost per cycle per unit time?

Solution: The pdf is.

$$f(t) = \lambda^2 t e^{-\lambda t} \quad \text{for } t > 0, \ \lambda > 0$$

The reliability function

$$R(t) = \int\limits_t^\infty f(x)dx = \int\limits_t^\infty \lambda^2 x e^{-\lambda x}dx = (1 + \lambda t)e^{-\lambda t}$$

The failure rate

$$r(t) = \frac{f(t)}{R(t)} = \frac{\lambda^2 t e^{-\lambda t}}{(1 + \lambda t)e^{-\lambda t}} = \frac{\lambda^2 t}{(1 + \lambda t)}. \tag{7.15}$$

From Eq. (7.12), the expected total cost per cycle per unit time is

$$E(T) = \frac{c_1 F(T) + c_2 R(T)}{\int\limits_0^T R(t)dt}.$$

The derivative of the function $E(T)$ is given by

$$\frac{\partial E(T)}{\partial T} = \frac{\left[c_1 f(T) - c_2 f(T)\right] \int_0^T R(t)dt - [c_1 F(T) + c_2 R(T)]R(T)}{\left(\int_0^T R(t)dt\right)^2}.$$

Setting the above equation to 0 we obtain the following:

$$\frac{\partial E(T)}{\partial T} = 0 \Leftrightarrow (c_1 - c_2)f(T) \int_0^T R(t)dt - [c_1 F(T) + c_2 R(T)]R(T) \equiv 0.$$

$$(c_1 - c_2)\frac{f(T)}{R(T)} \int_0^T R(t)dt = [c_1 F(T) + c_2 R(T)]$$

$$(c_1 - c_2)r(T) \int_0^T R(t)dt - (c_1 - c_2)F(T) = c_2$$

$$(c_1 - c_2)\left\{ r(T) \int_0^T R(t)dt - F(T) \right\} = c_2$$

or,

$$r(T) \int_0^T R(t)dt - F(T) = \frac{c_2}{(c_1 - c_2)}. \tag{7.16}$$

Let

$$G(T) = r(T) \int_0^T R(t)dt - F(T) \quad \text{and} \quad A_1 = \frac{c_2}{c_1 - c_2}. \tag{7.17}$$

That is, $G(T) = A_1$. Since $r(T)$ is increasing and $G(0) = 0$ we can show that $G(T)$ is increasing in T.

(a) If

$$r(\infty) > A \quad \text{where} \quad A = \frac{c_1}{\mu(c_1 - c_2)} \quad \text{and} \quad \mu = \int_0^\infty R(t)dt. \tag{7.18}$$

then $G(\infty) > A_1$. There exists a finite value T^* where

$$T^* = G^{-1}\left(\frac{c_2}{(c_1 - c_2)}\right). \tag{7.19}$$

(b)　If

$$r(\infty) \leq A \quad \text{then} \quad G(\infty) \leq \frac{c_2}{(c_1 - c_2)} \tag{7.20}$$

then the optimum replacement time is $T^* = \infty$. This implies that a unit will not be replaced until it fails.

We now calculate

$$\int_0^T R(t)dt = \int_0^T (1 + \lambda t)\, e^{-\lambda t} dt = \frac{2}{\lambda}\left(1 - e^{-\lambda T}\right) - Te^{-\lambda T}. \tag{7.21}$$

Then

$$\begin{aligned} G(T) &= r(T) \int_0^T R(t)dt - F(T) \\ &= \frac{\lambda^2 T}{(1 + \lambda T)}\left[\frac{2}{\lambda}\left(1 - e^{-\lambda T}\right) - Te^{-\lambda T}\right] - \left[1 - (1 + \lambda T)e^{-\lambda T}\right] \end{aligned} \tag{7.22}$$

Given $c_1 = 10$, $c_2 = 1$, and $\lambda = 2$, then

$$\frac{c_2}{c_1 - c_2} = \frac{1}{10 - 1} = \frac{1}{9}$$

From Eq. (7.22),

$$\begin{aligned} G(T) &= \frac{\lambda^2 T}{(1 + \lambda T)}\left[\frac{2}{\lambda}\left(1 - e^{-\lambda T}\right) - Te^{-\lambda T}\right] - \left[1 - (1 + \lambda T)e^{-\lambda T}\right] \\ &= \frac{4T}{(1 + 2T)}\left[\frac{2}{2}\left(1 - e^{-2T}\right) - Te^{-2T}\right] - \left[1 - (1 + 2T)e^{-2T}\right] \\ &= \frac{4T}{(1 + 2T)}\left[1 - e^{-2T} - Te^{-2T}\right] - \left[1 - (1 + 2T)e^{-2T}\right]. \end{aligned}$$

Here we can find T^* such as

$$G(T^*) = \frac{1}{9} = 0.1111$$

The expected total cost per cycle per unit time, from Eq. (7.12), is:

$$E(T) = \frac{c_1 F(T) + c_2 R(T)}{\int_0^T R(t)dt}.$$

$$= \frac{c_1 \left[1 - (1 + \lambda T)\, e^{-\lambda T}\right] + c_2 \left[(1 + \lambda T)\, e^{-\lambda T}\right]}{\left[\frac{2}{\lambda}\left(1 - e^{-\lambda T}\right) - Te^{-\lambda T}\right]}$$

$$= \frac{10\left[1 - (1 + 2T)\, e^{-2T}\right] + \left[(1 + 2T)\, e^{-2T}\right]}{\left[\left(1 - e^{-2T}\right) - Te^{-2T}\right]}$$

$$= \frac{10 - 9(1 + 2T)\, e^{-2T}}{1 - (1 + T)\, e^{-2T}}.$$

T	G(T)	E(T)
0.3	0.0930	7.3186
0.32	0.102	7.2939
0.33	0.1065	7.2883
0.335	0.1088	7.2870
0.34	0.1111	7.2865
0.35	0.1156	7.2881
0.5	0.1839	7.5375

Thus, the optimal planned replacement time T^* that minimizes the expected total cost per cycle per unit time is:

$$T^* = 0.34 \text{ and } E(T^*) = 7.2865.$$

7.2.2 Block Replacement

Consider that a unit begins to operate at time $t = 0$ and when it fails, it is discovered instantly and replaced immediately by a new one. Under this block policy, a unit is replaced at periodic times $kT (k = 1, 2, \cdots)$ independent of its age. Suppose that each unit has a failure time distribution $F(t)$ with finite mean μ. The expected total system cost per cycle $E_c(T)$ is given by

$$E_c(T) = c_1 E[N_1(T)] + c_2 E[N_2(T)] = c_1 M(T) + c_2 \qquad (7.23)$$

where $M(T) = E(N_1(T))$ is differential and the expected number of failed units per cycle. The expected total system cost per unit time for an infinite time span under block replacement policy is defined as

$$C(T) = \frac{c_1 M(T) + c_2}{T}. \tag{7.24}$$

This indicates that there will be one planned replacement per period at a cost of c_2 and the expected number of failures with corrective replacement per period where each corrective replacement has a cost of c_1.

Theorem 7.2 Given c_1 and c_2. There exists a finite optimum planned replacement time T^* that minimizes the expected total system cost per unit time $C(T)$:

$$T^* = D^{-1}\left(\frac{c_2}{c_1}\right). \tag{7.25}$$

the resulting expected total system cost is $C(T^*)$ where $D(T) = Tm(T) - M(T)$ and $m(t) \equiv dM(t)/dt$.

Similarly from Theorem 7.1, we can obtain the optimum planned replacement time T^* given in Eq. (7.25) which minimizes the expected cost per unit time $C(T)$ by differentiating the function $C(T)$ with respect to T and setting it equal to zero, we obtain

$$Tm(T) - M(T) = \frac{c_2}{c_1}$$

where $m(t) \equiv dM(t)/dt$. The results can immediately follow.

7.2.3 Periodic Replacement Policy

For some systems, we only need to perform minimal repair at each failure, and make the planned replacement or preventive maintenance at periodic times.

Consider a periodic replacement policy as follows: A unit is replaced periodically at periodic times kT $(k = 1, 2, \cdots)$. After each failure, only minimal repair is made so that the failure rate remains undisturbed by any repair of failures between successive replacements (Barlow and Proschan, 1965). This policy is commonly used with computers and airplanes. Specifically, a new unit begins to operate at $t = 0$, and when it fails, only minimal repair is made. That is, the failure rate of a unit remains undisturbed by repair of failures. Further, a unit is replaced at periodic times kT $(k = 1, 2, \cdots)$ independent of its age, and any units are as good as new after replacement. It is assumed that the repair and replacement times are negligible. Suppose that the failure times of each unit are independent, and have a cdf $F(t)$ and the failure rate $r(t) \equiv f(t)/R(t)$ where f is a probability density function density and $R(t)$ is the reliability function. The failures of a unit occur that follow a nonhomogeneous Poisson process with a mean-value function H(t) where $H(t) \equiv \int_0^t r(u)du$ and $R(t) = e^{-H(t)}$.

Consider one cycle with a constant time T from the planned replacement to the next one. Then, since the expected number of failures during one cycle is $E(N_1(T)) = M(T)$, the expected total system cost per cycle is

$$E_c(T) = c_1E(N_1(T)) + c_2E(N_2(T)) = c_1H(T) + c_2, \tag{7.26}$$

where c_1 is the cost of each minimal repair.

Therefore the expected total system cost per unit of time for an infinite time span is

$$C(T) \equiv \frac{1}{T}[c_1H(T) + c_2]. \tag{7.27}$$

If a unit is not replaced forever, *i.e.*, $T = \infty$, then $\lim_{T \to \infty} R(T)/T = r(\infty)$, which may be possibly infinite, and $C(\infty) = c_1r(\infty)$.

Given c_1 and c_2. There exists a finite optimum replacement time T^* such that

$$Tr(T) - H(T) = \frac{c_2}{c_1} \tag{7.28}$$

that minimizes the expected total system cost per unit time $C(T)$ as given in Eq. (7.27).

Differentiating the function $C(T)$ in Eq. (7.27) with respect to T and setting it equal to zero, we have.

$$Tr(T) - H(T) = \frac{c_2}{c_1}$$

If the cost of minimal repair depends on the age x of a unit and is given by $c_1(x)$, the expected total system cost per unit time can be defined as

$$C(T) = \frac{1}{T}[\int_0^T c_1(x)r(x)dx + c_2]. \tag{7.29}$$

One can obtain the optimum replacement policy T that minimizes the expected total system cost $C(T)$ by taking a derivative of the function $C(T)$ with respect to T given the function $c_1(x)$.

7.2.4 Replacement Models with Two Types of Units

In practice, many systems are consisted of vital and non-vital parts or essential and non-essential components. If vital parts fail then a system becomes dangerous or suffers a high cost. It would be wise to make the planned replacement or overhaul at suitable times. We may classify into two types of failures; partial and total failures, slight and serious failures, or simply faults and failures.

Consider a system consists of unit 1 and unit 2 which operate independently, where unit 1 corresponds to non-vital parts and unit 2 to vital parts. It is assumed that unit 1 is replaced always together with unit 2. Unit i has a failure time distribution $F_i(t)$, failure rate $r_i(t)$ and cumulative hazard $H_i(t)(i = 1, 2)$, i.e., $R_i(t) = \exp[-H_i(t)]$ and $H_i(t) = \int_0^t r_i(u)du$. Then, we consider the following four replacement policies which combine age, block and periodic replacements:

(a)   Unit 2 is replaced at failure or time T, whichever occurs first, and when unit 1 fails between replacements, it is replaced by a new unit. Then, the expected total system cost per unit time is

$$C(T) = \frac{c_1 \int_0^T f_1(t)R_2(t)dt + c_2F_2(T) + c_3}{\int_0^T R_2(t)dt} \qquad (7.30)$$

where c_1 is a cost of replacement for a failed unit 1, c_2 is an additional replacement for a failed unit 2, and c_3 is a cost of replacement for units 1 and 2.

(b)   In case (a), when unit 1 fails between replacements, it undergoes only minimal repair. Then, the expected total system cost per unit time is

$$C(T) = \frac{c_1 \int_0^T r_1(t)R_2(t)dt + c_2F_2(T) + c_3}{\int_0^T R_2(t)dt}, \qquad (7.31)$$

where c_1 is a cost of minimal repair for failed unit 1, and c_2 and c_3 are the same costs as case (a).

(c)   Unit 2 is replaced at periodic times $kT(k = 1, 2, \cdots)$ and undergoes only minimal repair at failures between planned replacements, and when unit 1 fails between replacements, it is replaced by a new unit. Then, the expected total system cost per unit time is

$$C(T) = \frac{1}{T}[c_1H_1(T) + c_2H_2(T) + c_3], \qquad (7.32)$$

where c_2 is a cost of minimal repair for a failed unit 2, and c_1 and c_3 are the same costs as case (a).

(d)   In case (c), when unit 1 fails between replacements, it also undergoes minimal repair. Then, the expected total system cost per unit time is

$$C(T) = \frac{1}{T}[c_1H_1(T) + c_2H_2(T) + c_3], \qquad (7.33)$$

where c_1 is a cost of minimal repair for a failed unit 1, and c_2 and c_3 are the same costs as case (c).

7.3 Non-repairable Degraded System Modeling

Maintenance has evolved from simple model that deals with machinery breakdowns, to time-based preventive maintenance, to today's condition-based maintenance. It is of great importance to avoid the failure of a system during its actual operating; especially, when such failure is dangerous and costly. This section discusses a relia-bility model and examines the problem of developing maintenance cost models for determining the optimal maintenance policies of non-repairable degraded systems with competing failure processes. The material in this section are based on Li and Pham (2005a).

Notation

C_c	Cost per CM action
C_p	Cost per PM action
C_m	Loss per unit idle time
C_i	Cost per inspection
$Y(t)$	Degradation process
$Y_i(t)$	Degradation process i, i = 1, 2
$D(t)$	Cumulative shock damage value up to time t
S	Critical value for shock damage
$C(t)$	Cumulative maintenance cost up to time t.
$E[C_1]$	Average total maintenance cost during a cycle
$E[W_1]$	Mean cycle length
$E[N_1]$	Mean number of inspections during a cycle
$E[\xi]$	Mean idle time during a cycle
$\{I_i\}_{i \in N}$	Inspection sequence
$\{U_i\}_{i \in N}$	Inter-inspection sequence
P_{i+1}	Probability that there are a total of $(i + 1)$ inspections in a renewal cycle
P_p	Probability that a renewal cycle ends by a PM action
P_c	Probability that a renewal cycle ends by a CM action (P_c = 1- P_p)

Consider that:

- The system has the state space $\Omega_U = \{M, \ldots, 1, 0, F\}$ and it starts at state M at time $t = 0;$
- System fails either due to degradation ($Y(t) > G$) or catastrophic failure $\left(D(t) = \sum_{i=1}^{N_2(t)} X_i > S \right)$. System may either goes from state i to the next degraded state $i-1$ or directly goes to catastrophic failure state F, $i = M, .0.1;$
- No repair or maintenance is performed on the system; and
- The two processes $Y(t)$ and $D(t)$ are independent.

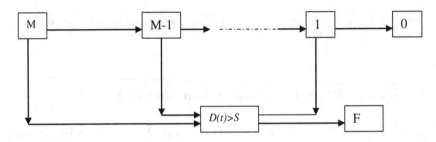

Fig. 7.1 Flow diagram of the system with two competing failure processes (Li and Pham, 2005a, b)

Figure 7.1 illustrates the case where systems are subject to two failure competing processes: degradation process $Y(t)$ and the random shocks process $D(t)$ and whichever process occurred first would cause the system to failure.

Suppose that the operating condition of the system at any time point could be classified into one of a finite number of the states, say $\Omega_U = \{M, \ldots, 1, 0, F\}$. A one-to-one relationship between the element of $\Omega = \{M, \ldots, 1, 0\}$ and its corresponding interval is defined as follows:

$$
\begin{aligned}
&\text{State M if}\quad && Y(t) \in [0, W_M] \\
&\text{State } M - 1 \text{ if} && Y(t) \in (W_M, W_{M-1}] \\
&\quad\vdots \\
&\text{State } i && Y(t) \in (W_{i+1}, W_i] \\
&\text{State 1} && Y(t) \in (W_2, W_1] \\
&\text{State 0} && Y(t) > W_1
\end{aligned}
$$

Let $P_i(t)$ be a probability that the value of $Y(t)$ will fall within a pre-defined interval corresponding to state i and $D(t) \leq S$. From state i, the system will make a direct transition to state $(i\text{-}1)$ due to gradual degradation or to state F due to a random shock (Fig. 7.1). The reliability function is defined as:

$$
R_M(t) = \sum_{i=1}^{M} P_i(t) = P(Y(t) \leq G, D(t) \leq S) \tag{7.34}
$$

where $P_i(t)$ is the probability of being in state i. Let T be the time to failure of the system. Then T can be defined as: $T = \inf\{t > 0 : Y(t) > G \text{ or } D(t) > S\}$. The mean time to failure is given by:

$$
E[T] = \int_0^\infty P(Y(t) \leq G, D(t) \leq S) dt
$$

$$= \int_0^\infty P(Y(t) \le G) \sum_{j=0}^\infty \frac{(\lambda_2 t)^j e^{-\lambda_2 t}}{j!} F_X^{(j)}(S) dt \qquad (7.35)$$

or, equivalently,

$$E[T] = \sum_{j=0}^\infty \frac{F_X^{(j)}(S)}{j!} \int_0^\infty P(Y(t) \le G)(\lambda_2 t)^j e^{-\lambda_2 t} dt \qquad (7.36)$$

Let $F_G(t) = P\{Y(t) \le G\}$, then $f_G(t) = \frac{d}{dt} F_G(t)$. The pdf of the time to failure, $f_T(t)$ can be easily obtained:

$$f_T(t) = -\frac{d}{dt}[P(Y(t) \le G)P(D(t) \le S)]$$

$$= -\sum_{j=0}^\infty \frac{F_X^{(j)}(S)}{j!} \frac{d}{dt}\Big[P(Y(t) \le G)(\lambda_2 t)^j e^{-\lambda_2 t}\Big]$$

After simplifications, we have

$$f_T(t) = -\sum_{j=1}^\infty \frac{F_X^{(j)}(S)}{j!}$$
$$\Big[f_G(t)(\lambda_2 t)^j e^{-\lambda_2 t} + F_G(t)j\lambda_2(\lambda_2 t)^{j-1} e^{-\lambda_2 t} - \lambda_2 F_G(t)(\lambda_2 t)^j e^{-\lambda_2 t}\Big] \quad (7.37)$$

Assume that the degradation process is described as the function $Y(t) = W\frac{e^{Bt}}{A+e^{Bt}}$ where the two random variables A and B are independent, and that A follows a uniform distribution with parameter interval $[0,a]$ and B follows exponential distribution with parameter $\beta > 0$. In short, $A \sim U[0, a]$, $a > 0$ and $B \sim Exp(\beta)$, $\beta > 0$.
The probability for the system of being in state M is as follows:

$$P_M(t) = P(Y(t) \le W_M, D(t) \le S)$$

$$= \left\{ \int_{\forall A} P\left(B < \frac{1}{t} \ln \frac{u_1 A}{1-u_1} \Big| A = x\right) f_A(x) dx \right\} P(D(t) \le S)$$

$$= \left\{ 1 - \frac{1}{a}\left(\frac{1-u_1}{u_1}\right)^{\frac{\beta}{t}} \left(\frac{t}{t-\beta}\right)\left(a^{1-\frac{\beta}{t}} - 1\right) \right\} e^{-\lambda_2 t} \sum_{j=0}^\infty \frac{(\lambda_2 t)^j}{j!} F_X^{(j)}(S)$$
$$(7.38)$$

Then the probability for the system of being in state i can be calculated as follows:

$$P_i(t) = P(W_{i+1} < W\frac{e^{Bt}}{A+e^{Bt}} \le W_i, D(t) \le S)$$

$$= \left\{ \int_0^a P\left(\frac{1}{t} \ln \frac{u_{i-1}A}{1-u_{i-1}} < B \le \frac{1}{t} \ln \frac{u_i A}{1-u_i} \Big| A = x \right) f_A(x) dx \right\} e^{-\lambda_2 t} \sum_{j=1}^{\infty} \frac{(\lambda_2 t)^j}{j!} F_X^{(j)}(S)$$

$$= \left\{ \frac{1}{a} \left(\frac{t}{t-\beta} \right) \left(a^{1-\frac{\beta}{t}} \right) \left[\left(\frac{1-u_i}{u_i} \right)^{\frac{\beta}{t}} - \left(\frac{1-u_{i-1}}{u_{i-1}} \right)^{\frac{\beta}{t}} \right] \right\} e^{-\lambda_2 t} \sum_{j=0}^{\infty} \frac{(\lambda_2 t)^j}{j!} F_X^{(j)}(S) \qquad (7.39)$$

where $\mu_i = \frac{W_i}{W}$, $i = M\text{-}1,..,1$.

Similarly, the probability for the system of being in state 0 is as follows:

$$P_0(t) = P(Y(t) = W \frac{e^{Bt}}{A+e^{Bt}} > G, D(t) \le S)$$

$$= \left\{ \frac{1}{a} \left(\frac{1-u_M}{u_M} \right)^{\frac{\beta}{t}} \left(\frac{t}{t-\beta} \right) \left(a^{1-\frac{\beta}{t}} \right) \right\} e^{-\lambda_2 t} \sum_{j=0}^{\infty} \frac{(\lambda_2 t)^j}{j!} F_X^{(j)}(S)$$

The probability for a catastrophic failure state F is given by:

$$P_F(t) = P(Y(t) = W \frac{e^{Bt}}{A+e^{Bt}} \le G, D(t) > S)$$

$$= \left\{ 1 - \frac{1}{a} \left(\frac{1-u_1}{u_1} \right)^{\frac{\beta}{t}} \left(\frac{t}{t-\beta} \right) \left(a^{1-\frac{\beta}{t}} \right) \right\} \left\{ 1 - e^{-\lambda_2 t} \sum_{j=0}^{\infty} \frac{(\lambda_2 t)^j}{j!} F_X^{(j)}(S) \right\}$$

Hence, the reliability $R_M(t)$ is given by:

$$R_M(t) = \sum_{i=1}^{M} P_i(t)$$

$$= \left\{ 1 - \frac{1}{a} \left(\frac{1-u_M}{u_M a} \right)^{\frac{\beta}{t}} \left(\frac{t}{t-\beta} \right) \left(a^{1-\frac{\beta}{t}} \right) \right\} \left\{ e^{-\lambda_2 t} \sum_{j=0}^{\infty} \frac{(\lambda_2 t)^j}{j!} F_X^{(j)}(S) \right\}$$

$$(7.40)$$

Example 7.3 Assume.

$Y(t) = W \frac{e^{Bt}}{A+e^{Bt}}$ where $A \sim U[0,5]$ and $B \sim Exp(10)$; and critical values for the degradation and the shock damage are: $G = 500$ and $S = 200$, respectively. The random shocks function: $D(t) = \sum_{i=1}^{N_2(t)} X_i$ where $X_i \sim Exp(0.3)$ and $X_i's$ are i.i.d. Figure 7.2 shows the reliability of the system using Eq. (7.40) for $\lambda_2 = .12$ and $\lambda_2 = .20$.

Fig. 7.2 Reliability $R_M(t)$ versus time t (Li and Pham, 2005a, b)

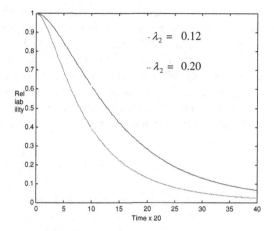

7.4 Inspection-Maintenance Repairable Degraded System Modeling

The system is assumed to be periodically inspected at times $\{I, 2I, \ldots, nI, \ldots\}$ and that the state of the system can only be detected by inspection. After a PM or CM action the system will store it back to as-good-as-new state. Assuming that the degradation $\{Y(t)\}_{t \geq 0}$ and random shock $\{D(t)\}_{t \geq 0}$ are independent, and a CM action is more costly than a PM and a PM costs much more than an inspection. In other words, $C_c > C_p > C_i$.

From Sect. 7.3, T is defined as the time-to-failure $T = \inf\{t > 0 : Y(t) > G \text{ or } D(t) > S\}$ where G is the critical value for $\{Y(t)\}_{t \geq 0}$ and S is the threshold level for $\{D(t)\}_{t \geq 0}$ (Li and Pham, 2005a, b).

The two threshold values L and G (G is fixed) effectively divide the system state into three zones as shown in Fig. 7.3. They are: Doing nothing zone when $Y(t) \leq L$ and $D(t) \leq S$; PM zone when $L < Y(t) \leq G$ and $D(t) \leq S$; and CM zone $Y(t) > G$ or $D(t) > S$. The maintenance action will be performed when either of the following situations occurs:

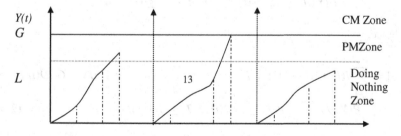

Fig. 7.3 The evolution of the system (Li and Pham, 2005a, b)

The current inspection reveals that the system condition falls into PM zone, however this state is not found at previous inspection. At the inspection time iI, the system falls into PM zone which means $\{Y((i-1)I) \leq L, D((i-1)I) \leq S\}$ $\cap\{L < Y(iI) \leq G, D(iI) \leq S\}$. Then PM action is performed and it will take a random time R_1. When the system fails at T, a CM action is taken immediately and would take a random of time R_2. Note that after a PM or a CM action is performed, the system is renewed and end of the cycle.

The average long-run maintenance cost per unit time can be defined as follows:

$$EC(L, I) = \frac{E[C_1]}{E[W_1]}. \tag{7.41}$$

The expected total maintenance cost during a cycle $E[C_1]$ is defined as:

$$E[C_1] = C_i E[N_I] + C_p E[R_1] P_p + C_c E[R_2] P_c \tag{7.42}$$

Note that there is a probability P_p that the cycle will end by a PM action and it will take on the average $E[R_1]$ amount of times to complete a PM action with a corresponding cost $C_p E[R_1] P_p$. Similarly, if a cycle ends by a CM action with probability P_c, it will take on the average $E[R_2]$ amount of times to complete a CM action with corresponding cost $C_c E[R_2] P_c$. We next discuss the analytical analysis of $E[C_1]$.

Calculate $E[N_I]$.

Let $E[N_I]$ denote the expected number of inspections during a cycle. Then

$$E[N_I] = \sum_{i=1}^{\infty} (i) P\{N_I = i\}$$

where $P\{N_I = i\}$ is the probability that there are a total of i inspections occurred in a renewal cycle. It can be shown that

$$P(N_I = i) = P(Y((i-1)I) \leq L, D((i-1)I) \leq S) \, P(L < Y(iI) \leq G, D(iI) \leq S)$$
$$+ P\{Y(iI) \leq L, D(iI) \leq S\} P\{iI < T \leq (i+1)I) \tag{7.43}$$

Hence,

$$E[N_I] = \sum_{i=1}^{\infty} i(P(Y((i-1)I) \leq L, D((i-1)I) \leq S)P(L < Y(iI) \leq G, D(iI) \leq S)$$
$$+ P(Y(iI) \leq L, D(iI) \leq S)P(iI < T \leq (i+1)I)) \tag{7.44}$$

Assume $Y(t) = A + Bg(t)$ where $A \sim N(\mu_A, \sigma_A^2)$, $B \sim N(\mu_B, \sigma_B^2)$, and A and B are independent. We now calculate the probabilities $P(Y((i-1)I) \le L, D((i-1)I) \le S)$ and $P(L < Y(iI) \le G, D(iI) \le S)$. Given $g(t) = t$.
$D(t) = \sum_{i=0}^{N(t)} X_i$ where $X_i's$ are i.i.d. and $N(t) \sim Possion(\lambda)$.
Then

$$P(Y((i-1)I) \le L, D((i-1)I) \le S)$$

$$= P(A + B(i-1)I \le L)P(D((i-1)I) = \sum^{N(i-1)I} X_i \le S)$$

$$= \Phi\left(\frac{L - (\mu_A + \mu_B(i-1)I)}{\sqrt{\sigma_A^2 + \sigma_B^2((i-1)I)^2}}\right) e^{-\lambda(i-1)I} \sum_{j=0}^{\infty} \frac{(\lambda(i-1)I)^j}{j!} F_X^{(j)}(S)$$

and

$$P(L < Y(iI) \le G, D(iI) \le S)$$

$$= \left\{ \Phi\left(\frac{G - (\mu_A + \mu_B iI)}{\sqrt{\sigma_A^2 + \sigma_B^2(iI)^2}}\right) - \Phi\left(\frac{L - (\mu_A + \mu_B iI)}{\sqrt{\sigma_A^2 + \sigma_B^2(iI)^2}}\right) \right\} e^{-\lambda iI} \sum_{j=0}^{\infty} \frac{(\lambda iI)^j}{j!} F_X^{(j)}(S)$$

Since T is $T = \inf\{t > 0 : Y(t) > G \text{ or } D(t) > S\}$, we have:

$$P(iI < T \le (i+1)I) = P(Y(iI) \le L, Y((i+1)I) > G)P(D((i+1)I) \le S)$$
$$+ P(Y((i+1)I) \le L) \, P(D(iI) \le S, D((i+1)I) > S)$$
$$(7.45)$$

In Eq. (7.45), since $Y(iI)$ and $Y((i+1)I)$ are not independent, we need to obtain the joint p.d.f $f_{Y(iI),Y((i+1)I)}(y_1, y_2)$ in order to compute $P(Y(iI) \le L, Y((i+1)I) > G)$.

Assume that $Y(t) = A + Bg(t)$ where $A > 0$ and $B > 0$ are two independent random variables, $g(t)$ is an increasing function of time t and $A \sim f_A(a), B \sim f_B(b)$. Let

$$\begin{cases} y_1 = a + bg(iI) \\ y_2 = a + bg((i+1)I) \end{cases} \tag{7.46}$$

After simultaneously solving the above equations in terms of y_1 and y_2, we obtain:

$$a = \frac{y_1 g((i+1)I) - y_2 g(iI)}{g((i+1)I) - g(iI)} = h_1(y_1, y_2)$$

$$b = \frac{y_2 - y_1}{g((i+1)I) - g(iI)} = h_2(y_1, y_2)$$

Then the random vector $(Y(iI), Y((i+1)I))$ has a joint continuous p.d.f as follows

$$f_{Y(iI),Y((i+1)I)}(y_1, y_2) = |J| f_A(h_1(y_1, y_2)) f_B(h_2(y_1, y_2)) \qquad (7.47)$$

where the *Jacobian* J is given by

$$J = \begin{vmatrix} \dfrac{\partial h_1}{\partial y_1} & \dfrac{\partial h_1}{\partial y_2} \\[2mm] \dfrac{\partial h_2}{\partial y_1} & \dfrac{\partial h_2}{\partial y_2} \end{vmatrix} = \left| \dfrac{1}{g(iI) - g((i+1)I)} \right|. \qquad (7.48)$$

Note that $D(iI)$ and $D(I_{i+1})$ are independent (Li and Pham, 2005a, b), therefore,

$$P(D(iI) \leq S, D((i+1)I) > S) = P(D(iI) \leq S)P(D((i+1)I) > S). \qquad (7.49)$$

Calculate P_p

Note that either a PM or CM action will end a renewal cycle. In other words, PM and CM these two events are mutually exclusive at renewal time point. As a consequence, $P_p + P_c = 1$. The probability P_p can be obtained as follows:

$$P_p = P(\text{PM ending a cycle})$$

$$= \sum_{i=1}^{\infty} P(Y(i-1)I) \leq L, L < Y(iI) \leq G) P(D(iI) \leq S) \qquad (7.50)$$

Expected Cycle Length Analysis

Since the renewal cycle ends either by a PM action with probability P_p or a CM action with probability P_c, the mean cycle length $E[W_1]$ is calculated as follows:

$$E[W_1] = \sum_{i=1}^{\infty} E[(iI + R_1)I_{PM \text{ occur in } ((i-1)I, iI]}] + E[(T + R_2)1_{CM \text{ occur}}]$$

$$= \left\{ \sum_{i=1}^{\infty} iIP(Y((i-1)I) \leq L, D((i-1)I) \leq S) \right.$$

$$P(L < Y(iI) \leq G, D(iI) \leq S)\} + E[R_1]P_p$$

$$+ (E[T] + E[R_2])P_c \qquad (7.51)$$

where $I_{PM \text{ occurs in}((i-1)I, iI]}$ and $I_{CM \text{ occurs}}$ are the indicator functions. The mean time to failure, $E[T]$ is given by:

$$E[T] = \int_0^\infty P\{T > t\}dt$$

$$= \int_0^\infty P\{Y(t) \le G, D(t) \le S\}dt$$

$$= \int_0^\infty P\{Y(t) \le G\} \sum_{j=0}^\infty \frac{(\lambda_2 t)^j e^{-\lambda_2 t}}{j!} F_X^{(j)}(S)dt \qquad (7.52)$$

or, equivalently, that

$$E[T] = \sum_{j=0}^\infty \frac{F_X^{(j)}(S)}{j!} \int_0^\infty P\{Y(t) \le G\}(\lambda_2 t)^j e^{-\lambda_2 t}dt \qquad (7.53)$$

The expression $E[T]$ would depend on the probability $P\{Y(t) \le G\}$ and sometimes it cannot easy obtain a closed-form.

Optimization Maintenance Cost Rate Policy

We determine the optimal inspection time I and PM threshold L such that the long-run average maintenance cost rate $EC(L, I)$ is minimized. In other words, we wish to minimize the following objective function (Li and Pham, 2005a):

$$EC(L, I)$$

$$= \frac{\sum_{i=1}^\infty iP\{Y(I_{i-1}) \le L, D(I_{i-1}) \le S\}P\{L < Y(I_i) \le G, D(I_i) \le S\}}{\left\{\sum_{i=1}^\infty I_i P\{Y(I_{i-1}) \le L, D(I_{i-1}) \le S\}P\{L < Y(I_i) \le G, D(I_i) \le S\}\right\} + E[R_1]P_p + E[R_2]P_c}$$

$$+ \frac{\sum_{i=1}^\infty iV_i\{P\{Y(I_i) \le L, Y(I_{i+1}) > G\}P\{D(I_{i+1}) \le S\} + P\{Y(I_{i+1}) \le L\}P\{D(I_i) \le S, D(I_{i+1}) > S\}\}}{\left\{\sum_{i=1}^\infty I_i P\{Y(I_{i-1}) \le L, D(I_{i-1}) \le S\}P\{L < Y(I_i) \le G, D(I_i) \le S\}\right\} + E[R_1]P_p + E[R_2]P_c}$$

$$+ \frac{C_p E[R_1] \sum_{i=1}^\infty P\{Y(I_{i-1}) \le L, D(I_{i-1}) \le S\}P\{L < Y(I_i) \le G, D(I_i) \le S\}}{\left\{\sum_{i=1}^\infty I_i P\{Y(I_{i-1}) \le L, D(I_{i-1}) \le S\}P\{L < Y(I_i) \le G, D(I_i) \le S\}\right\} + E[R_1]P_p + E[R_2]P_c}$$

$$+ \frac{C_c E[R_2]\left\{1 - \sum_{i=1}^\infty P\{Y(I_{i-1}) \le L, D(I_{i-1}) \le S\}P\{L < Y(I_i) \le G, D(I_i) \le S\}\right\}}{\left\{\sum_{i=1}^\infty I_i P\{Y(I_{i-1}) \le L, D(I_{i-1}) \le S\}P\{L < Y(I_i) \le G, D(I_i) \le S\}\right\} + E[R_1]P_p + E[R_2]P_c} \qquad (7.54)$$

where $I_{i-1} = (i-1)I, I_i = iI, I_{i+1} = (i+1)I$ and $V_i = P\{Y(iI) \le L, D(iI) \le S\}$.
The above complex objective function is a nonlinear optimization problem. Li and Pham (2005a, b) discussed a step-by-step algorithm based on the Nelder-Mead downhill simplex method.

Example 7.4 Assume that the degradation process is described by $Y(t) = A + Bg(t)$ where A and B are independent, $A \sim U(0, 4)$, $B \sim Exp(-0.3t)$, respectively, and

$g(t) = \sqrt{t}e^{0.005t}$. Assume that the random shock damage is described by $D(t) = \sum_{i=1}^{N(t)} X_i$ where X_i follows the exponential distribution, i.e., $X_i{\sim}Exp(-0.04t)$ and $N(t){\sim}Poisson(0.1)$. Given $G = 50$, $S = 100$, $C_i = 900/\text{inspection}$, $C_c = 5600/CM$, $C_p = 3000/PM$, $R_1{\sim}Exp(-.1t)$, and $R_2{\sim}Exp(-.04t)$. We now determine both the values of I and L that minimizes the average total system cost per unit time $EC(I,L)$.

From Eq. (7.54), the optimal values are $I^* = 37.5$, $L^* = 38$ and the corresponding cost value is $EC^*(I, L) = 440.7$. See Li and Pham (2005a, b) for the detailed calculations.

7.5 Warranty Concepts

A warranty is a contract under which the manufacturers of a product and/or service agree to repair, replace, or provide service when a product fails or the service does not meet intended requirements (Park and Pham, 2010a, 2010b, 2012a, 2012b, 2012c, 2016). These agreements exist because of the uncertainty present in the delivery of products or services, especially in a competitive environment. Warranties are important factors in both the consumers and manufacturers' decision making. A warranty can be the deciding factor on which item a consumer chooses to purchase when different products have similar functions and prices. The length and type of warranty is often thought of as a reflection of the reliability of a product as well as the company's reputation.

Warranty types are dependent on the kind of product that it protects. For larger or more expensive products with many components, it may be cheaper to repair the product rather than replacing it. These items are called repairable products. Other warranties simply replace an entire product because the cost to repair it is either close to or exceeds its original price. These products are considered non-repairable. The following are the most common types used in warranties:

Ordinary Free Replacement - Under this policy, when an item fails before a warranty expires it is replaced at no cost to the consumer. The new item is then covered for the remainder of the warranty length. This is the most common type of a warranty and often applies to cars and kitchen appliances.

Unlimited Free Replacement - This policy is the same as the ordinary free replacement policy but each replacement item carries a new identical warranty. This type of warranty is often used for electronic appliances with high early failure rates and usually has a shorter length because of it.

Pro-rata Warranty - The third most common policy takes into account how much an item is used. If the product fails before the end of the warranty length, then it is replaced at a cost that is discounted proportional to its use. Items that experience wear or aging, such as tires, are often covered under these warranties.

Different warranty models may include a combination of these three types as well as offering other incentives such as rebates, maintenance, or other services that can satisfy a customer and extend the life of their product. Bai and Pham (2006a,2006b,

2004, 2005), Wang and Pham (2006), Pham (2003), Wang and Pham (2010, 2011, 2012), and Murthy and Blischke (2006) can be served as good references of papers and books for further studies on maintenance and warranty topics.

7.6 Problems

1. Suppose that a system is restored to "as good as new" periodically at intervals of time T. So the system renews itself at time T, $2T$, ... Define system reliability under preventive maintenance (PM) as

$$R_M(t) = \text{Pr}\{\text{failure has not occurred by time t}\}$$

   In other words, the system survives to time t if and only if it survives every PM cycle $\{1,2,...,k\}$ and a further time $(t - kT)$.

(a) For a system with a failure time probability density function

$$f(t) = \frac{1}{\beta^2}te^{-\frac{t}{\beta}} \quad \text{for } t > 0, \ \beta > 0$$

   obtain the system reliability under PM and system mean time to first failure.
(b) Calculate the system mean time to first failure if PM is performed every 25 days and $\beta = 2$.

2. Suppose that a system is restored to "as good as new" periodically at intervals of time T. So the system renews itself at time T, $2T$, ... Define system reliability under preventive maintenance (PM) as

$$R_M(t) = \text{Pr}\{\text{failure has not occurred by time t}\}$$

   In other words, the system survives to time t if and only if it survives every PM cycle $\{1,2,...,k\}$ and a further time $(t - kT)$.

(a) For a system with a failure time probability density function

$$f(t) = \frac{1}{6}\lambda^3 t^2 e^{-\lambda t} \quad \text{for } t > 0, \ \lambda > 0$$

   obtain the system reliability under PM and system mean time to first failure.
(b) Calculate the system mean time to first failure if PM is performed every 50 days and $\lambda = 0.035$. (Hints: See Example 7.1)

3. Under an age replacement policy, the system is replaced at time T or at failure whichever occurs first. The costs are respectively c_p and c_f with $c_f > c_p$. For systems with a failure time probability density function

$$f(t) = \frac{1}{6}\lambda^3 t^2 e^{-\lambda t} \quad \text{for } t > 0, \ \lambda > 0$$

obtain the optimal planned replacement time T^* that minimizes the expected total cost per cycle per unit time. Given $c_p = 5$ and $c_f = 25$, what is the optimal planned replacement time T^* that minimizes the expected total cost per cycle per unit time?

References

Bai J, Pham H (2004) Discounted warranty cost for minimally repaired series systems. IEEE Trans Reliab 53(1):37–42

Bai J, Pham H (2005) Repair-limit risk-free warranty policies with imperfect repair. IEEE Trans Syst Man Cybernet (Part A) 35(6):765–772

Bai J, Pham H (2006a) Cost analysis on renewable full-service warranties for multi-component systems. Eur J Oper Res 168(2):492–508

Bai J, Pham H (2006b) Promotional warranty policies: analysis and perspectives. In: Pham H (ed) Springer handbook of engineering statistics. Springer, London

Barlow RE, Proshan F (1965) Mathematical theory of reliability. John Wiley & Sons, New York

Li W, Pham H (2005a) An inspection-maintenance model for systems with multiple competing processes. IEEE Trans Reliab 54(2):318–327

Li W, Pham H (2005b) Reliability modeling of multi-state degraded systems with multiple multi-competing failures and random shocks. IEEE Trans Reliab 54(2):297–303

Murthy DP, Blischke WR (2006) Warranty management and product manufacture. Springer

Park M, Pham H (2010a) Altered quasi-renewal concepts for modeling renewable warranty costs with imperfect repairs. Math Comput Model 52:1435–1450

Park M, Pham H (2010b) Warranty cost analyses using quasi-renewal processes for multi-component systems. IEEE Trans. on Syst Man Cybern Part A 40(6):1329–1340

Park M, Pham H (2012a) A new warranty policy with failure times and warranty servicing times. IEEE Trans Reliab 61(3):822–831

Park M, Pham H (2012b) A generalized block replacement policy for a k-out-of-n system with respect to threshold number of failed components and risk costs. IEEE Trans Syst Man Cybern Part A 42(2):453–463

Park M, Pham H (2012c) Warranty cost analysis for k-out-of-n systems with 2-D warranty. IEEE Trans Syst Man Cybern Part A 42(4):947–957

Park M, Pham H (2016) Cost models for age replacement policies and block replacement policies under warranty. Appl Math Model 40(9):5689–5702

Pham H, Wang HZ (1996) Imperfect maintenance. Eur J Oper Res 94:425–438

Pham H (2003) Handbook of reliability engineering. Springer, London

Wang HZ, Pham H (2006) Availability and maintenance of series systems subject to imperfect repair and correlated failure and repair. Eur J Operat Res

Wang Y, Pham H (2010) A dependent competing risk model with multiple degradation processes and random shocks using constant copulas. In: Proceedings of the 16th international conference on reliability and quality in design. Washington, DC

Wang Y, Pham H (2011) Analyzing the effects of air pollution and mortality by generalized additive models with robust principal components. Int J Syst Assur Eng Manag 2(3)

Wang Y, Pham H (2012) Modeling the dependent competing risks with multiple degradation processes and random shock using time-varying copulas. IEEE Trans Reliab 61(1)

Chapter 8
Statistical Machine Learning and Its Applications

This chapter aims to discuss some ongoing statistical machine learning methods and its applications by first providing a brief review of basic matrix calculations that commonly used in machine learning algorithms and computations. The chapter will then discuss the concept of singular value decomposition (SVD) and its applications of SVD in the recommender systems such as movie review ratings, and finally discuss the linear regression models.

8.1 Introduction to Linear Algebra

8.1.1 Definitions and Matrix Calculations

A matrix A of order $m \times n$ or m by n matrix is a rectangular array of mn elements having m rows and n columns. Denoting by a_{ij}, the element at the (i, j)th position, we can write the matrix A as

$$\begin{bmatrix} a_{11} & a_{12} & \dots & a_{1n} \\ a_{21} & a_{22} & \dots & a_{2n} \\ a_{m1} & a_{m2} & \dots & a_{mn} \end{bmatrix}$$

A matrix having only one row is called a row vector. A matrix having only one column is called a column vector. If the number of rows m and columns n are equal, the matrix is called *square matrix* of order $n \times n$.

If $A = (a_{jk})$ is an $m \times n$ matrix while $B = (b_{jk})$ is an $n \times p$ matrix then we define the product $A \cdot B$ as the matrix $C = (c_{jk})$ where

$$c_{jk} = \sum_{i=1}^{r} a_{ji} b_{ik} \tag{8.1}$$

© The Author(s), under exclusive license to Springer Nature Switzerland AG 2022
H. Pham, *Statistical Reliability Engineering*, Springer Series in Reliability Engineering,
https://doi.org/10.1007/978-3-030-76904-8_8

and where C is of order $m \times p$.

The sum A + B of two matrices is defined only if they have the same number of rows and the same number of columns. The product A · B of two matrices A of order $m \times n$ and B of order $n \times r$ is defined only if the number of columns of A is the same as the number of rows of B. Thus,

- Associative laws: $(A + B) + C = A + (B + C)$; $(A \cdot B)C = A(B \cdot C)$
- Distributive law: $A \cdot (B + C) = A \cdot B + A \cdot C$

In general, $A \cdot B \neq B \cdot A$

Example 8.1 Let

$$A = \begin{pmatrix} 2 & 1 & 4 \\ -3 & 0 & 2 \end{pmatrix}, \quad B = \begin{pmatrix} 3 & 5 \\ 2 & -1 \\ 4 & 2 \end{pmatrix}$$

then

$$A \cdot B = \begin{pmatrix} 24 & 17 \\ -1 & -11 \end{pmatrix}$$

If $\mathbf{A} = (a_{jk})$ is an $m \times n$ matrix, the transpose of $\mathbf{A}$ is denoted by $\mathbf{A}^T$ and $\mathbf{A}^T = (a_{kj})$. Note that

$$(\mathbf{A} + \mathbf{B})^T = \mathbf{A}^T + \mathbf{B}^T$$
$$(\mathbf{A} \cdot \mathbf{B})^T = \mathbf{B}^T \cdot \mathbf{A}^T$$
$$\left(\mathbf{A}^T\right)^T = \mathbf{A} \tag{8.2}$$

Definition 8.1 A square matrix in which all elements of the main diagonal are equal to 1 while all other elements are zero is called the *unit matrix*, and is denoted by **I**. That is,

$$\mathbf{A} \cdot \mathbf{I} = \mathbf{I} \cdot \mathbf{A} = \mathbf{A} \tag{8.3}$$

Definition 8.2 A matrix for which all elements a_{jk} of a matrix **A** are zero for $j \neq k$ is called a *diagonal matrix*.

Definition 8.3 If for a given square matrix **A** there exists a matrix **B** such that $\mathbf{A} \cdot \mathbf{B} = \mathbf{I}$ then **B** is called an *inverse* of **A** and is denoted by $\mathbf{A}^{-1}$.

Definition 8.4 The *determinant*, defined only for a square matrix, is a function of the elements of the matrix. The determinant of $A = ((a_{ij}))$, denoted by $|A|$ or $\det(\mathbf{A})$, is defined as

$$|A| = \sum \pm a_{1i_1} a_{2i_2} \dots a_{ni_n} \tag{8.4}$$

where the summation extends over all the $n!$ permutations i_1, i_2,..., i_n of the integers 1, 2,..., n. The sign before a summand is $+(-)$ if an even (odd) number of interchanges is required to permute 1, 2,..., n into i_1, i_2,..., i_n

A square matrix A such that $\det(\mathbf{A}) = 0$ is called a singular matrix. If $\det(\mathbf{A}) \neq 0$ then **A** is a non-singular matrix.

Every non-singular matrix **A** has an inverse matrix by $\mathbf{A}^{-1}$ such as

$$\mathbf{AA}^{-1} = \mathbf{A}^{-1}\mathbf{A} = \mathbf{I}. \tag{8.5}$$

A minor of a matrix **A** is a matrix obtained from **A** by any one of the following operations: (i) crossing out only certain rows of **A**; (ii) crossing out only certain columns of **A**; (iii) crossing out certain rows and columns of **A**.

Definition 8.5 The rank of a matrix is defined as the largest integer r for which it has a non-singular square minor with r rows.

Definition 8.6 Let v_1, v_2,...,v_n represent row vectors (or column vectors) of a square matrix **A** of order n. Then $\det(\mathbf{A}) = 0$ if and only if there exist constants λ_1, λ_2,...,λ_n, not all zero such that

$$\lambda_1 v_1 + \lambda_2 v_2 + \dots + \lambda_n v_n = 0 \tag{8.6}$$

where **0** is the zero row matrix. If Condition (8.6) is satisfied, we say that the vectors λ_1, λ_2,...,λ_n, are *linear dependent*. Otherwise, they are *linearly independent*.

Definition 8.7 A real matrix **A** is called an *orthogonal matrix* if its transpose is the same as its inverse, i.e., if $\mathbf{A}^T = \mathbf{A}^{-1}$ or $\mathbf{A}^T\mathbf{A} = \mathbf{I}$ where $\mathbf{A}^T$ is the transpose of A and I is the identify matrix.

In other words, an orthogonal matrix is always invertible and $\mathbf{A}^T = \mathbf{A}^{-1}$. In element form,

$$\left(a^{-1}\right)_{ij} = a_{ji}.$$

Example 8.2 The following matrices

$$A = \begin{bmatrix} \frac{1}{\sqrt{2}} & \frac{1}{\sqrt{2}} \\ \frac{1}{\sqrt{2}} & -\frac{1}{\sqrt{2}} \end{bmatrix}$$

and

$$B = \begin{bmatrix} \frac{2}{3} & -\frac{2}{3} & \frac{1}{3} \\ \frac{1}{3} & \frac{2}{3} & \frac{2}{3} \\ \frac{2}{3} & \frac{1}{3} & -\frac{2}{3} \end{bmatrix}$$

are orthogonal matrices, i.e., $A^T A = I$, and $B^T B = I$.

Orthogonal vectors: Two column vectors A and B are called orthogonal if $A^T B = 0$.

The rows of an orthogonal matrix are an orthogonal basis. That is, each row has length one and are mutually perpendicular. Similarly, the columns of an orthogonal matrix are also an orthogonal basis.

Lemma 8.1 *The determinant of an orthogonal matrix A is either 1 or −1.*

If A is orthogonal matrix, then $A^T = A^{-1}$. So, $\det(A^T A) = \det(I) = 1$.
Since $\det(A^T) = \det(A)$, we have $\det(A^T A) = \det(A^T) \cdot \det(A) = (\det(A))^2 = 1$. This implies that $\det(A) = \pm 1$.

Eigenvalues: Let $A = (a_{jk})$ be an $n \times n$ matrix and X a column vector. The equation $AX = \lambda X$ where λ is a number can be written as

$$\begin{pmatrix} a_{11} & a_{12} & \dots & a_{1n} \\ a_{21} & a_{22} & \dots & a_{2n} \\ a_{n1} & a_{n2} & \dots & a_{nn} \end{pmatrix} \begin{pmatrix} x_1 \\ x_2 \\ \dots \\ x_n \end{pmatrix} = \lambda \begin{pmatrix} x_1 \\ x_2 \\ \dots \\ x_n \end{pmatrix} \tag{8.7}$$

or

$$\begin{cases} (a_{11} - \lambda)x_1 + a_{12}x_2 + \dots + a_{1n}x_n = 0 \\ a_{21}x_1 + (a_{22} - \lambda)x_2 + \dots + a_{2n}x_n = 0 \\ \dots \\ a_{n1}x_1 + a_{n2}x_2 + \dots + (a_{nn} - \lambda)x_n = 0 \end{cases} \tag{8.8}$$

The Eq. (8.8) will have non-trivial solution if and only if

$$\begin{vmatrix} a_{11} - \lambda & a_{12} & \dots & a_{1n} \\ a_{21} & a_{22} - \lambda & \dots & a_{2n} \\ a_{n1} & a_{n2} & \dots & a_{nn} - \lambda \end{vmatrix} = 0 \tag{8.9}$$

which is a polynomial equation of degree n in λ. The roots of this polynomial equation are called eigenvalues of the matrix A. Corresponding to each eigenvalue there will be a solution X which is called an eigenvector. The Eq. (8.9) can be written as

$$\det(A - \lambda I) = 0 \tag{8.10}$$

and the equation in λ is called the characteristic equation. Eigenvalues play an important role in situations where the matrix is a transformation from one vector space onto itself.

Example 8.3 Find the eigenvalues and eigenvectors of the matrix A

$$A = \begin{pmatrix} 5 & 7 & -5 \\ 0 & 4 & -1 \\ 2 & 8 & -3 \end{pmatrix}$$

If

$$X = \begin{pmatrix} x_1 \\ x_2 \\ x_3 \end{pmatrix}$$

we obtain the equation $A\mathbf{X} = \lambda\mathbf{X}$. That is,

$$\begin{pmatrix} 5 & 7 & -5 \\ 0 & 4 & -1 \\ 2 & 8 & -3 \end{pmatrix} \begin{bmatrix} x_1 \\ x_2 \\ x_3 \end{bmatrix} = \lambda \begin{bmatrix} x_1 \\ x_2 \\ x_3 \end{bmatrix}$$

We can obtain

$$\begin{pmatrix} 5x_1 + 7x_2 - 5x_3 \\ 4x_2 - x_3 \\ 2x_1 + 8x_2 - 3x_3 \end{pmatrix} = \begin{bmatrix} \lambda x_1 \\ \lambda x_2 \\ \lambda x_3 \end{bmatrix}$$

This implies

$$\begin{cases} (5 - \lambda)x_1 + 7x_2 - 5x_3 = 0 \\ (4 - \lambda)x_2 - x_3 = 0 \\ 2x_1 + 8x_2 - (3 + \lambda)x_3 = 0 \end{cases} \tag{8.11}$$

This system will have non-trivial solutions if

$$\begin{vmatrix} (5 - \lambda) & 7 & -5 \\ & (4 - \lambda) & -1 \\ 2 & 8 & -(3 + \lambda) \end{vmatrix} = 0$$

Expansion of this determinant yields

$$\lambda^3 - 6\lambda^2 + 11\lambda - 6 = 0$$

or

$$(\lambda - 1)(\lambda - 2)(\lambda - 3) = 0.$$

Thus, the eigenvalues are:

$$\lambda = 1; \quad \lambda = 2; \quad \lambda = 3.$$

When $\lambda = 1$, Eq. (8.11) yields

$$\begin{cases} 4x_1 + 7x_2 - 5x_3 = 0 \\ 3x_2 - x_3 = 0 \\ 2x_1 + 8x_2 - 4x_3 = 0 \end{cases}$$

Solving for x_1, x_2 and x_3, we obtain $x_1 = 2\,x_2$, $x_3 = 3\,x_2$. Thus, an eigenvector is

$$\begin{pmatrix} x_1 \\ x_2 \\ x_3 \end{pmatrix} = \begin{pmatrix} 2x_2 \\ x_2 \\ 3x_2 \end{pmatrix} = \begin{pmatrix} 2 \\ 1 \\ 3 \end{pmatrix}$$

Similarly, for $\lambda = 2$, we obtain $x_1 = x_2$, $x_3 = 2\,x_2$. Thus, an eigenvector is

$$\begin{pmatrix} x_1 \\ x_2 \\ x_3 \end{pmatrix} = \begin{pmatrix} x_2 \\ x_2 \\ 2x_2 \end{pmatrix} = \begin{pmatrix} 1 \\ 1 \\ 2 \end{pmatrix}$$

Also, for $\lambda = 3$, we obtain $x_1 = -x_2$, $x_3 = x_2$ then an eigenvector is

$$\begin{pmatrix} x_1 \\ x_2 \\ x_3 \end{pmatrix} = \begin{pmatrix} -x_2 \\ x_2 \\ x_2 \end{pmatrix} = \begin{pmatrix} -1 \\ 1 \\ 1 \end{pmatrix}.$$

The unit eigenvectors have the property that they have length 1, i.e., the sum of the squares of their elements = 1. Thus the above eigenvectors become respectively:

$$\begin{pmatrix} \dfrac{2}{\sqrt{14}} \\ \dfrac{1}{\sqrt{14}} \\ \dfrac{3}{\sqrt{14}} \end{pmatrix}, \quad \begin{pmatrix} \dfrac{1}{\sqrt{6}} \\ \dfrac{1}{\sqrt{6}} \\ \dfrac{2}{\sqrt{6}} \end{pmatrix}, \quad \begin{pmatrix} \dfrac{-1}{\sqrt{3}} \\ \dfrac{1}{\sqrt{3}} \\ \dfrac{1}{\sqrt{3}} \end{pmatrix}$$

8.2 Singular Value Decomposition

The singular value decomposition (SVD) is a factorization of a real or complex matrix. It allows an exact representation of any matrix and can be used to form a subspace spanned by the columns of the matrix. SVD can be used to eliminate the less important data in the matrix to produce a low-dimensional approximation.

Definition 8.8 Let A be an $m \times n$ matrix and r be the rank of A, where $r \leq \min(m,n)$. The singular-value decomposition of A is a way to factor A in the form as

$$A = UDV^T \qquad (8.12)$$

where

U is an $m \times r$ column-orthonormal matrix, which is each of its columns is a unit vector and the dot product of any two columns is 0;
D is a diagonal matrix where the diagonal entries are called the singular values of A;
V^T is an $r \times n$ row-orthonormal matrix which is each of its rows is a unit vector and the dot product of any two columns is 0.

Here,

$$U^T U = I \text{ and } V^T V = I.$$

Let u_i and v_i be the orthonormal column ith of matrix U and column ith of V, respectively. Let $d_1, d_2, ..., d_r$ be the singular values of A where $d_1 \geq d_2 \geq, ..., \geq d_r > 0$. The singular values are the square root of positive eigenvalues. Singular values play an important role where the matrix is a transformation from one vector space to a different vector space, possibly with a different dimension.

Note that matrix U hold all the eigenvectors u_i of AA^T, matrix V hold all the eigenvectors v_i of $A^T A$, and D hold the square roots of all eigenvalues of $A^T A$. Thus, the matrix A can be written as

$$A = \begin{bmatrix} u_1 & u_2 & ... & u_r \end{bmatrix} \begin{bmatrix} d_1 & 0 & ... & 0 \\ 0 & d_2 & ... & 0 \\ 0 & 0 & ... & d_r \end{bmatrix} \begin{pmatrix} v_1^T \\ v_2^T \\ ... \\ v_r^T \end{pmatrix}$$

$$= u_1 d_1 v_1^T + u_2 d_2 v_2^T + ... + u_r d_r v_r^T$$

$$= \sum_{i=1}^{r} u_i d_i v_i^T . \qquad (8.13)$$

Since u_i and v_i have unit length, the most dominant factor in determining the significance of each term is the singular value d_i where $d_1 \geq d_2 \geq, ..., \geq d_r > 0$.

If the eigenvalues become too small, we can ignore the remaining terms

$$\left(+ \sum_{i=j}^{r} u_i d_i v_i^T \right).$$

Note that the singular values are the square root of positive eigenvalues of $A^T A$.

Lemma 8.2 Let $A = UDV^T$ be a singular value decomposition of an $m \times n$ matrix of rank r. Then

(a)

$$AV = UD \tag{8.14}$$

In other words, since V is orthogonal, we have

$$AV = \left(UDV^T \right) V = UD \left(V^T V \right) = UD$$

(b)

$$M = UDV^T = u_1 d_1 v_1^T + u_2 d_2 v_2^T + \ldots + u_r d_r v_r^T = \sum_{i=1}^{r} u_i d_i v_i^T$$

Example 8.4 Given matrix A

$$A = \begin{bmatrix} 3 & 2 & 2 \\ 2 & 3 & -2 \end{bmatrix}$$

We can obtain the following results

$$AA^T = \begin{bmatrix} 17 & 8 \\ 8 & 17 \end{bmatrix}$$

and the eigenvalues and the corresponding unit eigenvectors of matrix AA^T are

$$\lambda_1 = 25 \qquad \lambda_2 = 9$$

$$u_1 = \begin{pmatrix} \dfrac{1}{\sqrt{2}} \\ \dfrac{1}{\sqrt{2}} \end{pmatrix} \quad u_2 = \begin{pmatrix} \dfrac{1}{\sqrt{2}} \\ -\dfrac{1}{\sqrt{2}} \end{pmatrix}.$$

Similarly, we obtain

$$A^T A = \begin{bmatrix} 13 & 12 & 2 \\ 12 & 13 & -2 \\ 2 & -2 & 8 \end{bmatrix}$$

and the eigenvalues and the corresponding unit eigenvectors of matrix $A^T A$ are

$$\lambda_1 = 25 \qquad \lambda_2 = 9 \qquad \lambda_3 = 0$$

$$v_1 = \begin{pmatrix} \frac{1}{\sqrt{2}} \\ \frac{1}{\sqrt{2}} \\ 0 \end{pmatrix} \qquad v_2 = \begin{pmatrix} \frac{1}{\sqrt{18}} \\ -\frac{1}{\sqrt{18}} \\ \frac{4}{\sqrt{18}} \end{pmatrix} \qquad v_2 = \begin{pmatrix} \frac{2}{3} \\ -\frac{2}{3} \\ -\frac{1}{3} \end{pmatrix}$$

The rank of matrix A is: rank(A) = 2.

Note that the singular values are the square root of positive eigenvalues, i.e., 5 and 3. Therefore, the singular value decomposition matrix can be written as

$$A = U D V^T$$

$$= \underbrace{\begin{bmatrix} \frac{1}{\sqrt{2}} & \frac{1}{\sqrt{2}} \\ \frac{1}{\sqrt{2}} & -\frac{1}{\sqrt{2}} \end{bmatrix}}_{\equiv U} \underbrace{\begin{bmatrix} 5 & 0 & 0 \\ 0 & 3 & 0 \end{bmatrix}}_{\equiv D} \underbrace{\begin{bmatrix} \frac{1}{\sqrt{2}} & \frac{1}{\sqrt{2}} & 0 \\ \frac{1}{\sqrt{18}} & -\frac{1}{\sqrt{18}} & \frac{4}{\sqrt{18}} \\ \frac{2}{3} & -\frac{2}{3} & -\frac{1}{3} \end{bmatrix}}_{\equiv V^T}$$

Since 5 and 3 are the two positive singular values, therefore, we can rewrite the singular value decomposition matrix as follows:

$$A = U D V^T$$

$$= \underbrace{\begin{bmatrix} \frac{1}{\sqrt{2}} & \frac{1}{\sqrt{2}} \\ \frac{1}{\sqrt{2}} & -\frac{1}{\sqrt{2}} \end{bmatrix}}_{\equiv U} \underbrace{\begin{bmatrix} 5 & 0 \\ 0 & 3 \end{bmatrix}}_{\equiv D} \underbrace{\begin{bmatrix} \frac{1}{\sqrt{2}} & \frac{1}{\sqrt{2}} & 0 \\ \frac{1}{\sqrt{18}} & -\frac{1}{\sqrt{18}} & \frac{4}{\sqrt{18}} \end{bmatrix}}_{\equiv V^T}$$

Example 8.5 Let $A = \begin{bmatrix} 1 & 1 \\ 1 & 0 \end{bmatrix}$.

Note that A is a symmetric matrix. We can obtain

$$AA^T = A^T A = \begin{bmatrix} 1 & 1 \\ 1 & 0 \end{bmatrix}\begin{bmatrix} 1 & 1 \\ 1 & 0 \end{bmatrix} = \begin{bmatrix} 2 & 1 \\ 1 & 1 \end{bmatrix}$$

Rank(A) = 2. We can obtain the following results:

The two singular values are: $d_1 = 1.618034$ and $d_2 = 0.618034$ and the singular value decomposition matrix is given by:

$$A = UDV^T$$

$$= \begin{bmatrix} 0.850651 & 0.525731 \\ 0.525731 & -0.850651 \end{bmatrix} \begin{bmatrix} 1.618034 & 0 \\ 0 & 0.618034 \end{bmatrix} \begin{bmatrix} 0.850651 & 0.525731 \\ -0.525731 & 0.850651 \end{bmatrix}$$
$$\quad\quad \equiv U \quad\quad\quad\quad \equiv D \quad\quad\quad\quad \equiv V^T$$

Example 8.6 Find the singular values and the singular value decomposition of matrix A given as

$$A = \begin{pmatrix} 5 & 7 & -5 \\ 0 & 4 & -1 \\ 2 & 8 & -3 \end{pmatrix}$$

From Example 8.3, the eigenvalues of the matrix A are: $\lambda = 1$; $\lambda = 2$; $\lambda = 3$. Here, let us find the singular values of matrix A. Follow the same procedures from Example 8.4, we can obtain the following results:

Rank(A) = 3.

The three singular values are:

$$d_1 = 13.546948, \quad d_2 = 3.075623, \quad \text{and } d_3 = 0.144005$$

and the singular value decomposition of matrix A can be written as:

$$A = UDV^T$$

$$\equiv U \quad\quad\quad\quad\quad\quad\quad\quad\quad \equiv D$$
$$= \begin{bmatrix} 0.719021 & -0.660027 & 0.217655 \\ 0.276509 & 0.559003 & 0.781702 \\ 0.637614 & 0.501877 & -0.584438 \end{bmatrix} \begin{bmatrix} 13.546948 & 0 & 0 \\ 0 & 3.075623 & \\ 0 & 0 & 0.144005 \end{bmatrix}$$

$$\begin{bmatrix} 0.359515 & 0.829714 & -0.426993 \\ -0.746639 & 0.530248 & 0.401707 \\ -0.559714 & -0.174390 & -0.810129 \end{bmatrix}$$
$$\equiv V^T$$

8.2.1 Applications

Consider the following ratings of ten customers who provided the rating scores of eight movies as shown in Table 8.1. We now discuss an application of singular value decomposition to illustrate how to apply it in the marketing recommendation strategies based on the users' movie reviewing scores. Table 8.1 shows the ratings of 10 customers to each movie of all 8 movies. For example, the columns in Table

Table 8.1 customer ratings versus movie

User/movie	1	2	3	4	5	6	7	8
John	3	4	3	1	3	0	0	0
Andy	3	4	5	3	4	0	0	0
Cindy	4	3	4	4	5	0	0	0
Dan	5	4	3	4	3	0	0	0
Debbie	3	4	3	4	4	0	0	0
Frank	4	3	3	3	4	0	0	0
Good	0	0	0	0	0	4	4	4
Hanna	0	0	0	0	0	4	3	4
Ing	0	0	0	0	0	5	4	4
Jan	0	0	0	0	0	5	4	4

8.1 are the ratings of each movie. Each row represents the ratings of the movies per customer.

From the definition of the singular-value decomposition, we can obtain the singular-value decomposition of the rating-movie matrix based on the data in Table 8.1 as follows:

$$
\begin{bmatrix}
3\,4\,3\,1\,3\,0\,0\,0 \\
3\,4\,5\,3\,4\,0\,0\,0 \\
4\,3\,4\,4\,5\,0\,0\,0 \\
5\,4\,3\,4\,3\,0\,0\,0 \\
3\,4\,3\,4\,4\,0\,0\,0 \\
4\,3\,3\,3\,4\,0\,0\,0 \\
0\,0\,0\,0\,0\,4\,4\,4 \\
0\,0\,0\,0\,0\,4\,3\,4 \\
0\,0\,0\,0\,0\,5\,4\,4 \\
0\,0\,0\,0\,0\,5\,4\,4
\end{bmatrix}
=
\begin{bmatrix}
0.322182 & 0 \\
0.432183 & 0 \\
0.455123 & 0 \\
0.430322 & 0 \\
0.408184 & 0 \\
0.387930 & 0 \\
0 & 0.485776 \\
0 & 0.448523 \\
0 & 0.530495 \\
0 & 0.530495
\end{bmatrix}
\begin{bmatrix}
19.68259 & 0 \\
0 & 14.21969
\end{bmatrix}
$$

$$
\begin{bmatrix}
0.457839 & 0.452209 & 0.438319 & 0.404268 & 0.479932 & 0 & 0 & 0 \\
0 & 0 & 0 & 0 & 0 & 0.635889 & 0.529732 & 0.561275
\end{bmatrix}
$$

where, from the definition, the matrices U, D and V are,

$$U = \begin{bmatrix} 0.322182 & 0 \\ 0.432183 & 0 \\ 0.455123 & 0 \\ 0.430322 & 0 \\ 0.408184 & 0 \\ 0.387930 & 0 \\ 0 & 0.485776 \\ 0 & 0.448523 \\ 0 & 0.530495 \\ 0 & 0.530495 \end{bmatrix} \qquad D = \begin{bmatrix} 19.68259 & 0 \\ 0 & 14.21969 \end{bmatrix}$$

and

$$V = \begin{bmatrix} 0.457839 & 0 \\ 0.452209 & 0 \\ 0.438319 & 0 \\ 0.404268 & 0 \\ 0.479932 & 0 \\ 0 & 0.635889 \\ 0 & 0.529732 \\ 0 & 0.561275 \end{bmatrix}$$

respectively.

Suppose there is a new customer, say Amanda, who only watches the movie #2 and rate this movie with 3. We can use the results above to recommend other movies to Amanda by calculating the following:

$$(0, 3, 0, 0, 0, 0, 0, 0)VV^{T}$$

More specifically,

$$(0, 3, 0, 0, 0, 0, 0, 0) \, V \, V^{T} = (0, 3, 0, 0, 0, 0, 0, 0) \begin{bmatrix} 0.457839 & 0 \\ 0.452209 & 0 \\ 0.438319 & 0 \\ 0.404268 & 0 \\ 0.479932 & 0 \\ 0 & 0.635889 \\ 0 & 0.529732 \\ 0 & 0.561275 \end{bmatrix}$$

$$\begin{bmatrix} 0.457839 & 0.452209 & 0.438319 & 0.404268 & 0.479932 & 0 & 0 & 0 \\ 0 & 0 & 0 & 0 & 0 & 0.635889 & 0.529732 & 0.561275 \end{bmatrix}$$

$$= \begin{pmatrix} 0.621117 & 0.613478 & 0.594635 & 0.548440 & 0.651088 & 0 & 0 & 0 \end{pmatrix}.$$

The result indicates that Amanda would like movies 1, 2, 3, 4 and 5, but not movie 6, 7, and 8.

8.3 Linear Regression Models

We are often interested in whether there is a correlation between two variables, populations or processes. These correlations may give us information about the underlying processes of the problem under considerations. Given two sets of samples, X and Y, we usually ask the question, "Are the two variables X and Y correlated?" In other words, can Y be modeled as a linear function of X?

We consider the modeling between the dependent Y and one independent variable X. When there is only one independent variable in the linear regression model, the model is called as simple linear regression model. When there are more than one independent vairables in the models, then the linear model is termed as the multiple linear regression model. The linear regression model is applied if we want to model a numeric response variable and its dependency on at least one numeric factor variable. Consider a simple linear regression model

$$y = \alpha + \beta x + \varepsilon \tag{8.15}$$

where

$y = $ response or dependent variable

$x = $ independent or predictor or explanatory variable

$\varepsilon = $ random error, assumed to be normally distributed with mean 0 and standard deviation σ

$\beta = $ slope parameter

$\alpha = $ intercept.

Note that the error term ε represents the difference between the true and observed value of Y. For a given value of X, the mean of the random variable Y equals the deterministic part of the model with

$$\mu_y = E(Y) = \alpha + \beta x \tag{8.16}$$

and

$$\text{Var}(Y) = \sigma^2.$$

The parameters α, β and σ^2 are generally unknown in practice and ε is unobserved. The determination of the linear regression model depends on the estimation of parameters α, β and σ^2.

Assume n pairs of observations (x_i, y_i) for $i = 1, 2, \ldots, n$ are observed and are used to determine the unknown parameters α and β. Various methods of estimation can be used to determine the estimates of the parameters. The methods of least squares and maximum likelihood are the common methods of estimation.

8.3.1 Least Squares Estimation

Suppose a sample of n pairs of observations (x_i, y_i) for $i = 1, 2, \ldots, n$ are available where they are assumed to satisfy the simple linear regression model as follows:

$$y_i = \alpha + \beta x_i + \varepsilon_i \qquad i = 1, 2, \ldots, n \tag{8.17}$$

The principle of least squares estimates the parameters α and β by minimizing the sum of squares of difference between the observations and the line in the scatter diagram. We want to estimate the parameters of the model based on sample data. We will obtain the estimated regression line

$$\hat{y} = \hat{\alpha} + \hat{\beta}x \tag{8.18}$$

The sum of squared deviations for all n data points is

$$SS_e = \sum_{i=1}^{n} \varepsilon_i^2 = \sum_{i=1}^{n} (y_i - \alpha - \beta x_i)^2 \tag{8.19}$$

The derivatives of SS_e with respect to α and β are

$$\frac{\partial SS_e}{\partial \alpha} = -2 \sum_{i=1}^{n} (y_i - \alpha - \beta x_i) \tag{8.20}$$

and

$$\frac{\partial SS_e}{\partial \beta} = -2 \sum_{i=1}^{n} (y_i - \alpha - \beta x_i) x_i \tag{8.21}$$

respectively. The solutions of α and β are obtained by setting

$$\frac{\partial SS_e}{\partial \alpha} = 0 \quad \text{and} \quad \frac{\partial SS_e}{\partial \beta} = 0 \tag{8.22}$$

and the solution of these two equations in (8.22) are called the least squares estimates (LSE) of α and β. Thus, the LSE of α and β are

$$\hat{\alpha} = \overline{y} - \hat{\beta}\overline{x} \tag{8.23}$$

$$\hat{\beta} = \frac{S_{XY}}{S_{XX}} \tag{8.24}$$

where

$$S_{XY} = \sum_{i=1}^{n} (x_i - \overline{x})(y_i - \overline{y})$$

$$S_{XX} = \sum_{i=1}^{n} (x_i - \overline{x})^2$$

$$\overline{x} = \frac{1}{n}\sum_{i=1}^{n} x_i, \qquad \overline{y} = \frac{1}{n}\sum_{i=1}^{n} y_i. \tag{8.25}$$

We also obtain

$$\frac{\partial^2 SS_e}{\partial \alpha^2} = 2n, \qquad \frac{\partial^2 SS_e}{\partial \beta^2} = 2\sum_{i=1}^{n} x_i^2, \qquad \frac{\partial^2 SS_e}{\partial \alpha \, \partial \beta} = 2n\,\overline{x} \tag{8.26}$$

The Hessian matrix which is the matrix of second order partial derivatives is given by

$$H^* = \begin{pmatrix} \frac{\partial^2 SS_e}{\partial \alpha^2} & \frac{\partial^2 SS_e}{\partial \alpha \, \partial \beta} \\ \frac{\partial^2 SS_e}{\partial \alpha \, \partial \beta} & \frac{\partial^2 SS_e}{\partial \beta^2} \end{pmatrix} = 2\begin{pmatrix} n & n\overline{x} \\ n\overline{x} & \sum_{i=1}^{n} x_i^2 \end{pmatrix}$$

The matrix H^* is positive definite if its determinant and the element in the first row and column of H^* are positive. The determinant of H is given by

$$|H^*| = 2\left(n\sum_{i=1}^{n} x_i^2 - n^2 \overline{x}^2 \right) = 2n\sum_{i=1}^{n} (x_i - \overline{x})^2 \geq 0. \tag{8.27}$$

Note that the predicted values are

$$\hat{y}_i = \hat{\alpha} + \hat{\beta}x \qquad \text{for } i = 1, 2, \ldots, n \tag{8.28}$$

The difference between the observed value y_i and the fitted (or predicted) value is called as a residual. The ith residual is defined as

$$e_i = y_i - \hat{y}_i = y_i - \hat{\alpha} - \hat{\beta}x_i \quad \text{for } i = 1, 2, \ldots, n \qquad (8.29)$$

We now determine the unbiased estimators. We have

$$E(\hat{\alpha}) = E\left(\overline{y} - \hat{\beta}\overline{x}\right) = E\left(\alpha + \beta\overline{x} + \overline{\varepsilon} - \hat{\beta}\overline{x}\right) = \alpha + \beta\overline{x} - \beta\overline{x} = \alpha. \qquad (8.30)$$

Thus, $\hat{\alpha}$ is an unbiased estimator of α. Similarly, we can show that $\hat{\beta}$ is an unbiased estimator of β (see Problem 1) where $\hat{\beta}$ is given in Eq. (8.24). That is

$$E\left(\hat{\beta}\right) = E\left(\frac{S_{XY}}{S_{XX}}\right) = \beta. \qquad (8.31)$$

Calculate the Variances.

From Eq. (8.24) and since $\hat{\alpha} = \overline{y} - \hat{\beta}\overline{x}$ and $\hat{\beta} = \frac{S_{XY}}{S_{XX}}$, thus

$$\hat{\beta} = \sum_{i=1}^{n} k_i y_i$$

where $k_i = \frac{x_i - \overline{x}}{S_{XX}}$. The variance of $\hat{\beta}$ is

$$Var\left(\hat{\beta}\right) = \sum_{i=1}^{n} k_i^2 \, Var(y_i) + \sum_{i=1}^{n}\sum_{i \neq j}^{n} k_i k_j Cov\left(y_i, y_j\right)$$

$$= \sigma^2 \frac{\sum_{i=1}^{n} (x_i - \overline{x})^2}{S_{XX}^2} = \sigma^2 \frac{S_{XX}}{S_{XX}^2} = \frac{\sigma^2}{S_{XX}} \qquad (8.32)$$

where $Cov(y_i, y_j) = 0$ as $y_1, y_2, \ldots, y_n$ are independent. The variance of $\hat{\alpha}$ is

$$Var(\hat{\alpha}) = Var(\overline{y}) + \overline{x}^2 Var\left(\hat{\beta}\right) - 2\overline{x}Cov\left(\hat{\beta}, \overline{y}\right) \qquad (8.33)$$

Now,

$$Cov\left(\hat{\beta}, \overline{y}\right) = E\left[\left(\overline{y} - E(\overline{y})\right)\left(\hat{\beta} - E\left(\hat{\beta}\right)\right)\right]$$

$$= E\left[\overline{\varepsilon}\left(\sum_{i=1}^{n} c_i y_i - \beta\right)\right]$$

$$= \frac{1}{n}E\left[\left(\sum_{i=1}^{n} \varepsilon_i\right)\left(\alpha\sum_{i=1}^{n} c_i + \beta\sum_{i=1}^{n} c_i x_i + \sum_{i=1}^{n} c_i\varepsilon_i\right) - \beta\sum_{i=1}^{n} \varepsilon_i\right]$$

$$= 0. \tag{8.34}$$

So,

$$Var(\widehat{\alpha}) = \sigma^2 \left(\frac{1}{n} + \frac{\bar{x}^2}{S_{XX}} \right) \tag{8.35}$$

The covariance between $\widehat{\alpha}$ and $\widehat{\beta}$ is

$$Cov(\widehat{\alpha}, \widehat{\beta}) = Cov(\bar{y}, \widehat{\beta}) - \bar{x}var(\widehat{\beta}) = -\frac{\bar{x}^2}{S_{XX}}\sigma^2. \tag{8.36}$$

Calculate the residual sum of squares.

The residual sum of squares is

$$SS_{res} = \sum_{i=1}^{n} e_i^2 = \sum_{i=1}^{n} (y_i - \widehat{y}_i)^2$$

$$= \sum_{i=1}^{n} \left(y_i - \widehat{\alpha} - \widehat{\beta}x_i \right)^2$$

$$= \sum_{i=1}^{n} \left[(y_i - \bar{y}) - \widehat{\beta}(x_i - \bar{x}) \right]^2$$

$$= \sum_{i=1}^{n} (y_i - \bar{y})^2 + \widehat{\beta}^2 \sum_{i=1}^{n} (x_i - \bar{x})^2 - 2\widehat{\beta} \sum_{i=1}^{n} (x_i - \bar{x})(y_i - \bar{y})$$

$$= S_{YY} + \widehat{\beta}^2 S_{XX}^2 - \widehat{\beta}^2 S_{XX}^2$$

$$= S_{YY} - \widehat{\beta}^2 S_{XX}^2 = S_{YY} - \left(\frac{S_{XY}}{S_{XX}} \right)^2 S_{XX}$$

$$= S_{YY} - \frac{S_{XY}^2}{S_{XX}} = S_{YY} - \widehat{\beta} S_{XY} \tag{8.37}$$

where

$$S_{YY} = \sum_{i=1}^{n} (y_i - \bar{y})^2, \quad \bar{y} = \frac{1}{n} \sum_{i=1}^{n} y_i.$$

The estimators of variances of $\widehat{\alpha}$ and $\widehat{\beta}$ are given by

$$V\widehat{a}r(\widehat{\alpha}) = S^2 \left(\frac{1}{n} + \frac{\bar{x}^2}{S_{XX}} \right) \tag{8.38}$$

where

$$S^2 = \frac{SS_{res}}{n-2}.$$

(8.39)

and

$$V\hat{a}r\left(\hat{\beta}\right) = \frac{S^2}{S_{XX}}$$

(8.40)

It should be noted that $\sum\limits_{i=1}^{n} \left(y_i - \hat{y}\right) = 0$ so $\sum\limits_{i=1}^{n} e_i = 0$ where e_i can be regarded as an estimate of unknown ε_i for $i = 1,2,...,n$. Also, SS_{res} has a chi-square distribution with $(n-2)$ degrees of freedom, i.e., $SS_{res} \sim \chi^2_{n-2}$. Thus,

$$E(SS_{res}) = (n-2)\sigma^2.$$

(8.41)

Hence, $s^2 = \frac{SS_{res}}{n-2}$ is an unbiased estimator of σ^2.

8.3.2 Maximum Likelihood Estimation

We assume that ε_i's for $i = 1,2,..., n$ are independent and identically distributed following a normal distribution $N(0, \sigma^2)$. We now obtain the parameter estimate of the linear regression model:

$$y_i = \alpha + \beta x_i + \varepsilon_i \qquad i = 1, 2, \ldots, n$$

(8.42)

The observation y_i are independently distributed with $N(\alpha+\beta x_i, \sigma^2)$ for all $i = 1,2,..., n$. The likelihood function of the given observations (x_i, y_i) is

$$L(\alpha, \beta) = \prod_{i=1}^{n} \frac{1}{\sqrt{2\pi}\sigma} e^{-\frac{1}{2\sigma^2}(y_i - \alpha - \beta x_i)^2}$$

(8.43)

The $\log L(\alpha,\beta)$ is given by

$$\ln L(\alpha, \beta) = -\frac{n}{2}\ln(2\pi) - \frac{n}{2}\ln(\sigma^2) - \frac{1}{2\sigma^2}\sum_{i=1}^{n}(y_i - \alpha - \beta x_i)^2.$$

(8.44)

The maximum likelihood of α, β, and σ^2 can be obtained by taking a partial differentiation of $\ln L(\alpha, \beta)$ with respect to α, β, and σ^2:

$$\frac{\partial \ln L(\alpha,\beta)}{\partial \alpha} = -\frac{1}{\sigma^2} \sum_{i=1}^{n} (y_i - \alpha - \beta x_i) \equiv 0$$

$$\frac{\partial \ln L(\alpha,\beta)}{\partial \beta} = -\frac{1}{\sigma^2} \sum_{i=1}^{n} (y_i - \alpha - \beta x_i) x_i \equiv 0$$

$$\frac{\partial \ln L(\alpha,\beta)}{\partial \sigma^2} = -\frac{n}{2\sigma^2} + \frac{1}{2\sigma^4} \sum_{i=1}^{n} (y_i - \alpha - \beta x_i)^2 \equiv 0.$$

The solution of the above equations give the MLEs of α, β, and σ^2 as

$$\widetilde{\alpha} = \overline{y} - \widetilde{\beta}\overline{x}$$

$$\widetilde{\beta} = \frac{\sum_{i=1}^{n} (x_i - \overline{x})(y_i - \overline{y})}{\sum_{i=1}^{n} (x_i - \overline{x})^2} = \frac{S_{XY}}{S_{XX}} \tag{8.45}$$

and

$$\widetilde{s}^2 = \frac{\sum_{i=1}^{n} \left(y_i - \widetilde{\alpha} - \widetilde{\beta} x_i\right)^2}{n}.$$

Note that the least squares and maximum likelihood estimates of α, β are identical. The least squares and MLE of σ^2 are different. The LSE of σ^2 is

$$s^2 = \frac{1}{n-2} \sum_{i=1}^{n} (y_i - \overline{y})^2 \tag{8.46}$$

so that it is related to MLE as

$$\widetilde{s}^2 = \frac{n-2}{n} s^2. \tag{8.47}$$

Thus, $\widetilde{\alpha}$ and $\widetilde{\beta}$ are unbiased estimators of α and β where $\widetilde{s}^2$ is a biased estimate of σ^2.

8.4 Hypothesis Testing and Confidence Interval Estimation

8.4.1 For Intercept Coefficient

Case 1: When σ^2 is known

Suppose the null hypothesis under consideration is

$$H_0 : \quad \alpha = \alpha_0$$
$$H_1 : \quad \alpha \neq \alpha_0$$

where σ^2 is known, then using the result that

$$E\left(\hat{\alpha}\right) = \alpha, \quad \text{Var}\left(\hat{\alpha}\right) = \sigma^2\left(\frac{1}{n} + \frac{\bar{x}^2}{S_{xx}}\right) \tag{8.48}$$

where $\hat{\alpha} = \bar{y} - \hat{\beta}\bar{x}$, the following statistic

$$z_0 = \frac{\hat{\alpha} - \alpha_0}{\sqrt{\sigma^2\left(\frac{1}{n} + \frac{\bar{x}^2}{S_{xx}}\right)}} \tag{8.49}$$

has a $N(0,1)$ distribution when H_0 is true. Thus, we reject H_0 if $|z_0| > z_{\frac{\alpha_s}{2}}$ where $z_{\frac{\alpha_s}{2}}$ is the $\alpha_s/2$ percentage point on normal distribution. The $100(1-\alpha_s)\%$ confidence interval for α when σ^2 is known using the z_0 statistic from Eq. (8.49) as follows:

$$P\left(-z_{\frac{\alpha_s}{2}} < z_0 < z_{\frac{\alpha_s}{2}}\right) = 1 - \alpha_s$$

then

$$P\left(-z_{\alpha_s/2} < \frac{\hat{\alpha} - \alpha}{\sqrt{\sigma^2\left(\frac{1}{n} + \frac{\bar{x}^2}{S_{xx}}\right)}} < z_{\alpha_s/2}\right) = 1 - \alpha_s \tag{8.50}$$

After simplifications, we have

$$P\left(\hat{\alpha} - z_{\alpha_s/2}\sqrt{\sigma^2\left(\frac{1}{n} + \frac{\bar{x}^2}{S_{xx}}\right)} \leq \alpha \leq \hat{\alpha} + z_{\alpha_s/2}\sqrt{\sigma^2\left(\frac{1}{n} + \frac{\bar{x}^2}{S_{xx}}\right)}\right) = 1 - \alpha_s \tag{8.51}$$

So the $100(1-\alpha_s)\%$ confidence interval of α is

$$\left[\hat{\alpha} - z_{\alpha_s/2}\sqrt{\sigma^2\left(\frac{1}{n} + \frac{\bar{x}^2}{S_{xx}}\right)}, \hat{\alpha} + z_{\alpha_s/2}\sqrt{\sigma^2\left(\frac{1}{n} + \frac{\bar{x}^2}{S_{xx}}\right)}\right]. \tag{8.52}$$

Case 2: When σ^2 is unknown

When σ^2 is unknown, the following statistic

$$t_0 = \frac{\hat{\alpha} - \alpha_0}{\sqrt{\frac{SS_{res}}{n-2}\left(\frac{1}{n} + \frac{\bar{x}^2}{s_{xx}}\right)}} \tag{8.53}$$

where SS_{res} is in Eq. (8.37), which follows a t-distribution with $(n-2)$ degrees of freedom when H_0 is true. Thus, we reject H_0 if $|t_0| > t_{(n-2),\frac{\alpha_s}{2}}$ where $t_{(n-2),\frac{\alpha_s}{2}}$ is the $\alpha_s/2$ percentage point of the t-distribution with $(n-2)$ degrees of freedom. The $100(1-\alpha_s)\%$ confidence interval for α when σ^2 is unknown is given by:

$$P\left(-t_{(n-2),\frac{\alpha_s}{2}} < t_0 < t_{(n-2),\frac{\alpha_s}{2}}\right) = 1 - \alpha_s$$

then

$$P\left(-t_{(n-2),\alpha_s/2} < \frac{\hat{\alpha} - \alpha}{\sqrt{\frac{SS_{res}}{n-2}\left(\frac{1}{n} + \frac{\bar{x}^2}{s_{xx}}\right)}} < t_{(n-2),\alpha_s/2}\right) = 1 - \alpha_s$$

After simplifications, we have

$$P\left(\hat{\alpha} - t_{(n-2),\alpha_s/2}\sqrt{\frac{SS_{res}}{n-2}\left(\frac{1}{n} + \frac{\bar{x}^2}{s_{xx}}\right)} \le \alpha \le \hat{\alpha} + t_{(n-2),\alpha_s/2}\sqrt{\frac{SS_{res}}{n-2}\left(\frac{1}{n} + \frac{\bar{x}^2}{s_{xx}}\right)}\right)$$
$$= 1 - \alpha_s$$

The $100(1-\alpha_s)\%$ confidence interval of α is

$$\left[\hat{\alpha} - t_{(n-2),\alpha_s/2}\sqrt{\frac{SS_{res}}{n-2}\left(\frac{1}{n} + \frac{\bar{x}^2}{s_{xx}}\right)}, \hat{\alpha} + t_{(n-2),\alpha_s/2}\sqrt{\frac{SS_{res}}{n-2}\left(\frac{1}{n} + \frac{\bar{x}^2}{s_{xx}}\right)}\right]. \tag{8.54}$$

8.4.2 For Slope Coefficient

Case 1: When σ^2 is known

Consider a linear regression model

$$y_i = \alpha + \beta x_i + \varepsilon_i \qquad i = 1, 2, \ldots, n$$

It is assumed that ε_i's are independent and identically distributed and normally distributed with mean 0 and standard deviation σ. We develop a test for the null hypothesis under consideration for the slope coefficient as follows:

$$H_0 : \quad \beta = \beta_0$$
$$H_1 : \quad \beta \neq \beta_0$$

Assume σ^2 to be known, we know that

$$E\left(\widehat{\beta}\right) = \beta, \quad \text{Var}\left(\widehat{\beta}\right) = \frac{\sigma^2}{s_{xx}}$$

and $\widehat{\beta}$ is a linear combination of normally distributed y_i's. So, $\widehat{\beta} \sim N\left(\beta, \frac{\sigma^2}{s_{xx}}\right)$ and the following statistic

$$z_1 = \frac{\widehat{\beta} - \beta_0}{\sqrt{\frac{\sigma^2}{s_{xx}}}} \tag{8.55}$$

which is distributed as $N(0,1)$ when H_0 is true. Thus, we reject H_0 if $|z_1| > z_{\frac{\alpha_s}{2}}$ where $z_{\frac{\alpha_s}{2}}$ is the $\alpha_s/2$ percentage point on normal distribution. The $100(1-\alpha_s)\%$ confidence interval for β when σ^2 is known is as follows:

$$P\left(-z_{\frac{\alpha_s}{2}} < z_1 < z_{\frac{\alpha_s}{2}}\right) = 1 - \alpha_s$$

then

$$P\left(-z_{\alpha_s/2} < \frac{\widehat{\beta} - \beta}{\sqrt{\frac{\sigma^2}{s_{xx}}}} < z_{\alpha_s/2}\right) = 1 - \alpha_s$$

After simplifications, we have

$$P\left(\widehat{\beta} - z_{\alpha_s/2}\sqrt{\frac{\sigma^2}{s_{xx}}} \leq \beta \leq \widehat{\beta} + z_{\alpha_s/2}\sqrt{\frac{\sigma^2}{s_{xx}}}\right) = 1 - \alpha_s$$

So the $100(1-\alpha_s)\%$ confidence interval of β is

$$\left[\widehat{\beta} - z_{\alpha_s/2}\sqrt{\frac{\sigma^2}{s_{xx}}}, \widehat{\beta} + z_{\alpha_s/2}\sqrt{\frac{\sigma^2}{s_{xx}}}\right] \tag{8.56}$$

Case 2: When σ^2 is unknown

When σ^2 is unknown, we know that

$$\frac{SS_{res}}{\sigma^2} \sim \chi^2_{n-2} \text{ and } E\left(\frac{SS_{res}}{n-2}\right) = \sigma^2.$$

The following statistic

$$t_0 = \frac{\widehat{\beta} - \beta}{\sqrt{\frac{\sigma^2}{s_{xx}}}} = \frac{\widehat{\beta} - \beta}{\sqrt{\frac{SS_{res}}{(n-2)s_{xx}}}} \tag{8.57}$$

which follows a t-distribution with $(n-2)$ degrees of freedom when H_0 is true. Thus, we reject H_0 if $|t_0| > t_{(n-2),\frac{\alpha_s}{2}}$. The $100(1-\alpha_s)\%$ confidence interval for β when σ^2 is unknown is given by:

$$P\left(-t_{(n-2),\frac{\alpha_s}{2}} < t_0 < t_{(n-2),\frac{\alpha_s}{2}}\right) = 1 - \alpha_s$$

then

$$P\left(-t_{(n-2),\alpha_s/2} < \frac{\widehat{\beta} - \beta}{\sqrt{\frac{\widehat{\sigma}^2}{s_{xx}}}} < t_{(n-2),\alpha_s/2}\right) = 1 - \alpha_s$$

After simplifications, we have

$$P\left(\widehat{\beta} - t_{(n-2),\alpha_s/2}\sqrt{\frac{\widehat{\sigma}^2}{s_{xx}}} \le \beta \le \widehat{\beta} + t_{(n-2),\alpha_s/2}\sqrt{\frac{\widehat{\sigma}^2}{s_{xx}}}\right) = 1 - \alpha_s$$

The $100(1-\alpha_s)\%$ confidence interval of β is

$$\left[\widehat{\beta} - t_{(n-2),\alpha_s/2}\sqrt{\frac{SS_{res}}{(n-2)s_{xx}}}, \widehat{\beta} + t_{(n-2),\alpha_s/2}\sqrt{\frac{SS_{res}}{(n-2)s_{xx}}}\right] \tag{8.58}$$

8.5   Multiple Regression Models

In many applications where regression analysis is considred, more than one independent variable is needed in the regression model. A regression model that requires more than one indpendent variable is called a multiple regression model. When this model is linear in the coefficients, it is called a multiple linear regerssion model. In general, the dependent variable or response Y may be related to k independent variables, $x_1, x_2, \ldots, x_k$. The model

$$y = \beta_0 + \beta_1 x_1 + \beta_2 x_2 + \ldots + \beta_k x_k + \varepsilon \tag{8.59}$$

is called a multiple linear regression model with k independent variables. The parameters β_j are the regression coefficients for $j = 1, 2, \ldots, k$ and ε is the random error associated with the dependent variable y where $E(\varepsilon) = 0$ and $var(\varepsilon) = \sigma^2$. From Eq. (8.59), we can write the model in terms of the observations as

$$y_i = \beta_0 + \sum_{j=1}^{k} \beta_j x_{ij} + \varepsilon_i \qquad \text{for } i = 1, 2, \ldots, n \tag{8.60}$$

where ε_i are the random errors associated with y_i and that the $\{\varepsilon_i\}$ are uncorrelated random variables.

Similarly to the simple linear regression, the least squares function can be written as

$$L = \sum_{i=1}^{n} \varepsilon_i^2 = \sum_{i=1}^{n} \left(y_i - \beta_0 - \sum_{j=1}^{k} \beta_j x_{ij} \right)^2 \tag{8.61}$$

We can obtain the estimated coefficients β_0, β_j for $j = 1, 2, \ldots, k$ by minimizing the least squares function L with respect to $\beta_0, \beta_1, \beta_2, \ldots, \beta_k$. Differentiating L with respect to $\beta_0, \beta_1, \beta_2, \ldots, \beta_k$ and equating to zero, we generate the set of $(k+1)$ equations for $(k+1)$ unknown regression coefficients as follows:

$$n\hat{\beta}_0 + \hat{\beta}_1 \sum_{i=1}^{n} x_{i1} + \hat{\beta}_2 \sum_{i=1}^{n} x_{i2} + \ldots + \hat{\beta}_k \sum_{i=1}^{n} x_{ik} = \sum_{i=1}^{n} y_i$$

$$\hat{\beta}_0 \sum_{i=1}^{n} x_{i1} + \hat{\beta}_1 \sum_{i=1}^{n} x_{i1}^2 + \hat{\beta}_2 \sum_{i=1}^{n} x_{i1} x_{i2} + \ldots + \hat{\beta}_k \sum_{i=1}^{n} x_{i1} x_{ik} = \sum_{i=1}^{n} x_{i1} y_i$$

$$\ldots$$

$$\hat{\beta}_0 \sum_{i=1}^{n} x_{ik} + \hat{\beta}_1 \sum_{i=1}^{n} x_{ik} x_{i1} + \hat{\beta}_2 \sum_{i=1}^{n} x_{ik} x_{i2} + \ldots + \hat{\beta}_k \sum_{i=1}^{n} x_{ik}^2 = \sum_{i=1}^{n} x_{ik} y_i \tag{8.62}$$

In general, we can write the model in the matrix and vectors form as follows:

$$\underset{\sim}{y} = X\underset{\sim}{\beta} + \underset{\sim}{\varepsilon} \tag{8.63}$$

where

$$\underset{\sim}{y} = \begin{bmatrix} y_1 \\ y_2 \\ \dots \\ y_n \end{bmatrix}, \quad X = \begin{pmatrix} 1 & x_{11} & x_{12} & \dots & x_{1k} \\ 1 & x_{21} & x_{22} & \dots & x_{2k} \\ \dots & & & & \\ 1 & x_{n1} & x_{n2} & \dots & x_{nk} \end{pmatrix}, \quad \underset{\sim}{\beta} = \begin{bmatrix} \beta_0 \\ \beta_1 \\ \dots \\ \beta_k \end{bmatrix}, \quad \underset{\sim}{\varepsilon} = \begin{bmatrix} \varepsilon_1 \\ \varepsilon_2 \\ \dots \\ \varepsilon_n \end{bmatrix}$$

where

$\underset{\sim}{y}$ is an $(n \times 1)$ vector of the n observations,

X is an $nx(k+1)$ matrix based on the number of unknown coefficients and observations.

$\underset{\sim}{\beta}$ is a $(k+1) \times 1$ vector, and

$\underset{\sim}{\varepsilon}$ is an $(n \times 1)$ vector of random errors.

We can find the vector of unknown coefficients $\hat{\underset{\sim}{\beta}}$ using the least squares estimated method that minimizes

$$L = \sum_{i=1}^{n} \varepsilon_i^2 = \underset{\sim}{\varepsilon}'\underset{\sim}{\varepsilon} = \left(\underset{\sim}{y} - X\underset{\sim}{\beta} \right)' \left(\underset{\sim}{y} - X\underset{\sim}{\beta} \right). \tag{8.64}$$

or equivalently,

$$L = \underset{\sim}{y}'\underset{\sim}{y} - 2\underset{\sim}{\beta}'X'\underset{\sim}{y} + \underset{\sim}{\beta}'X'X\underset{\sim}{\beta} \tag{8.65}$$

Differentiating L with respect to $\underset{\sim}{\beta}$ and equating them equal zero, we can obtain the least squares estimators of $\underset{\sim}{\beta}$ called $\hat{\underset{\sim}{\beta}}$, as follows:

$$\left. \frac{\partial L}{\partial \underset{\sim}{\beta}} \right|_{\hat{\underset{\sim}{\beta}}} = -2X'\underset{\sim}{y} + 2X'X\hat{\underset{\sim}{\beta}} = 0 \tag{8.66}$$

From Eq. (8.66), we obtain

$$X'X\hat{\underset{\sim}{\beta}} = X'\underset{\sim}{y} \tag{8.67}$$

or, equivalently, the least squares estimator of $\underset{\sim}{\beta}$ is

$$\hat{\underset{\sim}{\beta}} = \left(X'X\right)^{-1}X'\underset{\sim}{y} \tag{8.68}$$

It is easy to write out the system of equations in Eq. (8.67) as follows:

$$
\begin{bmatrix}
n & \sum\limits_{i=1}^{n} x_{i1} & \sum\limits_{i=1}^{n} x_{i2} & \cdots & \sum\limits_{i=1}^{n} x_{ik} \\
\sum\limits_{i=1}^{n} x_{i1} & \sum\limits_{i=1}^{n} x_{i1}^2 & \sum\limits_{i=1}^{n} x_{i1}x_{i2} & \cdots & \sum\limits_{i=1}^{n} x_{i1}x_{ik} \\
\cdots\cdots\cdots & & & & \\
\sum\limits_{i=1}^{n} x_{ik} & \sum\limits_{i=1}^{n} x_{ik}x_{i1} & \sum\limits_{i=1}^{n} x_{ik}x_{i2} & \cdots & \sum\limits_{i=1}^{n} x_{ik}^2
\end{bmatrix}
\begin{bmatrix}
\hat{\beta}_0 \\
\hat{\beta}_1 \\
\cdots \\
\hat{\beta}_k
\end{bmatrix}
=
\begin{bmatrix}
\sum\limits_{i=1}^{n} y_i \\
\sum\limits_{i=1}^{n} x_{i1}y_i \\
\cdots \\
\sum\limits_{i=1}^{n} x_{ik}y_i
\end{bmatrix}
\tag{8.69}
$$

We can obtain the following results:

1. 1. $\hat{\underset{\sim}{\beta}}$ is an unbiased estimator of $\underset{\sim}{\beta}$, that is,

$$E\left(\hat{\underset{\sim}{\beta}}\right) = \underset{\sim}{\beta}$$

2. 2. The sum of squares of the residuals (SS_e) is

$$
SS_e = \sum_{i=1}^{n} \left(y_i - \hat{y}_i\right)^2 = \sum_{i=1}^{n} \varepsilon_i^2 = \underset{\sim}{\varepsilon}'\underset{\sim}{\varepsilon}
$$
$$
= \underset{\sim}{y}'\underset{\sim}{y} - \hat{\underset{\sim}{\beta}}'X'\underset{\sim}{y} \tag{8.70}
$$

3. The covariance matrix of $\hat{\underset{\sim}{\beta}}$ is

$$\mathrm{Cov}\left(\hat{\underset{\sim}{\beta}}\right) = \left(X'X\right)^{-1}\sigma^2 \tag{8.71}$$

4. The mean square error (MSE) is

$$MSE = \frac{SS_e}{n-k-1} \tag{8.72}$$

8.5.1 Applications

In this section we illustrate an application of advertising budget products using the multiple regression. In this application, we use the advertising budget data set [Advertising] for modeling the multiple regression where the sales for a particular product is a dependent variable (Y) of multiple regression and the three different media channels such as TV (X_1), Radio (X_2), and Newspaper (X_3) are independent variables as shown in Table 8.2. The advertising dataset consists of the sales of a product in 200 different markets (200 rows), together with advertising budgets for the product in each of those markets for three different media channels: TV, Radio and Newspaper. The sales are in thousands of units and the budget is in thousands of dollars. Table 8.2 below shows a small portion of the advertising budget data set. Table 8.5 in the appendix shows the entire advertising budget data set. Figures 3.6 and 3.7 (in Chap. 3) present the data plot and the correction coefficients between the pairs of variables of the advertising budget data, respectively. It shows that the pair of Sales and TV advertising have the highest positive correlation.

We will fit the multiple linear regression model

$$y = \beta_0 + \beta_1 x_1 + \beta_2 x_2 + \beta_3 x_3 + \varepsilon$$

to this advertising data as shown in Table 8.2 where the variables X_1, X_2, X_3, and Y represent the TV, Radio, Newspaper advertising budgets, and the sales for a particular product. From Table 8.3, the estimated regression model is

$$y = 2.93889 + 0.04577 \, x_1 + 0.18853 \, x_2 - 0.00104 \, x_3$$

Table 8.2 Advertising budget data in 200 different markets

| Market | TV | Radio | Newspaper | Sales |
No.	(X_1)	(X_2)	(X_3)	(Y)
1	230.1	37.8	69.2	22.1
2	44.5	39.3	45.1	10.4
3	17.2	45.9	69.3	9.3
4	151.5	41.3	58.5	18.5
5	180.8	10.8	58.4	12.9
6	8.7	48.9	75.0	7.2
...	...	...	...	...
197	94.2	4.9	8.1	9.7
198	177.0	9.3	6.4	12.8
199	283.6	42.0	66.2	25.5
200	232.1	8.6	8.7	13.4

Table 8.3 Coefficient results of the multiple regression model

Coefficients:	Estimate	Std. Error	t value	Pr(>|t|)
(Intercept)	2.93889	0.311908	9.422	2e−16 ***
TV	0.04577	0.001395	32.809	2e−16 ***
Radio	0.18853	0.008611	21.893	2e−16 ***
Newspaper	−0.00104	0.005871	−0.177	0.86

Residual standard error: 1.686 on 196 degrees of freedom
Multiple R-squared: 0.8972, Adjusted R-squared: 0.8956
F-statistic: 570.3 on 3 and 196 DF, p-value: 2.2e−16

From Table 8.3, it seems that there is no linear relationship between the newspaper budget (X_3) and the sales (Y) for the product. This implies that we can select the model with just two variables (TV and Radio) in the regression. Figures 8.1, 8.2, and 8.3 show the confidence intervals about the regression line of the independent variables TV, Radio, and Newspaper respectively and the Sales.

The final estimated regression model, as shown in Table 8.4, is given by

$$y = 2.92110 + 0.04575 \, x_1 + 0.18799 \, x_2$$

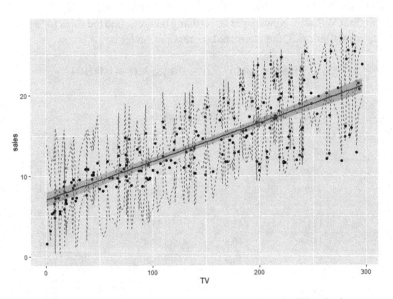

Fig. 8.1   The confidence interval about the regression line between the TV and sales

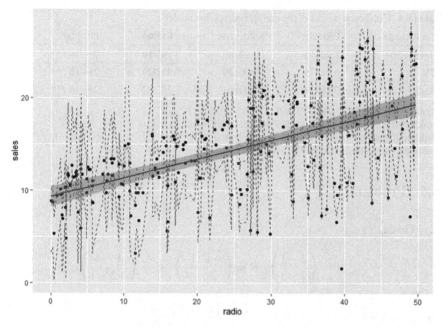

Fig. 8.2 The confidence interval about the regression line between the radio and sales

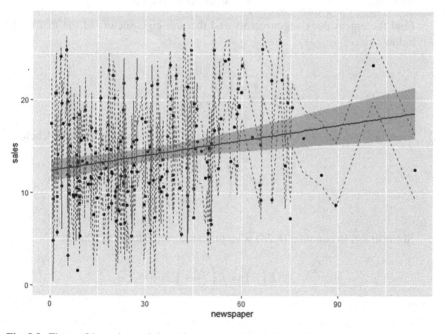

Fig. 8.3 The confidence interval about the regression line between the newspaper and sales

Table 8.4 Coefficient results of the multiple regression model

Coefficients	Estimate	Std. error	t value	Pr(>ltl)
(Intercept)	2.92110	0.29449	9.919	2e−16 ***
TV	0.04575	0.00139	32.909	2e−16 ***
Radio	0.18799	0.00804	23.382	2e−16 ***

Signif. codes: 0 '***' 0.001 '**' 0.01 '*' 0.05 '.' 0.1 ' ' 1
Residual standard error: 1.681 on 197 degrees of freedom
Multiple R-squared: 0.8972, Adjusted R-squared: 0.8962
F-statistic: 859.6 on 2 and 197 DF, p-value: <2.2e−16

8.6 Problems

1. Given

$$A = \begin{pmatrix} 2 & -1 \\ 4 & 3 \end{pmatrix} \quad B = \begin{pmatrix} -1 & 1 \\ 2 & -4 \end{pmatrix} \quad C = \begin{pmatrix} 1 & 4 \\ -2 & -1 \end{pmatrix}$$

using the associative law show that $(A \cdot B)C = A(B \cdot C)$.

2. Show that $A = \begin{pmatrix} \cos\theta & -\sin\theta \\ \sin\theta & \cos\theta \end{pmatrix}$ is an orthogonal matrix.

3. Find the eigenvalues, eigenvectors and the unit eigenvectors of the matrix A below

$$A = \begin{pmatrix} 2 & 0 & -2 \\ 0 & 4 & 0 \\ -2 & 0 & 5 \end{pmatrix}$$

4. Find the singular values and the singular value decomposition of matrix A below

$$A = \begin{pmatrix} 2 & 0 & -2 \\ 0 & 4 & 0 \\ -2 & 0 & 5 \end{pmatrix}$$

5. Show that $\widehat{\beta}$ in equation (8.24) is an unbiased estimator of β. That is

$$E\left(\widehat{\beta}\right) = E\left(\frac{S_{XY}}{S_{XX}}\right) = \beta.$$

Appendix

Table 8.5 shows the entire advertising budget data set with 200 different markets.

Table 8.5 Advertising budget data in 200 different markets

Market	TV	Radio	Newspaper	Sales
No.	(X_1)	(X_2)	(X_3)	(Y)
1	230.1	37.8	69.2	22.1
2	44.5	39.3	45.1	10.4
3	17.2	45.9	69.3	9.3
4	151.5	41.3	58.5	18.5
5	180.8	10.8	58.4	12.9
6	8.7	48.9	75.0	7.2
7	57.5	32.8	23.5	11.8
8	120.2	19.6	11.6	13.2
9	8.6	2.1	1.0	4.8
10	199.8	2.6	21.2	10.6
11	66.1	5.8	24.2	8.6
12	214.7	24.0	4.0	17.4
13	23.8	35.1	65.9	9.2
14	97.5	7.6	7.2	9.7
15	204.1	32.9	46.0	19.0
16	195.4	47.7	52.9	22.4
17	67.8	36.6	114.0	12.5
18	281.4	39.6	55.8	24.4
19	69.2	20.5	18.3	11.3
20	147.3	23.9	19.1	14.6
21	218.4	27.7	53.4	18.0
22	237.4	5.1	23.5	12.5
23	13.2	15.9	49.6	5.6
24	228.3	16.9	26.2	15.5
25	62.3	12.6	18.3	9.7

(continued)

Table 8.5 (continued)

Market	TV	Radio	Newspaper	Sales
No.	(X_1)	(X_2)	(X_3)	(Y)
26	262.9	3.5	19.5	12.0
27	142.9	29.3	12.6	15.0
28	240.1	16.7	22.9	15.9
29	248.8	27.1	22.9	18.9
30	70.6	16.0	40.8	10.5
31	292.9	28.3	43.2	21.4
32	112.9	17.4	38.6	11.9
33	97.2	1.5	30.0	9.6
34	265.6	20.0	0.3	17.4
35	95.7	1.4	7.4	9.5
36	290.7	4.1	8.5	12.8
37	266.9	43.8	5.0	25.4
38	74.7	49.4	45.7	14.7
39	43.1	26.7	35.1	10.1
40	228.0	37.7	32.0	21.5
41	202.5	22.3	31.6	16.6
42	177.0	33.4	38.7	17.1
43	293.6	27.7	1.8	20.7
44	206.9	8.4	26.4	12.9
45	25.1	25.7	43.3	8.5
46	175.1	22.5	31.5	14.9
47	89.7	9.9	35.7	10.6
48	239.9	41.5	18.5	23.2
49	227.2	15.8	49.9	14.8
50	66.9	11.7	36.8	9.7
51	199.8	3.1	34.6	11.4
52	100.4	9.6	3.6	10.7
53	216.4	41.7	39.6	22.6
54	182.6	46.2	58.7	21.2
55	262.7	28.8	15.9	20.2
56	198.9	49.4	60.0	23.7
57	7.3	28.1	41.4	5.5

(continued)

Table 8.5 (continued)

Market	TV	Radio	Newspaper	Sales
No.	(X_1)	(X_2)	(X_3)	(Y)
58	136.2	19.2	16.6	13.2
59	210.8	49.6	37.7	23.8
60	210.7	29.5	9.3	18.4
61	53.5	2.0	21.4	8.1
62	261.3	42.7	54.7	24.2
63	239.3	15.5	27.3	15.7
64	102.7	29.6	8.4	14.0
65	131.1	42.8	28.9	18.0
66	69.0	9.3	0.9	9.3
67	31.5	24.6	2.2	9.5
68	139.3	14.5	10.2	13.4
69	237.4	27.5	11.0	18.9
70	216.8	43.9	27.2	22.3
71	199.1	30.6	38.7	18.3
72	109.8	14.3	31.7	12.4
73	26.8	33.0	19.3	8.8
74	129.4	5.7	31.3	11.0
75	213.4	24.6	13.1	17.0
76	16.9	43.7	89.4	8.7
77	27.5	1.6	20.7	6.9
78	120.5	28.5	14.2	14.2
79	5.4	29.9	9.4	5.3
80	116.0	7.7	23.1	11.0
81	76.4	26.7	22.3	11.8
82	239.8	4.1	36.9	12.3
83	75.3	20.3	32.5	11.3
84	68.4	44.5	35.6	13.6
85	213.5	43.0	33.8	21.7
86	193.2	18.4	65.7	15.2
87	76.3	27.5	16.0	12.0
88	110.7	40.6	63.2	16.0
89	88.3	25.5	73.4	12.9

(continued)

Table 8.5 (continued)

Market	TV	Radio	Newspaper	Sales
No.	(X_1)	(X_2)	(X_3)	(Y)
90	109.8	47.8	51.4	16.7
91	134.3	4.9	9.3	11.2
92	28.6	1.5	33.0	7.3
93	217.7	33.5	59.0	19.4
94	250.9	36.5	72.3	22.2
95	107.4	14.0	10.9	11.5
96	163.3	31.6	52.9	16.9
97	197.6	3.5	5.9	11.7
98	184.9	21.0	22.0	15.5
99	289.7	42.3	51.2	25.4
100	135.2	41.7	45.9	17.2
101	222.4	4.3	49.8	11.7
102	296.4	36.3	100.9	23.8
103	280.2	10.1	21.4	14.8
104	187.9	17.2	17.9	14.7
105	238.2	34.3	5.3	20.7
106	137.9	46.4	59.0	19.2
107	25.0	11.0	29.7	7.2
108	90.4	0.3	23.2	8.7
109	13.1	0.4	25.6	5.3
110	255.4	26.9	5.5	19.8
111	225.8	8.2	56.5	13.4
112	241.7	38.0	23.2	21.8
113	175.7	15.4	2.4	14.1
114	209.6	20.6	10.7	15.9
115	78.2	46.8	34.5	14.6
116	75.1	35.0	52.7	12.6
117	139.2	14.3	25.6	12.2
118	76.4	0.8	14.8	9.4
119	125.7	36.9	79.2	15.9
120	19.4	16.0	22.3	6.6
121	141.3	26.8	46.2	15.5
122	18.8	21.7	50.4	7.0

(continued)

Table 8.5 (continued)

Market	TV	Radio	Newspaper	Sales
No.	(X_1)	(X_2)	(X_3)	(Y)
123	224.0	2.4	15.6	11.6
124	123.1	4.6	12.4	15.2
125	229.5	32.3	74.2	19.7
126	87.2	11.8	25.9	10.6
127	7.8	38.9	50.6	6.6
128	80.2	0.0	9.2	8.8
129	220.3	49.0	3.2	24.7
130	59.6	12.0	43.1	9.7
131	0.7	39.6	8.7	1.6
132	265.2	2.9	43.0	12.7
133	8.4	27.2	2.1	5.7
134	219.8	33.5	45.1	19.6
135	36.9	38.6	65.6	10.8
136	48.3	47.0	8.5	11.6
137	25.6	39.0	9.3	9.5
138	273.7	28.9	59.7	20.8
139	43.0	25.9	20.5	9.6
140	184.9	43.9	1.7	20.7
141	73.4	17.0	12.9	10.9
142	193.7	35.4	75.6	19.2
143	220.5	33.2	37.9	20.1
144	104.6	5.7	34.4	10.4
145	96.2	14.8	38.9	11.4
146	140.3	1.9	9.0	10.3
147	240.1	7.3	8.7	13.2
148	243.2	49.0	44.3	25.4
149	38.0	40.3	11.9	10.9
150	44.7	25.8	20.6	10.1
151	280.7	13.9	37.0	16.1
152	121.0	8.4	48.7	11.6
153	197.6	23.3	14.2	16.6
154	171.3	39.7	37.7	19.0
155	187.8	21.1	9.5	15.6
156	4.1	11.6	5.7	3.2
157	93.9	43.5	50.5	15.3
158	149.8	1.3	24.3	10.1

(continued)

Table 8.5 (continued)

Market	TV	Radio	Newspaper	Sales
No.	(X_1)	(X_2)	(X_3)	(Y)
159	11.7	36.9	45.2	7.3
160	131.7	18.4	34.6	12.9
161	172.5	18.1	30.7	14.4
162	85.7	35.8	49.3	13.3
163	188.4	18.1	25.6	14.9
164	163.5	36.8	7.4	18.0
165	117.2	14.7	5.4	11.9
166	234.5	3.4	84.8	11.9
167	17.9	37.6	21.6	8.0
168	206.8	5.2	19.4	12.2
169	215.4	23.6	57.6	17.1
170	284.3	10.6	6.4	15.0
171	50.0	11.6	18.4	8.4
172	164.5	20.9	47.4	14.5
173	19.6	20.1	17.0	7.6
174	168.4	7.1	12.8	11.7
175	222.4	3.4	13.1	11.5
176	276.9	48.9	41.8	27.0
177	248.4	30.2	20.3	20.2
178	170.2	7.8	35.2	11.7
179	276.7	2.3	23.7	11.8
180	165.6	10.0	17.6	12.6
181	156.6	2.6	8.3	10.5
182	218.5	5.4	27.4	12.2
183	56.2	5.7	29.7	8.7
184	287.6	43.0	71.8	26.2
185	253.8	21.3	30.0	17.6
186	205.0	45.1	19.6	22.6
187	139.5	2.1	26.6	10.3
188	191.1	28.7	18.2	17.3
189	286.0	13.9	3.7	15.9
190	18.7	12.1	23.4	6.7
191	39.5	41.1	5.8	10.8
192	75.5	10.8	6.0	9.9

(continued)

Table 8.5 (continued)

Market No.	TV (X_1)	Radio (X_2)	Newspaper (X_3)	Sales (Y)
193	17.2	4.1	31.6	5.9
194	166.8	42.0	3.6	19.6
195	149.7	35.6	6.0	17.3
196	38.2	3.7	13.8	7.6
197	94.2	4.9	8.1	9.7
198	177.0	9.3	6.4	12.8
199	283.6	42.0	66.2	25.5
200	232.1	8.6	8.7	13.4

Solutions to Selected Problems

Chapter 1

1. (i) $P\{1 \text{ boy}\} = \sum_{i=0}^{3} P\{1 \text{ boy}|i\}P\{i\} = 0.3938.$

    (ii) Let A denote the event that there are two children in the family and B the event that the family has one boy.

    $$P\{A|B\} = \frac{P\{A \cap B\}}{P\{B\}} = \frac{(0.5)(0.25)}{0.3938} = 0.317$$

2. We obtain $P\{A \cap B \cap C\} = \frac{1}{12}$ $P\{A\}P\{B\}P\{C\} = \frac{1}{24}.$
    Events A, B and C are not independent.

3. The probability of the event E that both units selected will be of the same category is

    $$P\{E\} = \frac{13}{119} + \frac{19}{714} + \frac{59}{238} = \frac{274}{714} = 0.383$$

4. n = 7 trials.

5. It is straightforward to show

    $$P\{A \cup B \cup C\} = P\{A\} + P\{B\} + P\{C\} - P\{A \cap B\}$$
    $$- P\{A \cap C\} - P\{B \cap C\} + P\{A \cap B \cap C\}$$

6. Let A, B and C be subsets of the sample space S defined by
    S = {1,2,3,4,5,6} A = {1,3,5}; B = {1,2,3}; C = {2,4,6}.
    Then (i) $A \cap (B \cap C) = \phi$
    (ii) $A \cup (B \cup C) = S$

7. $\frac{63}{64}$

8. (i) 0.495 (ii) 0.36 (iii) 0.115 (iv) 0.74

© The Editor(s) (if applicable) and The Author(s), under exclusive license 465
to Springer Nature Switzerland AG 2022
H. Pham, *Statistical Reliability Engineering*, Springer Series in Reliability Engineering,
https://doi.org/10.1007/978-3-030-76904-8

9. The probability of obtaining at least two heads given that the first toss resulted in a head is: 0.75

10. $p = \dfrac{\binom{100}{50}\binom{100}{50}}{\binom{200}{100}} = 0.1132$

16. (a) $\lambda = 0.00031$
    (b) P(T >800 | T >600)=0.9399
    (c) $P(T \le 900 \mid T \ge 720) = 0.0543$

17. (a) $a = \frac{b}{2}$.
    (b) $R(t = 6000) = 0.1429$.
    (c) The warranty period should be no more than 111.1 h.

18. (a) $R(t) = (t+1)e^{-t}$
    (b) The system MTTF is $MTTF = 2$.

19. (a) $\lambda \le 0.0342$.
    (b) $P(T \le 3 | T \ge 1.5) = 0.0015$.

20. (a) $a = \frac{6}{b^3}$.
    (b) $P(k < T < l) = \frac{l-k}{b^3}[k(3b - 2k - l) + l(3b - 2l - k)]$ for $0 \le k \le l \le b$.
    (c) $MTTF = \frac{b}{2}$.
    (d) $P(T > t + s | T > t) = \dfrac{1 - 3\left(\frac{t+s}{b}\right)^2 + 2\left(\frac{t+s}{b}\right)^3}{1 - 3\left(\frac{t}{b}\right)^2 + 2\left(\frac{t}{b}\right)^3}$ for $s + t \le b$.

21. $\int\limits_0^\infty R_1(t)dt \ge \int\limits_0^\infty R_2(t)dt$.

23. (a) $Mean = 13.863$, Variance $= 7.457$
    (b) $P(T \le 2 | T > 1) = \frac{F(2) - F(1)}{1 - F(1)} = 0.5568$

24. The failure rate function of this communication device is

$$h(t) = \begin{cases} 0 & t \le t_0 \\ \dfrac{4(t-t_0)}{(t_1 - t_0)^2 - 2(t - t_0)^2} & t_0 \le t \le \frac{t_0 + t_1}{2} \\ \dfrac{2}{(t_1 - t)} & \frac{t_0 + t_1}{2} \le t \le t_1 \\ \text{undefined} & t_1 \le t \end{cases}$$

26. $f(t) = \frac{1}{3}e^{-t} + 4e^t - \frac{7}{3}e^{2t}$.

27. (a) True (b) True (c) False (d) True (e) False.

28. $P(7, 4) = \frac{7!}{(7-4)!} = 840$ ways

29. $\Pr(\text{Mr.Best second}) = \frac{(n-1)!}{n!} = \frac{1}{n}$.

30. (a) 210 ways (b) 329 ways (c) 301 ways (d) 120 ways (e) 108.

31. 226,800 ways.

32. a. The reliability function, R(t):

$$R(t) = \left(1 - \frac{t}{a}\right)^3.$$

b. The failure rate function, h (t), is given by

$$h(t) = \frac{3}{a - t}.$$

c. The expected life of the component, MTTF, is given by

$$MTTF = \frac{a}{4}.$$

33. (a) $F(1) = 0.0902$
    (b) The probability of the motor's lasting at least 3 years: $R(t = 3) = 0.5578$.
    (c) The maximum number of days for which the motor can be warranted is 195.3 days.
34. 2025.
35. (a) $a = 0.602$
    (b) $a = 166.7$ gallons.
36. The expected value of the weekly costs is: 1.6975.
37. (a) 0.3996 (b) 0.594 (c) 2.
38. The expected value of the smallest of the three claims is 1125.
39. The expected value of the smallest of the five claims is 1071.4.
40. The probability that the system functions for at least 5 months is 0.2873.
41. The expected value of the smallest of the four claims is $83.333
42. (a) The expected cost of the project is: $E(C(Y)) = 30,000\left(\frac{1}{p}\right) - 5,000$
    (b) 0.01.
43. 0.084
44. (a) 0.81141 (b) $375.
45. The expected completion time is: 1.213
46. 3
47. (a) Calculate the expected length of the meeting, $E(X)$. (b) 48.333

Chapter 2

10. By induction, we can conclude the pdf of Erlang distribution is $f(t) = \frac{\beta^n t^{n-1}}{(n-1)!} e^{-t\beta}$
12. $Var(X) = \frac{1}{\lambda^2}\left(1 - 2\lambda a e^{-\lambda a} - e^{-2\lambda a}\right)$.
13. Upon differentiation the the failure rate h(t) and set $h'(t_0) = 0$, we obtain

$$t_0 = \left(\frac{1 - \alpha}{\alpha \ln a}\right)^{\frac{1}{\alpha}}.$$

and it can be shown that the function h(t) is initially decreasing and then increasing in t.
14. Upon differentiation of the failure rate function, we obtain

$$h'(t) = \alpha \ln a \left(t^{\alpha-2} a^{t^{\alpha}} \right) \left[(\alpha - 1) + \alpha t^{\alpha} \ln a \right] \geq 0 \text{ for } \alpha \geq 1.$$

Thus, h(t) is increasing in t. This implies the loglog distribution is IFR.

15. 0.9298
16. 0.758
17. Let X be a random variable that represents the number of good fuses in the sample of size n.

$$P[X = x] = \frac{\binom{k}{x}\binom{N-k}{n-x}}{\binom{N}{n}} \quad x = 0, 1, 2, \ldots, k$$

18. 0.1046
19. The expected number of failures during the interval (0,t) is λt.
20. It can show directly using Property 3 (Chi-square distribution).
21. Let S_N = the number of developed plants. $E(S_N) = p\lambda$ and $Var(S_N) = \lambda p$.
22. Let n = the minimum number of components required to ensure a successful mission. $n = 6$.
23. Let X be the number of claims received by a week.

$$P(X \leq 2) = \sum_{x=0}^{2} \frac{(10)^x e^{-10}}{x!} = 0.$$

24. Let X be a geometric random variable that represent the number of accidents in a given month with parameter p. $P(X > 2) = p^3 = (0.1)^3 = 0.001$.
25. Let X be the number of present for work and Y the number of absentee. This implies that $n = X + Y$.

$$P(Y \geq 2) = \sum_{y=2}^{20} \binom{20}{y}(0.05)^y(0.95)^{20-y} = 0.000001.$$

26. Let X denote the number of failures before n successes; p is the probability of success at each trial.

$$P(X = x) = \binom{n+x-1}{x} p^n (1 - p)^x \quad \text{for } x = 0, 1, 2, \ldots$$

27. Let Y be the number of trials up to and including the 5th success. $P(15 \leq Y \leq 20) = 0.2405$.

28. Let X and Y denote independent random variables with binomial distribution $X \sim b(m,p)$ and $Y \sim b(n,p)$, respectively. $\dfrac{P(X=k)P(Y=s-k)}{P(X+Y=s)} = \dfrac{\binom{m}{k}\binom{n}{s-k}}{\binom{m+n}{s}}$.

29. Let X denote the number of girls in the selected number of children.

$$P(X = 2) = \frac{\binom{10}{2}\binom{12}{4}}{\binom{22}{6}} = 0.2986$$

and

$$P(X = 3) = \frac{\binom{10}{3}\binom{12}{3}}{\binom{22}{6}} = 0.3538.$$

30. (a) 0.6065 (b) 0.8187.
31. 0.488
32. 0.10
33. (a) $P = p^k + k(1-p)p^{k-1} + \frac{k(k-1)}{2}(1-p)^2 p^{k-2}$
    (b) 0.9298092.
34. The probability that he shall reach a false conclusion is 0.057.
35. (a) $P(T_2 > T_1) = \int\limits_{t_2=0}^{\infty} F_1(t_2)f_2(t_2)dt_2$.
    (b) $P(T_2 > T_1) = \frac{\lambda_1}{\lambda_1+\lambda_2}$.
    (c) $P(T_2 > T_1) = 0.625$.
36. 0.990909
37. 0.420175

Chapter 3

1. see (Pham 2000a, page 330)
2. Let p is the parameter to be estimated. The maximum likelihood estimator, $\hat{p}$, is

$$\hat{p} = \frac{\sum_{i=1}^{n} x_i}{n}.$$

6. $\max\{x_i\} - d \leq \hat{c} \leq \min\{x_i\} + d \quad \text{for } i = 1, 2, \ldots, n$
7. (a) $\hat{c} = \dfrac{\max\limits_{1 \leq x_i \leq n}\{x_i\} + \min\limits_{1 \leq x_i \leq n}\{x_i\}}{2} \cdot \hat{d} = \dfrac{\max\limits_{1 \leq x_i \leq n}\{x_i\} - \min\limits_{1 \leq x_i \leq n}\{x_i\}}{2}.$

   (b) $\hat{c} = 27.5; \quad \hat{d} = 7.5$
8. $\hat{d} = \max\limits_{1 \leq i \leq n}\{|x_i|\}.$
9. The number of customers who enter the restaurant per day were not the same on the five days at the 5% level of significance.
10. We cannot reject the null hypothesis.
12. The results of current year are consistent with the results from last year at the 5% level of significance.
16. We can obtain the maximum likelihood estimates

$$\hat{p}_i = \frac{x_i}{n} \quad \text{for } i = 1, 2, \ldots, k$$

17. The MLE of p is $\hat{p} = \frac{k}{n}$.
18. (a) 0.1819
19. (a) $\hat{\lambda} = -1 - \dfrac{n}{\sum\limits_{i=1}^{n} \ln(t_i)}$.

   (b) $\hat{\lambda} = 0.4739$
20. $E(\hat{\beta}) = (\alpha + \beta) - \left(\alpha + \frac{\beta}{n}\right) = \beta\left(1 - \frac{1}{n}\right)$ and $E(\hat{\alpha}) = \alpha + \frac{\beta}{n}$.

   (c) $\hat{\alpha} = 2.4$ and $\hat{\beta} = 1.2$
21. The number of customers who enter the restaurant per day were not the same on the five days at the 5% level of significance.
22. The maximum likelihood estimate of θ is $\hat{\theta} = 0.5$.
23. (a) $\hat{\theta} = 3$. (b) $\hat{\theta} = 3.116$
24. (a) 22.32(b) 14.35
25. (a) 8.486 (b) 7.4
26. **(a)** $\hat{\theta} = \frac{2\bar{x}-1}{1-\bar{x}}$ (b) $\hat{\theta} = \frac{2(0.6)-1}{1-0.6} = 0.5$
27. The maximum likelihood estimate of $\lambda = 0.012$.
28. $\hat{\theta} = 10.54675$
29. The sampling is stopped at the 16th observation and that the lot would be rejected.
30. The sampling is stopped at the 6th observation and that the lot would be rejected.
31. (a) $\hat{\lambda} = \frac{2}{\bar{X}}$. (b) $\hat{\lambda} = \frac{2}{\bar{X}}$.

32. (a) $\hat{\lambda} = \frac{3}{\bar{X}}$. (b) $\hat{\lambda} = 0.1595745$

33. We reject the null hypothesis. That is, there is sufficient evidence, at the 0.10 level, to conclude that facility conditions and pricing policy are dependent of one another.

Chapter 4

1. Letting $r_i(t)$ denote the failure rate of the ith component, then $r(t) = \sum\limits_{i=1}^{n} r_i(u)$ is the system failure rate.

2. The reliability and MTTF of a 2-out-of-4 system are as follows:

$$R(t) = e^{-(\lambda_1+\lambda_2)t} + e^{-(\lambda_1+\lambda_3)t} + e^{-(\lambda_1+\lambda_4)t} + e^{-(\lambda_2+\lambda_3)t} + e^{-(\lambda_2+\lambda_4)t} + e^{-(\lambda_3+\lambda_4)t}$$
$$- 2e^{-(\lambda_1+\lambda_2+\lambda_3)t} - 2e^{-(\lambda_1+\lambda_2+\lambda_4)t} - 2e^{-(\lambda_1+\lambda_3+\lambda_4)t}$$
$$- 2e^{-(\lambda_2+\lambda_3+\lambda_4)t} + 3e^{-(\lambda_1+\lambda_2+\lambda_3+\lambda_4)t}$$

and

$$MTTF = \frac{1}{\lambda_1+\lambda_2} + \frac{1}{\lambda_1+\lambda_3} + \frac{1}{\lambda_1+\lambda_4} + \frac{1}{\lambda_2+\lambda_3} + \frac{1}{\lambda_2+\lambda_4} + \frac{1}{\lambda_3+\lambda_4}$$
$$- \frac{2}{\lambda_1+\lambda_2+\lambda_3} - \frac{2}{\lambda_1+\lambda_2+\lambda_4} - \frac{2}{\lambda_1+\lambda_3+\lambda_4} - \frac{2}{\lambda_2+\lambda_3+\lambda_4}$$
$$+ \frac{3}{\lambda_1+\lambda_2+\lambda_3+\lambda_4}.$$

3. The reliability and the MTTF function of the high-voltage load-sharing system are given as follows:

$$R(t) = e^{-(\lambda_A+\lambda_B+\lambda_P)t} + \frac{\lambda_A\, e^{-(\lambda_P+\lambda'_B)t}}{(\lambda_A + \lambda_B + \lambda_C - \lambda'_B)}\left(1 - e^{-(\lambda_A+\lambda_B+\lambda_C-\lambda'_B)t}\right)$$
$$+ \frac{\lambda_B\, e^{-(\lambda_P+\lambda'_A)t}}{(\lambda_A + \lambda_B + \lambda_C - \lambda'_A)}\left(1 - e^{-(\lambda_A+\lambda_B+\lambda_C-\lambda'_A)t}\right)$$

The system MTTF is

$$MTTF = \frac{1}{\lambda_A + \lambda_B + \lambda_P} + \frac{\lambda_A}{(\lambda_A + \lambda_B + \lambda_C - \lambda'_B)}$$
$$\left(\frac{1}{\lambda_P + \lambda'_B} - \frac{1}{\lambda_A + \lambda_B + \lambda_C + \lambda_P}\right)$$

$$+ \frac{\lambda_B}{\left(\lambda_A + \lambda_B + \lambda_C - \lambda_A'\right)} \left(\frac{1}{\lambda_P + \lambda_A'} - \frac{1}{\lambda_A + \lambda_B + \lambda_C + \lambda_P} \right)$$

4. The system reliability is 0.99975
5. (a)

$$R = p^k + k(1-p)p^{k-1} + \frac{k(k-1)}{2}(1-p)^2 p^{k-2}$$

(b) R = 0.9298.
6. (a) The reliability of the device is

$$R = \left[1 - (1-p_1)^m - mp_1(1-p_1)^{m-1} \right] \left[1 - (1-p_2)^n - np_2(1-p_2)^{n-1} \right]$$

(b) R = 0.9992.
8. (a) The reliability of the system is

$$R(t) = (1 + 2\lambda t)e^{-2\lambda t}.$$

(b) The MTTF of the system is

$$MTTF = \frac{1}{\lambda}.$$

(c) $R(t = 10) = 0.9953$. System $MTTF = 200$ hours.
9. $R_{system} = 0.97847$
10. (a) 0.9999685 (b) 0.9992

Chapter 5

1. (*Theorem 4*) Show that the statistic T_k is the UMVUE of R^k where σ_x and σ_y are known, $k < \min \{m, a\}$, and $T_k =$

$$\begin{cases} \frac{(a-k)(k\sigma_x + m\sigma_y)\sigma_y^{k-1}}{ma(\sigma_x+\sigma_y)^k} e^{\frac{(Y_{(1)}-X_{(1)})k}{\sigma_y}} I(Y_{(1)} < X_{(1)}) \\ + \sum_{i=0}^{k} (-1)^i \binom{k}{i} \frac{(m-i)(a\sigma_x + i\sigma_y)\sigma_x^{i-1}}{ma(\sigma_x+\sigma_y)^i} e^{-\frac{i[Y_{(1)}-X_{(1)}]}{\sigma_x}} I(Y_{(1)} \geq X_{(1)}). \end{cases}$$

Chapter 6

1. Using Eqs. (42) and (43)), one can obtain the results.
2. Table below presents the results.
   Reliability and MTTF for varying time t.

Time	Sensor reliability	Channel reliability	System reliability

(continued)

(continued)

Time	Sensor reliability	Channel reliability	System reliability
10	0.9851	0.9947	1.0000
20	0.9704	0.9889	1.0000
30	0.9506	0.9826	0.9999
40	0.9418	0.9758	0.9999
MTTF	666.6	628.6	546.2

3. (a) Let the random variable X be the amount of time that the system functions is in the good state. The probability of having a replacement because of a system failure is given by:

$$\sum_{n=0}^{\infty} P\{nT < X \le [(n+1)T - a]\} = \sum_{n=0}^{\infty} \left(e^{-\lambda nT} - e^{-\lambda[(n+1)T-a]} \right)$$

$$= \left(1 - e^{-\lambda(T-a)} \right) \sum_{n=0}^{\infty} e^{-\lambda nT}$$

$$= \frac{\left(1 - e^{-\lambda(T-a)} \right)}{\left(1 - e^{-\lambda T} \right)}$$

(b) The time between two replacements is equal to nT with probability

$$P\{(n - 1)T < X \le nT\} \quad \text{for} \quad n = 1, 2, \dots$$

Hence,

$$E(\text{Time between two replacements}) = \sum_{n=1}^{\infty} nT \left(e^{-\lambda(n-1)T} - e^{-\lambda nT} \right)$$

$$= \frac{T}{\left(1 - e^{-\lambda T} \right)}.$$

5. (a) $MTTF = \frac{\mu + 20\lambda_a + 19\lambda_b}{380\lambda_a \lambda_b}$.
   (b) MTTF = 683.5526 h.

6. $MTTF = \frac{\mu + 15\lambda_a + 14\lambda_b}{210\lambda_a \lambda_b}$.

   (b) MTTF = 2141.25 h.

8. Show that the renewal function, $M(t)$, can be written as follows:

$$M(t) = F(t) + \int\limits_0^t M(t-s)dF(s)$$

where $F(t)$ is the distribution function of the inter-arrival time or the renewal period.

Chapter 7

1. From example 1 in Chap. 7, substitute $\lambda = \frac{1}{\beta}$, we then obtain the results (based on example 1).
2. Follow the same procedure from example 1 in Chap. 7, one can easily obtain the results.
3. Follow the same procedure from example 2 in Chap. 7, one can easily obtain the results.

Chapter 8

1. $(AB)C = \begin{pmatrix} -16 & -22 \\ 18 & 16 \end{pmatrix} A(BC) = \begin{pmatrix} -16 & -22 \\ 18 & 16 \end{pmatrix}$
   Hence $(AB)\,C = A(BC)$.

2. $A^T A = \begin{pmatrix} \cos\theta & \sin\theta \\ -\sin\theta & \cos\theta \end{pmatrix} \begin{pmatrix} \cos\theta & -\sin\theta \\ \sin\theta & \cos\theta \end{pmatrix} = \begin{pmatrix} 1 & 0 \\ 0 & 1 \end{pmatrix} = I$

   Thus A is an orthogonal matrix.
3. The eigenvalues are:

$$\lambda = 1; \quad \lambda = 4; \quad \lambda = 6.$$

For the eigenvalue $\lambda = 1$, the eigenvector can be obtained as

$$\begin{pmatrix} x_1 \\ x_2 \\ x_3 \end{pmatrix} = \begin{pmatrix} 2 \\ 0 \\ 1 \end{pmatrix}$$

Similarly, for $\lambda = 4$, we can obtain the eigenvector as

$$\begin{pmatrix} x_1 \\ x_2 \\ x_3 \end{pmatrix} = \begin{pmatrix} 0 \\ 1 \\ 0 \end{pmatrix}$$

Also, for $\lambda = 6$, the eigenvector is.

$$\begin{pmatrix} x_1 \\ x_2 \\ x_3 \end{pmatrix} = \begin{pmatrix} 1 \\ 0 \\ -2 \end{pmatrix}$$

The unit eigenvectors have the property that they have length 1, i. e., the sum of the squares of their elements = 1. Thus the above eigenvectors become respectively:

$$\begin{pmatrix} \dfrac{2}{\sqrt{5}} \\ 0 \\ \dfrac{1}{\sqrt{5}} \end{pmatrix}, \begin{pmatrix} 0 \\ 1 \\ 0 \end{pmatrix}, \begin{pmatrix} \dfrac{1}{\sqrt{5}} \\ 0 \\ \dfrac{-2}{\sqrt{5}} \end{pmatrix}.$$

4. From problem 3, the eigenvalues of the matrix A are: $\lambda = 1$; $\lambda = 4$; $\lambda = 6$. Note that matrix A is symmetric, so the eigenvalues and singular values are the same. The three singular values are: $d_1 = 6, d_2 = 4$, and $d_3 = 1$, and the singular value decomposition of matrix A can be written as:

$$A = UDV^T$$

$$= \begin{bmatrix} -0.447214 & 0 & 0.894427 \\ 0 & 1 & 0 \\ 0.894427 & 0 & 0.447214 \end{bmatrix} \begin{bmatrix} 6 & 0 & 0 \\ 0 & 4 & 0 \\ 0 & 0 & 1 \end{bmatrix} \begin{bmatrix} -0.447214 & 0 & 0.894427 \\ 0 & 1 & 0 \\ 0.894427 & 0 & 0.447214 \end{bmatrix}$$

$$\equiv U \qquad\qquad \equiv D \qquad\qquad \equiv V^T$$

5. $\hat{\beta} = \sum\limits_{i=1}^{n} k_i y_i$ where $k_i = \dfrac{x_i - \bar{x}}{S_{xx}}$

Note that

$$\sum_{i=1}^{n} k_i = 0, \qquad \sum_{i=1}^{n} k_i x_i = 1.$$

Hence,

$$E\left(\hat{\beta}\right) = E\left(\frac{S_{XY}}{S_{XX}}\right) = \sum_{i=1}^{n} k_i E(y_i)$$

$$= \sum_{i=1}^{n} k_i (\alpha + \beta x_i) = \alpha \sum_{i=1}^{n} k_i + \beta \sum_{i=1}^{n} k_i x_i = \beta.$$

Therefore, $\hat{\beta}$ is an unbiased estimator of β.

Appendix A
Distribution Tables

See Tables A.1, A.2, A.3, A.4, A.5, A.6, A.7, A.8

Table A.1 Cumulative areas under the standard normal distribution

Z	0	1	2	3	4	5	6	7	8	9
−3.0	0.0013	0.0010	0.0007	0.0005	0.0003	0.0002	0.0002	0.0001	0.0001	0.0000
−2.9	0.0019	0.0018	0.0017	0.0017	0.0016	0.0016	0.0015	0.0015	0.0014	0.0014
−2.8	0.0026	0.0025	0.0024	0.0023	0.0023	0.0022	0.0021	0.0021	0.0020	0.0019
−2.7	0.0035	0.0034	0.0033	0.0032	0.0031	0.0030	0.0029	0.0028	0.0027	0.0026
−2.6	0.0047	0.0045	0.0044	0.0043	0.0041	0.0040	0.0039	0.0038	0.0037	0.0036
−2.5	0.0062	0.0060	0.0059	0.0057	0.0055	0.0054	0.0052	0.0051	0.0049	0.0048
−2.4	0.0082	0.0080	0.0078	0.0075	0.0073	0.0071	0.0069	0.0068	0.0066	0.0064
−2.3	0.0107	0.0104	0.0102	0.0099	0.0096	0.0094	0.0091	0.0089	0.0087	0.0084
−2.2	0.0139	0.0136	0.0132	0.0129	0.0126	0.0122	0.0119	0.0116	0.0113	0.0110
−2.1	0.0179	0.0174	0.0170	0.0166	0.0162	0.0158	0.0154	0.0150	0.0146	0.0143
−2.0	0.0228	0.0222	0.0217	0.0212	0.0207	0.0202	0.0197	0.0192	0.0188	0.0183
−1.9	0.0287	0.0281	0.0274	0.0268	0.0262	0.0256	0.0250	0.0244	0.0238	0.0233
−1.8	0.0359	0.0352	0.0344	0.0336	0.0329	0.0322	0.0314	0.0307	0.0300	0.0294
−1.7	0.0446	0.0436	0.0427	0.0418	0.0409	0.0401	0.0392	0.0384	0.0375	0.0367
−1.6	0.0548	0.0537	0.0526	0.0516	0.0505	0.0495	0.0485	0.0475	0.0465	0.0455
−1.5	0.0668	0.0655	0.0643	0.0630	0.0618	0.0606	0.0594	0.0582	0.0570	0.0559
−1.4	0.0808	0.0793	0.0778	0.0764	0.0749	0.0735	0.0722	0.0708	0.0694	0.0681
−1.3	0.0968	0.0951	0.0934	0.0918	0.0901	0.0885	0.0869	0.0853	0.0838	0.0823
−1.2	0.1151	0.1131	0.1112	0.1093	0.1075	0.1056	0.1038	0.1020	0.1003	0.0985
−1.1	0.1357	0.1335	0.1314	0.1292	0.1271	0.1251	0.1230	0.1210	0.1190	0.1170
−1.0	0.1587	0.1562	0.1539	0.1515	0.1492	0.1469	0.1446	0.1423	0.1401	0.1379
−0.9	0.1841	0.1814	0.1788	0.1762	0.1736	0.1711	0.1685	0.1660	0.1635	0.1611

(continued)

© The Editor(s) (if applicable) and The Author(s), under exclusive license
to Springer Nature Switzerland AG 2022
H. Pham, *Statistical Reliability Engineering*, Springer Series in Reliability Engineering,
https://doi.org/10.1007/978-3-030-76904-8

Table A.1 (continued)

Z	0	1	2	3	4	5	6	7	8	9
−0.8	0.2119	0.2090	0.2061	0.2033	0.2005	0.1977	0.1949	0.1922	0.1894	0.1867
−0.7	0.2420	0.2389	0.2358	0.2327	0.2297	0.2266	0.2236	0.2206	0.2177	0.2148
−0.6	0.2743	0.2709	0.2676	0.2643	0.2611	0.2578	0.2546	0.2514	0.2483	0.2451
−0.5	0.3085	0.3050	0.3015	0.2981	0.2946	0.2912	0.2877	0.2843	0.2810	0.2776
−0.4	0.3446	0.3409	0.3372	0.3336	0.3300	0.3264	0.3228	0.3192	0.3156	0.3121
−0.3	0.3821	0.3783	0.3745	0.3707	0.3669	0.3632	0.3594	0.3557	0.3520	0.3483
−0.2	0.4207	0.4168	0.4129	0.4090	0.4052	0.4013	0.3974	0.3936	0.3897	0.3859
−0.1	0.4602	0.4562	0.4522	0.4483	0.4443	0.4404	0.4364	0.4325	0.4286	0.4247
−0.0	0.5000	0.4960	0.4920	0.4880	0.4840	0.4801	0.4761	0.4721	0.4681	0.4641
0.0	0.5000	0.5040	0.5080	0.5120	0.5160	0.5199	0.5239	0.5279	0.5319	0.5359
0.1	0.5398	0.5438	0.5478	0.5517	0.5557	0.5596	0.5636	0.5675	0.5714	0.5753
0.2	0.5793	0.5832	0.5871	0.5910	0.5948	0.5987	0.6026	0.6064	0.6103	0.6141
0.3	0.6179	0.6217	0.6255	0.6293	0.6331	0.6368	0.6406	0.6443	0.6480	0.6517
0.4	0.6554	0.6591	0.6628	0.6664	0.6700	0.6736	0.6772	0.6808	0.6844	0.6879
0.5	0.6915	0.6950	0.6985	0.7019	0.7054	0.7088	0.7123	0.7157	0.7190	0.7224
0.6	0.7257	0.7291	0.7324	0.7357	0.7389	0.7422	0.7454	0.7486	0.7517	0.7549
0.7	0.7580	0.7611	0.7642	0.7673	0.7703	0.7734	0.7764	0.7794	0.7823	0.7852
0.8	0.7881	0.7910	0.7939	0.7967	0.7995	0.8023	0.8051	0.8078	0.8106	0.8133
0.9	0.8159	0.8186	0.8212	0.8238	0.8264	0.8289	0.8315	0.8340	0.8365	0.8389
1.0	0.8413	0.8438	0.8461	0.8485	0.8508	0.8531	0.8554	0.8577	0.8599	0.8621
1.1	0.8643	0.8665	0.8686	0.8708	0.8729	0.8749	0.8770	0.8790	0.8810	0.8830
1.2	0.8849	0.8869	0.8888	0.8907	0.8925	0.8944	0.8962	0.8980	0.8997	0.9015
1.3	0.9032	0.9049	0.9066	0.9082	0.9099	0.9115	0.9131	0.9147	0.9162	0.9177
1.4	0.9192	0.9207	0.9222	0.9236	0.9251	0.9265	0.9278	0.9292	0.9306	0.9319
1.5	0.9332	0.9345	0.9357	0.9370	0.9382	0.9394	0.9406	0.9418	0.9430	0.9441
1.6	0.9452	0.9463	0.9474	0.9484	0.9495	0.9505	0.9515	0.9525	0.9535	0.9545
1.7	0.9554	0.9564	0.9573	0.9582	0.9591	0.9599	0.9608	0.9616	0.9625	0.9633
1.8	0.9641	0.9648	0.9656	0.9664	0.9671	0.9678	0.9686	0.9693	0.9700	0.9706
1.9	0.9713	0.9719	0.9726	0.9732	0.9738	0.9744	0.9750	0.9756	0.9762	0.9767
2.0	0.9772	0.9778	0.9783	0.9788	0.9793	0.9798	0.9803	0.9808	0.9812	0.9817
2.1	0.9821	0.9826	0.9830	0.9834	0.9838	0.9842	0.9846	0.9850	0.9854	0.9857
2.2	0.9861	0.9864	0.9868	0.9871	0.9874	0.9878	0.9881	0.9884	0.9887	0.9890
2.3	0.9893	0.9896	0.9898	0.9901	0.9904	0.9906	0.9909	0.9911	0.9913	0.9916
2.4	0.9918	0.9920	0.9922	0.9925	0.9927	0.9929	0.9931	0.9932	0.9934	0.9936
2.5	0.9938	0.9940	0.9941	0.9943	0.9945	0.9946	0.9948	0.9949	0.9951	0.9952
2.6	0.9953	0.9955	0.9956	0.9957	0.9959	0.9960	0.9961	0.9962	0.9963	0.9964
2.7	0.9965	0.9966	0.9967	0.9968	0.9969	0.9970	0.9971	0.9972	0.9973	0.9974
2.8	0.9974	0.9975	0.9976	0.9977	0.9977	0.9978	0.9979	0.9979	0.9980	0.9981
2.9	0.9981	0.9982	0.9982	0.9983	0.9984	0.9984	0.9985	0.9985	0.9986	0.9986
3.0	0.9987	0.9990	0.9993	0.9995	0.9997	0.9998	0.9998	0.9999	0.9999	1.000

Table A.2 Percentage points of the t-distribution

$\nu\backslash\alpha$	0.100	0.050	0.025	0.01	0.005	0.001
1	3.078	6.314	12.706	31.821	63.657	318.310
2	1.886	2.920	4.303	6.965	9.925	23.326
3	1.638	2.353	3.182	4.541	5.841	10.213
4	1.533	2.132	2.776	3.747	4.604	7.173
5	1.476	2.015	2.571	3.365	4.032	5.893
6	1.440	1.943	2.447	3.143	3.707	5.208
7	1.415	1.895	2.365	2.998	3.499	4.785
8	1.397	1.860	2.306	2.896	3.355	4.501
9	1.383	1.833	2.262	2.821	3.250	4.297
10	1.372	1.812	2.228	2.764	3.169	4.144
11	1.363	1.796	2.201	2.718	3.106	4.025
12	1.356	1.782	2.179	2.681	3.055	3.930
13	1.350	1.771	2.160	2.650	3.012	3.852
14	1.345	1.761	2.145	2.624	2.977	3.787
15	1.341	1.753	2.131	2.602	2.947	3.733
16	1.337	1.746	2.120	2.583	2.921	3.686
17	1.333	1.740	2.110	2.567	2.898	3.646
18	1.330	1.734	2.101	2.552	2.878	3.610
19	1.328	1.729	2.093	2.539	2.861	3.579
20	1.325	1.725	2.086	2.528	2.845	3.552
21	1.323	1.721	2.080	2.518	2.831	3.527
22	1.321	1.717	2.074	2.508	2.819	3.505
23	1.319	1.714	2.069	2.500	2.807	3.485
24	1.318	1.711	2.064	2.492	2.797	3.467
25	1.316	1.708	2.060	2.485	2.787	3.450
26	1.315	1.706	2.056	2.479	2.779	3.435
27	1.314	1.703	2.052	2.473	2.771	3.421
28	1.313	1.701	2.048	2.467	2.763	3.408
29	1.311	1.699	2.045	2.462	2.756	3.396
30	1.310	1.697	2.042	2.457	2.750	3.385
40	1.303	1.684	2.021	2.423	2.704	3.307
60	1.296	1.671	2.000	2.390	2.660	3.232
120	1.289	1.658	1.980	2.358	2.617	3.160
∞	1.282	1.645	1.960	2.326	2.576	3.090

Table A.3 Percentage points of the chi-squared distribution

$\nu \backslash \chi_\alpha^2$	$\chi_{0.99}^2$	$\chi_{0.975}^2$	$\chi_{0.95}^2$	$\chi_{0.90}^2$	$\chi_{0.10}^2$	$\chi_{0.05}^2$	$\chi_{0.025}^2$	$\chi_{0.01}^2$
1	0	0.00	0.00	0.02	2.71	3.84	5.02	6.64
2	0.02	0.05	0.10	0.21	4.61	5.99	7.38	9.21
3	0.12	0.22	0.35	0.58	6.25	7.82	9.35	11.35
4	0.30	0.48	0.71	1.06	7.78	9.49	11.14	13.28
5	0.55	0.83	1.15	1.61	9.24	11.07	12.83	15.09
6	0.87	1.24	1.64	2.20	10.65	12.59	14.45	16.81
7	1.24	1.69	2.17	2.83	12.02	14.07	16.01	18.48
8	1.65	2.18	2.73	3.49	13.36	15.51	17.54	20.09
9	2.09	2.70	3.33	4.17	14.68	16.92	19.02	21.67
10	2.56	3.25	3.94	4.87	15.99	18.31	20.48	23.21
11	3.05	3.82	4.58	5.58	17.28	19.68	21.92	24.73
12	3.57	4.40	5.23	6.30	18.55	21.03	23.34	26.22
13	4.11	5.01	5.89	7.04	19.81	22.36	24.74	27.69
14	4.66	5.63	6.57	7.79	21.06	23.69	26.12	29.14
15	5.23	6.26	7.26	8.57	22.31	25.00	27.49	30.58
16	5.81	6.91	7.96	9.31	23.54	26.30	28.85	32.00
17	6.41	7.56	8.67	10.09	24.77	27.59	30.19	33.41
18	7.02	8.23	9.39	10.87	25.99	28.87	31.53	34.81
19	7.63	8.91	10.12	11.65	27.20	30.14	32.85	36.19
20	8.26	9.59	10.85	12.44	28.41	31.41	34.17	37.57
21	8.90	10.28	11.59	13.24	29.62	32.67	35.48	38.93
22	9.54	10.98	12.34	14.04	30.81	33.92	36.78	40.29
23	10.20	11.69	13.09	14.85	32.01	35.17	38.08	41.64
24	10.86	12.40	13.85	15.66	33.20	36.42	39.36	42.98
25	11.52	13.12	14.61	16.47	34.38	37.65	40.65	44.31
26	12.20	13.84	15.38	17.29	35.56	38.89	41.92	45.64
27	12.88	14.57	16.15	18.11	36.74	40.11	43.19	46.96
28	13.57	15.31	16.93	18.94	37.92	41.34	44.46	48.28
29	14.26	16.05	17.71	19.77	39.09	42.56	45.72	49.59
30	14.95	16.79	18.49	20.60	40.26	43.77	46.98	50.89
35	18.48	20.56	22.46	24.81	46.03	49.80	53.21	57.36
40	22.14	24.42	26.51	29.07	51.78	55.76	59.35	63.71
50	29.69	32.35	34.76	37.71	63.14	67.50	71.42	76.17
60	37.47	40.47	43.19	46.48	74.37	79.08	83.30	88.39
70	45.43	48.75	51.74	55.35	85.50	90.53	95.03	100.44
80	53.53	57.15	60.39	64.30	96.55	101.88	106.63	112.34
90	61.74	65.64	69.12	73.31	107.54	113.15	118.14	124.13
100	70.05	74.22	77.93	82.38	118.47	124.34	129.57	135.81
120	86.92	91.58	95.70	100.62	140.23	146.57	152.21	158.95

Table A.4 Critical values $d_{n,\alpha}$ of the maximum absolute difference between sample and population cumulative distribtuions for the Kolmogorov–Smirnov (KS) test (Smirnov 1948)

$n\backslash\alpha$	0.2	0.1	0.05	0.02	0.01
1	0.900	0.950	0.975	0.990	0.995
2	0.684	0.776	0.842	0.900	0.929
3	0.565	0.636	0.708	0.785	0.829
4	0.493	0.565	0.624	0.689	0.734
5	0.447	0.509	0.563	0.627	0.669
6	0.410	0.468	0.519	0.577	0.617
7	0.381	0.436	0.483	0.538	0.576
8	0.358	0.410	0.454	0.507	0.542
9	0.339	0.387	0.430	0.480	0.513
10	0.323	0.369	0.409	0.457	0.489
11	0.308	0.352	0.391	0.437	0.468
12	0.296	0.338	0.375	0.419	0.449
13	0.285	0.325	0.361	0.404	0.432
14	0.275	0.314	0.349	0.390	0.418
15	0.266	0.304	0.338	0.377	0.404
16	0.258	0.295	0.327	0.366	0.392
17	0.250	0.286	0.318	0.355	0.381
18	0.244	0.279	0.309	0.346	0.371
19	0.237	0.271	0.301	0.337	0.361
20	0.232	0.265	0.294	0.329	0.352
21	0.226	0.259	0.287	0.321	0.344
22	0.221	0.253	0.281	0.314	0.337
23	0.216	0.247	0.275	0.307	0.330
24	0.212	0.242	0.264	0.301	0.323
25	0.208	0.238	0.264	0.295	0.317
26	0.204	0.233	0.259	0.290	0.311
27	0.200	0.229	0.254	0.284	0.305
28	0.197	0.225	0.250	0.279	0.300
29	0.193	0.221	0.246	0.275	0.295
30	0.190	0.218	0.242	0.270	0.281
35	0.180	0.210	0.230	0.265	0.270
Over 35	$1.07/\sqrt{n}$	$1.22/\sqrt{n}$	$1.36/\sqrt{n}$	$1.56/\sqrt{n}$	$1.63/\sqrt{n}$

Table A.5 Range values d_n
for average value of w

n (number in sample)	d_n
2	1.128
3	1.693
4	2.059
5	2.326
6	2.534
7	2.704
8	2.847
9	2.970
10	3.078
11	3.173
12	3.258

Table A.6 Values of outlier
ratio for significant level
$\alpha = 0.05$

No. in sample	Critical value	No. in sample	Critical value
3	1.15	20	2.71
4	1.48	21	2.73
5	1.71	22	2.76
		23	2.78
6	1.89	24	2.80
7	2.02		
8	2.13	25	2.82
9	2.21	30	2.91
		32	2.98
10	2.29	40	3.04
11	2.36	45	3.09
12	2.41		
13	2.46	50	3.13
14	2.51	60	3.20
		70	3.26
15	2.55	80	3.31
16	2.59	90	3.35
17	2.62	100	3.38
18	2.65		
19	2.68		

Table A.7 Percentage points of the F-distribution—$F_{0.05}(v_1, v_2)$

$v_2 \backslash v_1$	1	2	3	4	5	6	7	8	9
1	161.4	199.5	215.7	224.6	230.2	234.0	236.8	238.9	240.5
2	18.51	19.00	19.16	19.25	19.30	19.33	19.35	19.37	19.38
3	10.13	9.55	9.28	9.12	9.01	8.94	8.89	8.85	8.81
4	7.71	6.94	6.59	6.39	6.26	6.16	6.09	6.04	6.00
5	6.61	5.79	5.41	5.19	5.05	4.95	4.88	4.82	4.77
6	5.99	5.14	4.76	4.53	4.39	4.28	4.21	4.15	4.10
7	5.59	4.74	4.35	4.12	3.97	3.87	3.79	3.73	3.68
8	5.32	4.46	4.07	3.84	3.69	3.58	3.50	3.44	3.39
9	5.12	4.26	3.86	3.63	3.48	3.37	3.29	3.23	3.18
10	4.96	4.10	3.71	3.48	3.33	3.22	3.14	3.07	3.02
11	4.84	3.98	3.59	3.36	3.20	3.09	3.01	2.95	2.90
12	4.75	3.89	3.49	3.26	3.11	3.00	2.91	2.85	2.80
13	4.67	3.81	3.41	3.18	3.03	2.92	2.83	2.77	2.71
14	4.60	3.74	3.34	3.11	2.96	2.85	2.76	2.70	2.65
15	4.54	3.68	3.29	3.06	2.90	2.79	2.71	2.64	2.59
16	4.49	3.63	3.24	3.01	2.85	2.74	2.66	2.59	2.54
17	4.45	3.59	3.20	2.96	2.81	2.70	2.61	2.55	2.49
18	4.41	3.55	3.16	2.93	2.77	2.66	2.58	2.51	2.46
19	4.38	3.52	3.13	2.90	2.74	2.63	2.54	2.48	2.42
20	4.35	3.49	3.10	2.87	2.71	2.60	2.51	2.45	2.39
21	4.32	3.47	3.07	2.84	2.68	2.57	2.49	2.42	2.37
22	4.30	3.44	3.05	2.82	2.66	2.55	2.46	2.40	2.34
23	4.28	3.42	3.03	2.80	2.64	2.53	2.44	2.37	2.32
24	4.26	3.40	3.01	2.78	2.62	2.51	2.42	2.36	2.30
25	4.24	3.39	2.99	2.76	2.60	2.49	2.40	2.34	2.28
26	4.23	3.37	2.98	2.74	2.59	2.47	2.39	2.32	2.27
27	4.21	3.35	2.96	2.73	2.57	2.46	2.37	2.31	2.25
28	4.20	3.34	2.95	2.71	2.56	2.45	2.36	2.29	2.24
29	4.18	3.33	2.93	2.70	2.55	2.43	2.35	2.28	2.22
30	4.17	3.32	2.92	2.69	2.53	2.42	2.33	2.27	2.21
40	4.08	3.23	2.84	2.61	2.45	2.34	2.25	2.18	2.12
60	4.00	3.15	2.76	2.53	2.37	2.25	2.17	2.10	2.04
120	3.92	3.07	2.68	2.45	2.29	2.17	2.09	2.02	1.96
∞	3.84	3.00	2.60	2.37	2.21	2.10	2.01	1.94	1.88

(continued)

Table A.7 (continued)

$\nu_2 \backslash \nu_1$	10	12	15	20	24	30	40	60	120	∞
1	241.9	243.9	245.9	248.0	249.1	250.1	251.1	252.2	253.3	254.3
2	19.40	19.41	19.43	19.45	19.45	19.46	19.47	19.48	19.49	19.50
3	8.79	8.74	8.70	8.66	8.64	8.62	8.59	8.57	8.55	8.53
4	5.96	5.91	5.86	5.80	5.77	5.75	5.72	5.69	5.66	5.63
5	4.74	4.68	4.62	4.56	4.53	4.50	4.46	4.43	4.40	4.36
6	4.06	4.00	3.94	3.87	3.84	3.81	3.77	3.74	3.70	3.67
7	3.64	3.57	3.51	3.44	3.41	3.38	3.34	3.30	3.27	3.23
8	3.35	3.28	3.22	3.15	3.12	3.08	3.04	3.01	2.97	2.93
9	3.14	3.07	3.01	2.94	2.90	2.86	2.83	2.79	2.75	2.71
10	2.98	2.91	2.85	2.77	2.74	2.70	2.66	2.62	2.58	2.54
11	2.85	2.79	2.72	2.65	2.61	2.57	2.53	2.49	2.45	2.40
12	2.75	2.69	2.62	2.54	2.51	2.47	2.43	2.38	2.34	2.30
13	2.67	2.60	2.53	2.46	2.42	2.38	2.34	2.30	2.25	2.21
14	2.60	2.53	2.46	2.39	2.35	2.31	2.27	2.22	2.18	2.13
15	2.54	2.48	2.40	2.33	2.29	2.25	2.20	2.16	2.11	2.07
16	2.49	2.42	2.35	2.28	2.24	2.19	2.15	2.11	2.06	2.01
17	2.45	2.38	2.31	2.23	2.19	2.15	2.10	2.06	2.01	1.96
18	2.41	2.34	2.27	2.19	2.15	2.11	2.06	2.02	1.97	1.92
19	2.38	2.31	2.23	2.16	2.11	2.07	2.03	1.98	1.93	1.88
20	2.35	2.28	2.20	2.12	2.08	2.04	1.99	1.95	1.90	1.84
21	2.32	2.25	2.18	2.10	2.05	2.01	1.96	1.92	1.87	1.81
22	2.30	2.23	2.15	2.07	2.03	1.98	1.94	1.89	1.84	1.78
23	2.27	2.20	2.13	2.05	2.01	1.96	1.91	1.86	1.81	1.76
24	2.25	2.18	2.11	2.03	1.98	1.94	1.89	1.84	1.79	1.73
25	2.24	2.16	2.09	2.01	1.96	1.92	1.87	1.82	1.77	1.71
26	2.22	2.15	2.07	1.99	1.95	1.90	1.85	1.80	1.75	1.69
27	2.20	2.13	2.06	1.97	1.93	1.88	1.84	1.79	1.73	1.67
28	2.19	2.12	2.04	1.96	1.91	1.87	1.82	1.77	1.71	1.65
29	2.18	2.10	2.03	1.94	1.90	1.85	1.81	1.75	1.70	1.64
30	2.16	2.09	2.01	1.93	1.89	1.84	1.79	1.74	1.68	1.62
40	2.08	2.00	1.92	1.84	1.79	1.74	1.69	1.64	1.58	1.51
60	1.99	1.92	1.84	1.75	1.70	1.65	1.59	1.53	1.47	1.39
120	1.91	1.83	1.75	1.66	1.61	1.55	1.50	1.43	1.35	1.25
∞	1.83	1.75	1.67	1.57	1.52	1.46	1.39	1.32	1.22	1.00

Table A.8 Percentage points of the F-distribution—$F_{0.01}(v_1, v_2)$

$v_2 \backslash v_1$	1	2	3	4	5	6	7	8	9
1	4052	4999.5	5403	5625	5764	5859	5928	5981	6022
2	98.50	99.00	99.17	99.25	99.30	99.33	99.36	99.37	99.39
3	34.12	30.82	29.46	28.71	28.24	27.91	27.67	27.49	27.35
4	21.20	18.00	16.69	15.98	15.52	15.21	14.98	14.80	14.66
5	16.26	13.27	12.06	11.39	10.97	10.67	10.46	10.29	10.16
6	13.75	10.92	9.78	9.15	8.75	8.47	8.26	8.10	7.98
7	12.25	9.55	8.45	7.85	7.46	7.19	6.99	6.84	6.72
8	11.26	8.65	7.59	7.01	6.63	6.37	6.18	6.03	5.91
9	10.56	8.02	6.99	6.42	6.06	5.80	5.61	5.47	5.35
10	10.04	7.56	6.55	5.99	5.64	5.39	5.20	5.06	4.94
11	9.65	7.21	6.22	5.67	5.32	5.07	4.89	4.74	4.63
12	9.33	6.93	5.95	5.41	5.06	4.82	4.64	4.50	4.39
13	9.07	6.70	5.74	5.21	4.86	4.62	4.44	4.30	4.19
14	8.86	6.51	5.56	5.04	4.69	4.46	4.28	4.14	4.03
15	8.68	6.36	5.42	4.89	4.56	4.32	4.14	4.00	3.89
16	8.53	6.23	5.29	4.77	4.44	4.20	4.03	3.89	3.78
17	8.40	6.11	5.18	4.67	4.34	4.10	3.93	3.79	3.68
18	8.29	6.01	5.09	4.58	4.25	4.01	3.84	3.71	3.60
19	8.18	5.93	5.01	4.50	4.17	3.94	3.77	3.63	3.52
20	8.10	5.85	4.94	4.43	4.10	3.87	3.70	3.56	3.46
21	8.02	5.78	4.87	4.37	4.04	3.81	3.64	3.51	3.40
22	7.95	5.72	4.82	4.31	3.99	3.76	3.59	3.45	3.35
23	7.88	5.66	4.76	4.26	3.94	3.71	3.54	3.41	3.30
24	7.82	5.61	4.72	4.22	3.90	3.67	3.50	3.36	3.26
25	7.77	5.57	4.68	4.18	3.85	3.63	3.46	3.32	3.22
26	7.72	5.53	4.64	4.14	3.82	3.59	3.42	3.29	3.18
27	7.68	5.49	4.60	4.11	3.78	3.56	3.39	3.26	3.15
28	7.64	5.45	4.57	4.07	3.75	3.53	3.36	3.23	3.12
29	7.60	5.42	4.54	4.04	3.73	3.50	3.33	3.20	3.09
30	7.56	5.39	4.51	4.02	3.70	3.47	3.30	3.17	3.07
40	7.31	5.18	4.31	3.83	3.51	3.29	3.12	2.99	2.89
60	7.08	4.98	4.13	3.65	3.34	3.12	2.95	2.82	2.72
120	6.85	4.79	3.95	3.48	3.17	2.96	2.79	2.66	2.56
∞	6.63	4.61	3.78	3.32	3.02	2.80	2.64	2.51	2.41

(continued)

Table A.8 (continued)

$\nu_2 \backslash \nu_1$	10	12	15	20	24	30	40	60	120	∞
1	6056	6106	6157	6209	6235	6261	6287	6313	6339	6366
2	99.40	99.42	99.43	99.45	99.46	99.47	99.47	99.48	99.49	99.50
3	27.23	27.05	26.87	26.69	26.60	26.50	26.41	26.32	26.22	26.13
4	14.55	14.37	14.20	14.02	13.93	13.84	13.75	13.65	13.56	13.46
5	10.05	9.89	9.72	9.55	9.47	9.38	9.29	9.20	9.11	9.02
6	7.87	7.72	7.56	7.40	7.31	7.23	7.14	7.06	6.97	6.88
7	6.62	6.47	6.31	6.16	6.07	5.99	5.91	5.82	5.74	5.65
8	5.81	5.67	5.52	5.36	5.28	5.20	5.12	5.03	4.95	4.86
9	5.26	5.11	4.96	4.81	4.73	4.65	4.57	4.48	4.40	4.31
10	4.85	4.71	4.56	4.41	4.33	4.25	4.17	4.08	4.00	3.91
11	4.54	4.40	4.25	4.10	4.02	3.94	3.86	3.78	3.69	3.60
12	4.30	4.16	4.01	3.86	3.78	3.70	3.62	3.54	3.45	3.36
13	4.10	3.96	3.82	3.66	3.59	3.51	3.43	3.34	3.25	3.17
14	3.94	3.80	3.66	3.51	3.43	3.35	3.27	3.18	3.09	3.00
15	3.80	3.67	3.52	3.37	3.29	3.21	3.13	3.05	2.96	2.87
16	3.69	3.55	3.41	3.26	3.18	3.10	3.02	2.93	2.84	2.75
17	3.59	3.46	3.31	3.16	3.08	3.00	2.92	2.83	2.75	2.65
18	3.51	3.37	3.23	3.08	3.00	2.92	2.84	2.75	2.66	2.57
19	3.43	3.30	3.15	3.00	2.92	2.84	2.76	2.67	2.58	2.49
20	3.37	3.23	3.09	2.94	2.86	2.78	2.69	2.61	2.52	2.42
21	3.31	3.17	3.03	2.88	2.80	2.72	2.64	2.55	2.46	2.36
22	3.26	3.12	2.98	2.83	2.75	2.67	2.58	2.50	2.40	2.31
23	3.21	3.07	2.93	2.78	2.70	2.62	2.54	2.45	2.35	2.26
24	3.17	3.03	2.89	2.74	2.66	2.58	2.49	2.40	2.31	2.21
25	3.13	2.99	2.85	2.70	2.62	2.54	2.45	2.36	2.27	2.17
26	3.09	2.96	2.81	2.66	2.58	2.50	2.42	2.33	2.23	2.13
27	3.06	2.93	2.78	2.63	2.55	2.47	2.38	2.29	2.20	2.10
28	3.03	2.90	2.75	2.60	2.52	2.44	2.35	2.26	2.17	2.06
29	3.00	2.87	2.73	2.57	2.49	2.41	2.33	2.23	2.14	2.03
30	2.98	2.84	2.70	2.55	2.47	2.39	2.30	2.21	2.11	2.01
40	2.80	2.66	2.52	2.37	2.29	2.20	2.11	2.02	1.92	1.80
60	2.63	2.50	2.35	2.20	2.12	2.03	1.94	1.84	1.73	1.60
120	2.47	2.34	2.19	2.03	1.95	1.86	1.76	1.66	1.53	1.38
∞	2.32	2.18	2.04	1.88	1.79	1.70	1.59	1.47	1.32	1.00

Appendix B
Laplace Transform

The Laplace transform of $f(t)$ is the function $f^*(s)$ where

$$f^*(s) = \int_0^\infty e^{-st} f(t) dt$$

Often the Laplace transform is denoted as $f^*(s)$ or $\mathcal{L}\left(f(t)\right)$ or $\mathcal{L}\left(f\right)$. A summary of some common Laplace transform functions is listed in Table B.1 below.

© The Editor(s) (if applicable) and The Author(s), under exclusive license to Springer Nature Switzerland AG 2022
H. Pham, *Statistical Reliability Engineering*, Springer Series in Reliability Engineering,
https://doi.org/10.1007/978-3-030-76904-8

Table B.1 List of common
Laplace transform functions

$f(t)$	$\ell\{f(t)\} = f^*(s)$
f(t)	$f^*(s) = \int\limits_0^\infty e^{-st} f(t)dt$
$\frac{\partial f(t)}{\partial t}$	$s\, f^*(s) - f(0)$
$\frac{\partial^2}{\partial t^2}[f(t)]$	$s^2 f^*(s) - s\, f(0) - \frac{\partial}{\partial t} f(0)$
$\frac{\partial^n}{\partial t^n}[f(t)]$	$s^n f^*(s) - s^{n-1} f(0) - \ldots - \frac{\partial^{n-1}}{\partial t^{n-1}} f(0)$
f(at)	$\frac{1}{a} f^*\left(\frac{s}{a}\right)$
1	$\frac{1}{s}$
t	$\frac{1}{s^2}$
a	$\frac{a}{s}$
e^{-at}	$\frac{1}{s+a}$
te^{at}	$\frac{1}{(s-a)^2}$
$(1+at)e^{at}$	$\frac{s}{(s-a)^2}$
$\frac{1}{a}e^{-\frac{t}{a}}$	$\frac{1}{(1+sa)}$
t^p for $p > -1$	$\frac{\Gamma(p+1)}{s^{p+1}}$ for $s > 0$
$t^n \quad n = 1, 2, 3, \ldots$	$\frac{n!}{s^{n+1}} \quad s > 0$
$\frac{1}{a}\left(1 - e^{-at}\right)$	$\frac{1}{s(s+a)}$
$\frac{1}{a}\left(e^{at} - 1\right)$	$\frac{1}{s(s-a)}$
$\frac{1}{a^2}\left(e^{-at} + at - 1\right)$	$\frac{1}{s^2(s+a)}$
$\frac{1}{b-a}\left(e^{-at} - e^{-bt}\right)$	$\frac{1}{(s+a)(s+b)} \quad a \neq b$
$\frac{(ae^{at} - be^{bt})}{a-b}$	$\frac{s}{(s-a)(s-a)} \quad a \neq b$
$\frac{\alpha^k t^{k-1} e^{-\alpha t}}{\Gamma(k)}$	$\left(\frac{\alpha}{\alpha+s}\right)^k$

Bibliography

Abouammoh, El-Neweihi (1984) Closure of the NBUE and DMRL classes under formation of parallel systems. Stat Probab Lett 4:223–225

Chatterjee S, Shukla A, Pham H (2019) Modeling and analysis of software fault detectability and removability with time variant fault exposure ratio, fault removal efficiency, and change point. J Risk Reliab 233(2):246–256

Pham H (2014) Reliability: models, statistical methods, and applications. Rutgers University, Lecture book

Pham H (2016) A generalized fault-detection software reliability model subject to random operating environments. Vietnam J Comput Sci 3(3):145–150

Pham T, Pham H (2019) A generalized software reliability model with stochastic fault-detection rate. Ann Oper Res 277(1):83–93. https://doi.org/10.1007/s10479-017-2486-3

Pham H (ed) (1992d) Fault-tolerant software systems: techniques and applications. IEEE Computer Society Press, Los Alamitos (CA)

Sharma M, Singh VB, Pham H (2018) Entropy based software reliability analysis of multi-version open source software. IEEE Trans Softw Eng 44(12):1207–1223. https://doi.org/10.1109/TSE.2017.2766070

Xie M, Tang Y, Goh TN (2002) A modified Weibull extension with bathtub-shaped failure rate function. Reliab Eng Syst Safety 76:279–285

Zhang Y, Pham H (2014) A dual-stochastic process model for surveillance systems with the uncertainty of operating environments subject to the incident arrival and system failure processes. Int J Perform Eng 10(2):133–142

Zhu M, Pham H (2018) A multi-release software reliability modeling for open source software incorporating dependent fault detection process. Ann Oper Res. 269(1):773–790, https://doi.org/10.1007/s10479-017-2556-6

Zhu M, Pham H (2018) A two-phase software reliability modeling involving with software fault dependency and imperfect fault removal. Comput Lang Syst Struct 53(2018):27–42

© The Editor(s) (if applicable) and The Author(s), under exclusive license to Springer Nature Switzerland AG 2022
H. Pham, *Statistical Reliability Engineering*, Springer Series in Reliability Engineering, https://doi.org/10.1007/978-3-030-76904-8

Index

Printed in the United States
by Baker & Taylor Publisher Services